METHODS IN MOLECULAR BIOLOGY™

Series Editor
John M. Walker
School of Life Sciences
University of Hertfordshire
Hatfield, Hertfordshire, AL10 9AB, UK

For further volumes:
http://www.springer.com/series/7651

Cancer Cytogenetics

Methods and Protocols

Second Edition

Edited by

Lynda J. Campbell

Victorian Cancer Cytogenetics Service, St. Vincent's Hospital Melbourne, Fitzroy, VIC, Australia

Humana Press

Editor
Lynda J. Campbell
Victorian Cancer Cytogenetics Service
St. Vincent's Hospital Melbourne
Fitzroy, VIC, Australia
lynda.campbell@svhm.org.au

ISSN 1064-3745 e-ISSN 1940-6029
ISBN 978-1-4939-5655-5 ISBN 978-1-61779-074-4 (eBook)
DOI 10.1007/978-1-61779-074-4
Springer New York Dordrecht Heidelberg London

Softcover re-print of the Hardcover 2nd edition 2011

Printed on acid-free paper

Humana Press is part of Springer Science+Business Media (www.springer.com)

Preface

This book comprises a series of chapters describing conventional cytogenetic analysis methods, fluorescence in situ hybridization and array methods. All of these techniques have been presented in the context of their use in the diagnosis and investigation of different varieties of cancer. The authors of the different chapters have provided both standard methods and also extensive notes to guide individuals who are new to these methods through the pitfalls that bedevil all such testing. The experience of the authors should enable all testing to be readily transferable. Inevitably, there are some areas of repetition as the methods for preparing chromosomes for analysis in the setting of myelodysplastic syndromes are similar to those methods used to analyse chronic myeloid leukaemia or acute myeloid leukaemia. However, every laboratory uses slight variations in their methods, and it is these differences that should prove useful to both novice and experienced cytogeneticists. Hopefully, readers will be able to either establish new techniques in their laboratories or will find the different variations of standard methods helpful in improving their results.

The pace of change in scientific and medical research condemns most text books to being out of date by the time they are published. However, while some of the data presented here concerning the standard testing algorithms for different disorders may change with time, many of the methods have already withstood the test of time and will continue to provide useful information to cytogeneticists for some time to come.

I would like to thank my colleagues from around the world and also closer to home for their generous contributions to this volume. The cancer cytogenetics community is not a large one, and so it is possible to establish wonderful, productive collaborations around the globe. I have been fortunate indeed to gather together such an experienced group of individuals to share their knowledge.

Lastly, I would like to thank the staff of the Victorian Cancer Cytogenetics Service who, despite impossibly large workloads, have both contributed directly to the contents of this volume and made it possible for me to spend time assembling the book. They are an extremely talented and hardworking group of individuals for whom I am extremely grateful.

Fitzroy, VIC
Melbourne, 2010

Lynda J. Campbell

Contents

Contributors

SHILANI ARULIAH • *Clinical Cytogenetics Laboratory, The Royal Marsden Hospital and The Institute of Cancer Research, Sutton, Surrey, UK*
ULRIKE BACHER • *Clinic for Stem Cell Transplantation, University Cancer Center Hamburg, Hamburg, Germany*
LYNDA J. CAMPBELL • *Victorian Cancer Cytogenetics Service, St. Vincent's Hospital Melbourne, Fitzroy, VIC, Australia*
K-JOHN J. CHEUNG • *Department of Pathology and Laboratory Medicine, British Columbia Cancer Agency, Vancouver, BC, Canada*
ILSE CHUDOBA • *Metasystems, Altlussheim, Germany*
RYAN C. DENLEY • *Department of Pathology, Molecular Diagnostic Service, Memorial Sloan-Kettering Cancer Center, New York, NY, USA*
CLAUDIA HAFERLACH • *MLL Munich Leukemia Laboratory, Munich, Germany*
CHRISTINE J. HARRISON • *Leukaemia Research Cytogenetics Group, Northern Institute for Cancer Research, Newcastle University, Newcastle, UK*
DOUGLAS E. HORSMAN • *Department of Pathology and Laboratory Medicine, British Columbia Cancer Agency, Vancouver, BC, Canada*
SURESH C. JHANWAR • *Department of Pathology, Molecular Diagnostic Service, Memorial Sloan-Kettering Cancer Center, New York, NY, USA*
ATSUSHI KODAMA • *Chromosome Unit, Central Laboratory, Tokyo Medical University, Tokyo, Japan*
ROBYN LUKEIS • *Cytogenetics Laboratory, Department of Haematology, SydPath, St. Vincent's Hospital, Darlinghurst, NSW, Australia*
RUTH N. MACKINNON • *Victorian Cancer Cytogenetics Service, St. Vincent's Hospital Melbourne, Fitzroy, VIC, Australia*
BRUCE R. MERCER • *Victorian Cancer Cytogenetics Service, St. Vincent's Hospital Melbourne, Fitzroy, VIC, Australia*
TOON MIN • *Clinical Cytogenetics Laboratory, The Royal Marsden Hospital and The Institute of Cancer Research, Sutton, Surrey, UK*
CHRISTINE M. MORRIS • *Cancer Genetics Research Group, Department of Pathology, University of Otago Christchurch School of Medicine and Health Services, Christchurch, New Zealand*
CHARLES G. MULLIGHAN • *Department of Pathology, St. Jude Children's Research Hospital, Memphis, TN, USA*
JUNKO H. OHYASHIKI • *First Department of Internal Medicine, Tokyo Medical University, Tokyo, Japan*
KAZUMA OHYASHIKI • *First Department of Internal Medicine, Tokyo Medical University, Tokyo, Japan*

KATHLEEN C. RAYEROUX • *Victorian Cancer Cytogenetics Service, St. Vincent's Hospital Melbourne, Fitzroy, VIC, Australia*
JEFFREY R. SAWYER • *Department of Pathology, Myeloma Institute for Research and Therapy, University of Arkansas for Medical Sciences, Little Rock, AR, USA*
CLAIRE SCHWAB • *Leukaemia Research Cytogenetics Group, Northern Institute for Cancer Research, Newcastle University, Newcastle, UK*
RASHMI R. SINGH • *Department of Pathology and Laboratory Medicine, British Columbia Cancer Agency, Vancouver, BC, Canada*
MARY SUTER • *Cytogenetics Laboratory, Department of Haematology, SydPath, St. Vincent's Hospital, Sydney, NSW, Australia*
JOHN SWANSBURY • *Clinical Cytogenetics Laboratory, The Royal Marsden Hospital and The Institute of Cancer Research, Sutton, Surrey, UK*
JOANNE S. WHITE • *Victorian Cancer Cytogenetics Service, St. Vincent's Hospital Melbourne, Fitzroy, VIC, Australia*
ADRIAN ZORDAN • *Victorian Cancer Cytogenetics Service, St. Vincent's Hospital Melbourne, Fitzroy, VIC, Australia*

Chapter 1

Introduction

Lynda J. Campbell

Like most cytogeneticists, I have been informed by well-meaning colleagues at regular intervals over the last couple of decades that cytogenetics is an old-fashioned science that will be superseded in the very near future by molecular biological techniques. Yet, my laboratory is busier than it has ever been, despite the introduction of quantitative polymerase chain reaction (PCR) to detect the common translocations and of array-based techniques to provide a more precise global view of the genome than conventional cytogenetics could ever achieve. The value of cytogenetic analysis, particularly of malignancies, lies in its ability to detect new and unsuspected abnormalities, at a sufficiently low power that the abnormalities can be analysed and interpreted without the need for sophisticated computer software. It is rather like a low flying aeroplane able to read the lie of the land and identify the major landmarks without being able to necessarily identify the types of trees or read the street signs.

Of course, without the advances in molecular biology, cytogenetics probably would not play the important role it currently undertakes in the management of cancer patients. Despite the labour-intensive nature of cytogenetic analysis and its reputation as "low-tech," cytogeneticists have absorbed new techniques into their armamentarium, as evidenced by the contents of this book. Molecular biological advances have enabled the development of fluorescence in situ hybridization (FISH), the importance of which is demonstrated by the presence of seven separate chapters in this volume covering the use of FISH in different settings. Cytogeneticists have embraced FISH as a vital adjunct to the analysis of chromosome abnormalities. FISH allows the identification of chromosomal rearrangements that cannot be fathomed by banding alone. FISH also provides a window into the genetic rearrangements within non-dividing cells, particularly valuable in disorders from which it may be difficult to obtain metaphase spreads.

Lynda J. Campbell (ed.), *Cancer Cytogenetics: Methods and Protocols*, Methods in Molecular Biology, vol. 730,
DOI 10.1007/978-1-61779-074-4_1, © Springer Science+Business Media, LLC 2011

Cancer cytogeneticists are also now looking towards comparative genomic hybridization (CGH) via arrays, in reality an adaptation of FISH testing, to identify gains and losses of parts of chromosomes more accurately, and at a resolution not possible by conventional cytogenetics. The clinical significance of the extraordinary amounts of information generated by array CGH and single nucleotide polymorphism (SNP) arrays has yet to be fully understood, but there is little doubt that these technologies will play a role in the management of patients in the future.

Another reason for the exponential growth in demand for cancer cytogenetic analysis in recent years has been the expansion in our knowledge of the significance of genetic aberrations underlying specific types of tumours. A decade ago, we were not routinely karyotyping all plasma cell myeloma cases or performing FISH studies with panels of probes on new cases of chronic lymphocytic leukaemia (CLL). Now, every cancer cytogenetics laboratory is being bombarded by myeloma and CLL cases, and the demand is, in many instances, outstripping the ability of laboratories to supply results in a timely fashion.

Despite all of the advances in the study of cytogenetics, skilled cytogeneticists remain the most crucial components of any successful laboratory. The automation available for harvesting, finding metaphase spreads, and capturing those spreads into a computer analysis program has not obviated the need for individuals with the expertise to recognise, at a glance, abnormalities in the banding pattern of individual chromosomes. The skill required by the cytogeneticist who seeks to ascertain specific abnormalities in short, often indistinctly banded chromosomes found in many cancer cells, is different to the skill required to analyse 850 band level chromosomes in an endeavour to identify subtle constitutional abnormalities, but not less critical. The following chapters represent a variety of methods from different laboratories around the globe to assist in that analysis.

Chapter 2

Evolution of Cytogenetic Methods in the Study of Cancer

Lynda J. Campbell

Abstract

Cytogenetic methods have not changed greatly over the last 50 years since Nowell and Hungerford's description of the Philadelphia chromosome but the clinical utility of these methods has evolved dramatically. The multicentre clinical studies that have identified major clinical applications for cytogenetic analysis in different cancers and the development of in situ hybridization have contributed to an explosion in cytogenetic testing for cancer patients.

Key words: Cytogenetic analysis, Fluorescence in situ hybridization, Translocations, Array comparative genomic hybridization, SNP arrays

1. Introduction

Cytogenetic analysis has become an integral part of the diagnosis and management of many malignancies. Theodor Boveri was the first to suggest that malignant tumours could be due to an abnormal chromosome constitution (1). His hypothesis stated that the cell of a malignant tumour has an abnormal chromosome constitution and that any event leading to an abnormal chromosome constitution will result in a malignant tumour. He also postulated the existence of enhancing or suppressing chromosomes, suggesting that malignant growth would result from loss of suppressing chromosomes or the predominance of enhancing chromosomes. Thus, prior to the concept of genes, Boveri foreshadowed the existence of oncogenes and tumour suppressor genes.

The term chromosome was first coined by Waldeyer in 1888 but it took nearly 70 years for the chromosome complement of a normal human cell to be reliably determined. The birth of cytogenetics is generally dated from Tjio and Levan's identification of

Lynda J. Campbell (ed.), *Cancer Cytogenetics: Methods and Protocols*, Methods in Molecular Biology, vol. 730,
DOI 10.1007/978-1-61779-074-4_2, © Springer Science+Business Media, LLC 2011

the true chromosome complement in human cells as it is from this time that abnormalities of chromosome number and subsequently chromosome structure were reported (2). Their discovery was made possible by a number of advances. In his delightful book, Hsu divided the study of human cytogenetics into four periods: the pre-hypotonic period (or Dark Ages), the period from 1952 to 1959 that included the discovery of hypotonic solution pre-treatment for cytological preparations, the third period (1959–1969) during which time chromosome abnormalities were linked to clinical syndromes, and the post-banding (modern) period (3).

During the pre-hypotonic era, chromosomes were studied in mouse and rat cancers and camera lucida drawings of metaphases suggested the presence of many more chromosomes than normal and of structural abnormalities within these chromosomes. The drawings were taken from squash preparations, a technique that was used to flatten metaphase spreads of chromosomes into a two-dimensional configuration, but still resulted in crowded overlapping aggregations of chromosomes that were very difficult to count. Colchicine, an extract of the autumn crocus, was used to arrest the cells in the metaphase stage of the cell cycle and increase the number of mitoses available for analysis.

The use of a hypotonic pre-treatment method was an enormous step forward in the production of analysable chromosome preparations. The chromosomes could now be separated and viewed individually. Counting chromosomes was simplified and gross structural abnormalities could be discerned. Hsu describes the discovery of the utility of a hypotonic pre-treatment as a laboratory accident, the perpetrator of which never owned up to the error; thus, a major discovery in the history of cytogenetics was apparently made by an unknown technician.

It was in this era, in 1956, that Tjio and Levan finally answered the question that had been plaguing investigators for more than 30 years, when they reported that there were 46 chromosomes in the human cell rather than 48. Subsequently, a number of researchers were able to identify chromosome abnormalities that appeared specific for clinical syndromes. Lejeune and his colleagues published the chromosomal nature of Down syndrome in 1959 (4). Their observation of an extra G group chromosome in patients with a specific congenital malformation syndrome showed for the first time that cytogenetic analysis could be used to diagnose a human condition. From this time, there was a stream of publications describing chromosome aneuploidies associated with other malformation syndromes. A number of the early, seminal papers in the field have been reproduced in Peter Harper's excellent study of the beginnings of human cytogenetics (5). The study of the constitutional karyotype was aided by the discovery that phytohemagglutinin (PHA) could induce peripheral blood lymphocytes to divide (6). This method was adopted by Moorhead et al. (7) for the study of human chromosomes and remains one of the mainstays of modern cytogenetic analysis.

The confirmation of the correct chromosome number in human cells led in the late 1950s and early 1960s to a flood of publications describing numerical and structural abnormalities of chromosomes. The resulting confusion in the literature made clear that there was a need for a common nomenclature to describe these rearrangements in a manner that was intelligible to other workers in the field. Thus, a small group met in Denver, CO, to establish a system of describing chromosome abnormalities. They published the results of their deliberations in a report entitled "A Proposed Standard System of Nomenclature of Human Mitotic Chromosomes," also known as the Denver Conference (1960), and this report has formed the basis for all subsequent nomenclature reports, now published as An International System for Human Cytogenetic Nomenclature (ISCN) (8).

The ability to create a banding pattern on human chromosomes understandably complicated the nomenclature. Meetings in Paris and Edinburgh proposed a basic system for designating chromosome regions and bands, resulting in a report of the Paris Conference (1971). Crucially, this report provided a way of describing structural rearrangements in terms of the band composition and the breakpoints involved in the rearrangement. The subsequent ISCN publications have been updated to encompass the various advances in the field, including fluorescence in situ hybridization (FISH) and arrays, but their core function remains the description of chromosome abnormalities in a manner that allows a cytogeneticist to interpret the report of a colleague immediately from anywhere in the world. Whilst the complexity of chromosome rearrangements in cancer cells frequently tests this system to the limit, as a scientific form of Esperanto, it has been spectacularly successful over the years.

The advent of banding enabled chromosomes to be individually identified and the normal homologues paired. Initially, banding patterns along the length of each chromosome were induced by preparations stained with quinacrine mustard and visualized via a fluorescence microscope (9) or depended upon a method whereby slides were incubated in warm saline or buffer solutions prior to staining by Giemsa. The initial Giemsa staining method required 3 days for completion. Seabright's rapid banding technique was therefore embraced as the whole procedure could be carried out at room temperature using air-dried slides and producing G-banded chromosomes ready for observation within 10 minutes (10). Once banding became available, there were numerous publications describing recurrent chromosome abnormalities that appeared to be found in specific tumour types.

2. Cancer Cytogenetics

The history of cancer cytogenetics is not a long one but it has been eventful and much knowledge has been accumulated in the 50 years since Peter Nowell and David Hungerford published their finding of a small marker chromosome in the chromosome complement of cells cultured from seven patients with chronic myeloid leukaemia (CML) (11). Nowell and Hungerford made their landmark discovery in 1960, only just beating another group from Edinburgh who had also noted the same marker chromosome in their CML patients (12).

Although the hypothesis that malignant cells were derived from normal tissue cells that had acquired an abnormal chromatin content was first proposed by Boveri, it was not until Nowell and Hungerford's description of the Philadelphia chromosome that a revolution in our understanding of the processes underlying the development of malignancy began. Nowell and Hungerford called the marker chromosome the Philadelphia chromosome 1, Ph^1, after the city in which they worked and the number 1 superscript signalled that they fully expected that there were many more cytogenetic markers of cancer to be discovered. This was an exciting finding and an exciting time that effectively launched the field of cancer cytogenetics but the following years were frustrating as abnormalities were observed in various cancers but the inability to identify specific chromosomes by any method other than their basic shape limited researchers' abilities to link abnormalities with different morphological subtypes of haematological or solid tumours. All this changed with the advent of banding techniques.

Banding allowed the chromosomes to be clearly distinguished from one another and, most importantly, revealed the nature of structural abnormalities: balanced translocations of material between chromosomes, deletion of part of a chromosome, duplication of another segment, or an inversion of a chromosome segment. In 1973, Janet Rowley reported that a reciprocal translocation between chromosomes 9 and 22 resulted in the Philadelphia chromosome (13). Since then, hundreds of rearrangements have been identified including not only translocations but also deletions and additions of part or all of chromosomes and also inversions of genetic material within chromosomes.

From the descriptions of chromosome rearrangements, long before the Human Genome Project shone a light on the location of genes strung along our chromosomes, the molecular biologists were able to discover critical genes at the breakpoints of translocations. Banding continues to allow us to identify new chromosome abnormalities in both haematological and solid tumours and these abnormalities provide the sign posts to the critical genetic changes that underlie the transformation of normal cells into cancer cells.

The first cytogenetic abnormality to have its genetic secrets unlocked was the 8;14 translocation which characterizes Burkitt lymphoma/leukaemia. Researchers identified that the translocation caused two genes, *MYC* on 8q24 and the immunoglobulin heavy chain gene, *IGH* on 14q32, to come together (14, 15). It is now known that two classes of translocations are found in malignancies. The first type is epitomized by the t(8;14) in Burkitt lymphoma, one gene which is already actively transcribed in the cell type, such as *IGH* in B lymphocytes, is juxtaposed to a gene such as *MYC* which is, by virtue of its resulting proximity to *IGH*, up-regulated. Other translocations such as the t(8;21) in acute myeloid leukaemia (AML), also described by Janet Rowley in 1973 (16), and the t(9;22) in CML, form fusion genes with a "new" gene product which incorporates part of the normal genes broken at the sites of translocation. In the case of t(8;21), the genes involved are the *RUNX1T1* (originally named ETO after "Eight Twenty-One") gene on 8q22 and the *RUNX1* (*AML1*) gene on 21q22 (17); the t(9;22) causes a fusion of the *BCR* gene on 22q11.2 and *ABL1* on 9q34 (18, 19). It is the altered function of these "new" fusion genes that appears to transform the cell, as shown by the development of CML-like disorders in mice into which *BCR–ABL1* constructs have been inserted (20).

It took time for the medical and scientific worlds to realize the importance of the cytogenetic discoveries of the 1960s and 1970s. It was necessary to convince clinicians that the chromosome changes being described in the marrow and peripheral blood of their patients with a variety of malignancies could provide valuable information about the type and prognosis of these disorders. Many important clinical correlations were either identified or confirmed in the International Workshops on Chromosomes in Leukaemia. These constituted gatherings of physicians and scientists from around the world who brought together case studies of chromosome analyses together with clinical and laboratory data relating to each case.

The first of these was held in Helsinki, Finland in August 1977 (21). Laboratories participated from Belgium, Finland, Sweden, England, Germany, and the USA and the participants reviewed the data of 223 patients with Ph^1-positive CML and 279 patients with acute non-lymphocytic leukaemia. A number of further workshops were held and information regarding the incidence and prognostic significance of rearrangements in CML, AML, acute lymphoblastic leukaemia (ALL), and myelodysplastic syndromes (MDS) provided by these workshops formed the basis for all future studies. Subsequently, national and multinational clinical trial groups have incorporated cytogenetic studies into their prospective trials and provided a wealth of data to show that cytogenetic analysis is of diagnostic and prognostic importance in most haematological malignancies and a number of solid tumours.

It has only been by the careful observation of chromosome abnormalities and their correlations with clinical features that true insights have been obtained as to the underlying genetic basis of malignancy.

Whilst the basic cytogenetic methods used in laboratories around the world today are very similar to those first described in the 1960s and 1970s, there are areas where improvements have been made. Mitogens were introduced into cultures to induce chronic lymphoid malignancy cells to divide in the late 1970s (22) but further refinements and combinations of mitogens are still being discovered. For example, Chapter 9 describes a recently discovered method that enables chromosome abnormalities to be identified in the majority of cases of chronic lymphocytic leukaemias (23).

3. Introduction of FISH Testing

The identification of the genes involved in chromosome translocations paralleled the development of *in situ* hybridization (ISH) and so allowed the most significant advancement in the field of cytogenetics to come into being. Early ISH studies used tritiated thymidine to label the DNA fragments that were used as probes (24). Slides were prepared by dropping fixed cytogenetic suspension onto the slide. After the subsequent application of probe, the slide was immersed in photographic emulsion, wrapped in foil, and stored away in a light proof box for up to 2 months. The localization of the probe was identified by "developing" the slide so that silver granules were deposited at the site of the tritiated thymidine emissions. In expert hands, this method worked well and many of the early gene localizations were made using this method. However, the necessity of performing most of the steps in total darkness and the time required for hybridization made this a most frustrating method as, after 2 months, it was entirely likely that the test had been unsuccessful and determining the reasons for failure after such a time period was extremely difficult.

The development of FISH was therefore very welcome indeed. FISH did not require total darkness for successful completion of testing and a result could be obtained in 24 h. The ease with which routine diagnostic laboratories could establish FISH techniques now enabled them to be used in the routine diagnostic setting for the first time. Moreover, FISH probes were developed that allowed the cytogeneticist to determine the presence or absence of extra copies of chromosomes, translocations, and deletions in non-dividing cells. Initially, the probes were home-grown with single colour fluorescent signals for gains and losses of chromosomes and the translocation probes produced only a single fusion signal. However, the increasing use of commercial FISH probes in the clinical setting ensured that the probe designs evolved. False-positive and false-negative results were reduced by designing probes with built-in controls or with a resulting signal pattern that could not be readily duplicated by accidental co-localization in a normal patient control slide.

The ability to identify chromosome rearrangements in non-dividing cells has proved particularly useful in the chronic lymphoid malignancies. FISH has been used with panels of probes to identify prognostic subgroups within CLL (25) and plasma cell myeloma (26). FISH has also identified cryptic translocations such as the t(12;21) in paediatric ALL (27) and t(4;14) in myeloma (28) and cryptic deletions such as the 4q12 deletion that results in a *PDGFRA–FIP1L1* fusion gene (29). Both translocations and deletions are invisible microscopically and could only be found by molecular methods.

Further refinements of FISH methods enabled the effective painting of each chromosome a different colour so that complex karyotypes could be elucidated (30, 31) and the combination of FISH probes with fluorescent-labelled antibodies to identify individual cell types has proved invaluable in identifying chromosome abnormalities in disorders with variable marrow infiltration such as myeloma (32). Such strides have been made in the last 20 years in the use of FISH in cancer diagnosis that it now seems inconceivable for cytogenetics laboratories not to use FISH routinely.

4. Array-Based Karyotyping Methods

One of the innovative uses for FISH testing that has evolved has been the development of comparative genomic hybridization (CGH). CGH involves the labelling of patient DNA and normal control DNA with green and red fluorochromes, respectively. The two DNAs are then allowed to compete for hybridization on a slide containing normal chromosome preparations. The concept relies on a computer "reading" each chromosome and assessing the proportion of green- and red-labelled DNA that has hybridized along the length of each chromosome. If there are no gains or losses of DNA in the patient sample there should be an equal proportion of patient and control DNA hybridized to each chromosome and an equal mixture of green and red fluorescence rendering each chromosome yellow. In the event that there is loss of part of a chromosome in the patient genome, there is a disproportionate amount of control DNA hybridizing to that chromosome and so it appears red. In contrast, an extra segment of DNA in the patient sample causes that segment of the normal chromosome to appear green. CGH has been used largely in the research setting and achieved only limited use in diagnostic laboratories.

It produced a picture of genetic abnormalities across the genome without the need to produce metaphase spreads but its major drawback was the resolution only allowed the detection of very large gains and losses of DNA.

However, the application of CGH to arrays of bacterial artificial chromosomes (BACs) or oligonucleotides dotted onto slides or "chips" has overcome the problem of resolution and made CGH an enormously powerful tool in cytogenetics. Array CGH is capable of mapping deletions or amplifications measured in kilobases rather than megabases. It is also possible to design arrays that target specific areas of interest or cover the entire genome. To date, these arrays are not capable of detecting balanced translocations but their ability to detect changes in copy number is extraordinary. For the cancer cytogeneticist, the challenge will be how to interpret the vast amount of information generated by these arrays. A leukaemia karyotype may appear to contain a simple chromosome abnormality but the array CGH applied to the same genome may uncover hundreds of sub-microscopic rearrangements. Only large clinical trials incorporating array data collection will enable us to determine what is important and what is not from these vast repositories of information.

Another refinement has been added to the use of arrays. It is now possible to detect uniparental disomy or, as those who work in the field of acquired abnormalities in cancer prefer, copy number neutral loss of heterozygosity (LOH) or acquired isodisomy. LOH without loss of one copy of a DNA segment refers to regions of cancer genomes where it appears that one chromosome has lost a region but has replaced it with a duplication of the identical segment from the other homologue. The regions of LOH can be identified by the use of single nucleotide polymorphism (SNP) arrays. SNP arrays detect the presence of thousands of polymorphisms along the length of each chromosome, and so a stretch of DNA without any variation observed between the two homologues indicates either that only one copy of the region is present or, if it is clear that there is no deletion, that duplication of one copy has occurred. The biological impetus for this action appears to be, in many instances, to achieve a doubling of a gene mutation such as *TET2* mutations in chronic myelomonocytic leukaemia (CMML). CMML usually has a normal karyotype but SNP arrays have shown copy number neutral LOH involving a region of the long arm of chromosome 4 in up to 35% of cases and most of these have been shown to carry homozygous *TET2* mutations (33). The power of these arrays appears likely to reveal many more genetic rearrangements in different cancers.

Ultimately, the challenge for cytogeneticists and for clinicians will be how to use the current and future technologies to best serve the needs of our patients. Conventional cytogenetics remains a powerful and affordable test that is integral to the management of patients with a wide variety of malignancies. FISH, too, has become an important tool both for diagnosis and to predict outcome in many cancers. The potential of the array technologies cannot be under-estimated but their role in the care of cancer patients remains to be defined. And thus, just as the 1960s and 1970s were exciting decades for cytogeneticists, so too will be the coming years as we cope with integrating existing and emerging technologies.

References

1. Sandberg AA. Before 1956: Some historical background to the study of chromosomes in human cancer and leukemia. Cancer Genet Cytogenet 1979;1:87–94.
2. Tjio JH, Levan A. The chromosome number of man. Hereditas 1956;42:1–6.
3. Hsu TC. Human and Mammalian Cytogenetics: An Historical Perspective. New York: Springer; 1979.
4. Lejeune J, Gautier M, Turpin R. Etude des chromosomes somatiques de neuf enfants mongoliens. C R Acad Sci 1959;248:1721–2.
5. Harper PS. First Years of Human Chromosomes. Bloxham, Oxfordshire: Scion Publishing Limited; 2006.
6. Nowell PC. Phytohemagglutinin: an initiator of mitosis in cultures of normal human leukocytes. Cancer Res 1960;20:462–6.
7. Moorhead PS, Nowell PC, Mellman WJ, Battips DM, Hungerford DA. Chromosome preparations of leukocytes cultured from human peripheral blood. Exp Cell Res 1960;20:613–6.
8. Shaffer LG, Slovak ML, Campbell LJ, eds. ISCN (2009). An International System for Human Cytogenetic Nomenclature. Basel: S. Karger; 2009.
9. Caspersson T, Farber S, Foley GE, et al. Chemical differentiation along metaphase chromosomes. Exp Cell Res 1968;49:219–22.
10. Seabright M. A rapid banding technique for human chromosomes. Lancet 1971;ii:971–2.
11. Nowell PC, Hungerford DA. Chromosome studies on normal and leukemic human leukocytes. J Natl Cancer Inst 1960;25:85–109.
12. Baikie AG, Court-Brown WM, Buckton KE, Harnden DG, Jacobs PA, Tough IM. A possible specific chromosome abnormality in human chronic myeloid leukaemia. Nature 1960;188:1165–6.
13. Rowley JD. A new consistent chromosomal abnormality in chronic myelogenous leukaemia identified by quinacrine fluorescence and Giemsa staining. Nature 1973;243:290–3.
14. Dalla-Favera R, Bregni M, Erikson J, Patterson D, Gallo RC, Croce CM. Human c-myc onc gene is located on the region of chromosome 8 that is translocated in Burkitt lymphoma cells. Proc Natl Acad Sci U S A 1982;79:7824–7.
15. Taub R, Kirsch I, Morton C, Lenoir G, Swan D, Tronick S, Aaronson S, Leder P. Translocation of the c-myc gene into the immunoglobulin heavy chain locus in human Burkitt lymphoma and murine plasmacytoma cells. Proc Natl Acad Sci U S A 1982;79:7837–41.
16. Rowley JD. Identification of a translocation with quinacrine fluorescence in a patient with acute leukemia. Ann Génét 1973;16:109–12.
17. Miyoshi H, Shimizu K, Kozu T, Maseki N, Kaneko Y, Ohki M. t(8;21) breakpoints on chromosome 21 in acute myeloid leukemia are clustered within a limited region of a single gene, AML1. Proc Natl Acad Sci U S A 1991;88:10431–4.
18. Heisterkamp N, Stephenson JR, Groffen J, Hansen PF, de Klein A, Bartram CR, Grosveld G. Localization of the c-abl oncogene adjacent to a translocation break point in chronic myelocytic leukaemia. Nature 1983;306:239–42.
19. Groffen J, Stephenson JR, Heisterkamp N, de Klein A, Bartram CR, Grosveld G. Philadelphia chromosomal breakpoints are clustered within a limited region, bcr, on chromosome 22. Cell 1984;36:93–9.
20. Daley GQ, Van Etten RA, Baltimore D. Induction of chronic myelogenous leukemia in mice by the P210bcr/abl gene of the Philadelphia chromosome. Science 1990;247:824–30.
21. Rowley JD, de la Chapelle A. General report on the First International Workshop on Chromosomes in Leukemia. Int J Cancer 1978;21:307–8.
22. Gahrton G, Robert K, Friberg K, Zech L, Bird A. Nonrandom chromosomal aberrations in chronic lymphocytic leukemia revealed by polyclonal B-cell-mitogen stimulation. Blood 1980;56:640–7.

23. Dicker F, Schnittger S, Haferlach T, Kern W, Schoch C. Immunostimulatory oligonucleotide-induced metaphase cytogenetics detect chromosomal aberrations in 80% of CLL patients: a study of 132 CLL cases with correlation to FISH, IgVH status, and CD38 expression. Blood 2006;108:3152–60.
24. Harper ME, Saunders GF. Localization of single copy DNA sequences of G-banded human chromosomes by in situ hybridization. Chromosoma 1981;83:431–9.
25. Dohner H, Stilgenbauer S, Benner A, et al. Genomic aberrations and survival in chronic lymphocytic leukemia. N Engl J Med 2000;343:1910–6.
26. Avet-Loiseau H. Role of genetics in prognostication in myeloma. Best Pract Res Clin Haematol 2007;20:625–35.
27. Romana S, Le Coniat M, Berger R. t(12;21): a new recurrent translocation in acute lymphoblastic leukemia. Genes Chromosomes Cancer 1994;9:186–91.
28. Chesi M, Nardini E, Brents LA, et al. Frequent translocation t(4;14)(p16.3;q32.3) in multiple myeloma is associated with increased expression and activating mutations of fibroblast growth factor receptor 3. Nat Genet 1997;16:260–4.
29. Gotlib J, Cools J, Malone JM III, Schrier SL, Gilliland DG, Coutre SE. The FIP1L1–PDGFR{alpha} fusion tyrosine kinase in hypereosinophilic syndrome and chronic eosinophilic leukemia: implications for diagnosis, classification, and management. Blood 2004;103:2879–91.
30. Schröck E, du Manoir S, Veldman T, Schoell B, Wienberg J, Ferguson-Smith MA, Ning Y, Ledbetter DH, Bar-Am I, Soenksen D, Garini Y, Ried T. Multicolor spectral karyotyping of human chromosomes. Science 1996;273:494–7.
31. Speicher MR, Ballard SG, Ward DC. Karyotyping human chromosomes by combinatorial multi-fluor FISH. Nat Genet 1996;12:368–75.
32. Ahmann GJ, Jalal SM, Juneau AL, et al. A novel three-color, clone-specific fluorescence in situ hybridization procedure for monoclonal gammopathies. Cancer Genet Cytogenet 1998;101:7–11.
33. Jankowska AM, Szpurka H, Tiu RV, et al. Loss of heterozygosity 4q24 and TET2 mutations associated with myelodysplastic/myeloproliferative neoplasms. Blood 2009;113:6403–10.

Chapter 3

Fluorescence In Situ Hybridization Methods and Troubleshooting Applied to Fixed Cell Suspensions

John Swansbury, Toon Min, and Shilani Aruliah

Abstract

A robust procedure for performing fluorescence in situ hybridization (FISH) is described, with tips for troubleshooting. FISH probes are now more reliable and there is a greater range commercially available. FISH is an essential part of the cytogeneticist's repertoire. It remains a powerful, complementary adjunct to karyotype studies, and knowledge of the underlying chromosome abnormalities can be essential for understanding the FISH signal patterns.

Key words: FISH, Technique, Troubleshooting, Interpretation

1. Introduction

Fluorescence in situ hybridization (FISH) is a process by which specific fragments of DNA are labeled with a fluorochrome (a dye that emits visible light when irradiated with ultraviolet light) and are then allowed to attach to a particular location on a chromosome. This chapter describes the use of FISH principally in the context of a hospital laboratory, providing a routine, diagnostic service. The FISH technique is now well established and generally robust and reliable. However, the correct interpretation of the results is not always obvious, and so this chapter also includes a description of some of the factors that need to be considered.

The chromosomes obtained in studies of malignancy are often of poor morphology, and tend to be involved in complex and subtle rearrangements; in such cases, it is often impossible to define the entire karyotype by simply using G-banding. There is a variety of other newer genetic techniques that are available. These include polymerase chain reaction assays that can detect specific

Lynda J. Campbell (ed.), *Cancer Cytogenetics: Methods and Protocols*, Methods in Molecular Biology, vol. 730,
DOI 10.1007/978-1-61779-074-4_3, © Springer Science+Business Media, LLC 2011

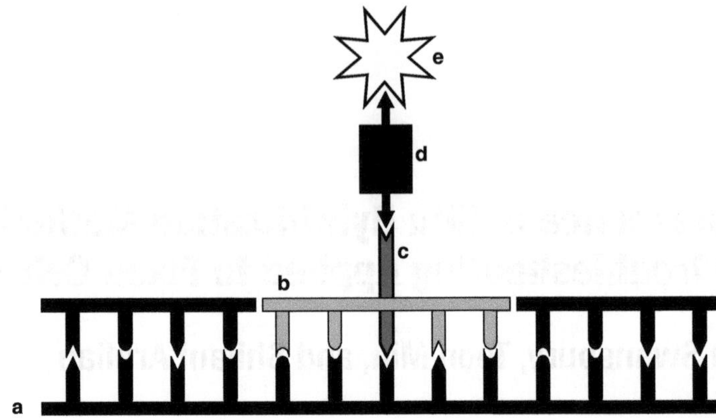

Fig. 1. Schematic illustration of the FISH principle: (**a**) represents the DNA in the chromosome being studied, i.e., the target DNA; (**b**) represents the manufactured DNA (i.e., the probe DNA), which precisely matches target DNA; incorporated into the probe DNA is a label (**c**), such as digoxigenin-11-dUTP; this provides an attachment for (**d**), an antibody (anti-digoxigenin), which in turn is attached to a fluorochrome (**e**) such as Texas Red.

gene rearrangements (including translocations) but not gains or losses of chromosome material, and array comparative genomic hybridization that can detect minute gains and losses but not balanced rearrangements. FISH has some of the attributes of both assays and forms a very powerful combination with karyotype studies. In particular, FISH can compensate for one of main restrictions of karyotype studies, the need for metaphase divisions: clinically useful information can often be obtained from FISH studies of interphase nuclei (1).

The basic principle of FISH is illustrated in Fig. 1. The impression is sometimes given that FISH studies are simply a matter of buying a kit with the right DNA probe, following the supplier's instructions, and reading a simple positive or negative result. In practice, getting a reliable result from a FISH study requires experience, time spent in testing to determine the precise local conditions needed for optimum hybridization, and time spent in assessing and scoring positive and negative controls to determine local baseline levels. The techniques described here will provide some useful guidelines about those aspects of the procedures that can be varied and those that are critical, and will provide advice about the origin and resolution of commonly encountered problems.

1.1. Choice of Probes

For both research and clinical applications, there is now a great variety of DNA probes that are commercially available. Suppliers include Abbott (Vysis), Cambio, Cytocell, Dako, Kreatech, and Poseidon (Stretton). Probes have been produced that identify whole chromosomes, parts of chromosomes, centromeres, telomeres, genes, and parts of genes. Most of these probes have been prepared to the exacting standards that are required for clinical use.

In a busy hematology cytogenetics laboratory, the use of commercially produced, standardized, directly labeled probes saves valuable time. However, there are also very many unlabelled probes available that have been produced in a research context, and which have not yet been adopted for commercial development. These are often freely available, subject to certain restrictions, on request to the laboratory that produced them. They will require labeling by nick translation (an enzymatic labeling system which synthesizes nucleic acids using nucleotides that have a hapten), and the attachment of a fluorochrome. The nick translation procedure was described in the previous edition of this book (2). The fluorochromes most commonly used are fluorescein, rhodamine, coumarin, Texas Red, SpectrumOrange, and SpectrumGreen.

Commercial probes are generally easy to use, as they are prepared to high standards of consistency and quality. However, they can vary in terms of ease of use, size, and the area of chromosome that is covered. Always read the marketing literature carefully, as there is a tendency not to provide adequate information about limitations. For example, some probes may not cover all of the genes being tested and some probes may not be contiguous, i.e., there may be gaps in the length of DNA being covered, which in interphase nuclei can sometimes give the appearance of the signal being split.

A single-color probe that spans a translocation breakpoint will result in an extra signal if the translocation is present, but it will give a very similar result if there is an extra copy of that part of the chromosome: signal size cannot be relied on to discriminate between these two situations. Also, if the same single-color probe was used in a situation when there had been a translocation followed by loss of one of the products, the resulting FISH picture would be very similar to normal, i.e. two signals present. For this reason, dual-color or triple-color probes are usually more informative.

Studies of translocations are usually done with one of two types of FISH probe. For common, regular translocations, it is best to use the "dual-fusion" type, in which each translocation partner is identified by a probe of different color, usually red or green. The normal configuration is therefore two red signals and two green signals in each nucleus. When there has been a translocation, one of each signal is split and the different halves join together; the resulting signal pattern is one red, one green, and two fused signals (see Fig. 2). Probes that produce a single fusion are not good for follow-up studies, since a small percentage of cells will happen to have normal red and green signals that are colocated just by chance.

For translocations that can involve multiple partners, a "break-apart" type of probe can be more useful. For example, the *MLL* gene at 11q23 has been found to fuse with over 50 different genes; it would not be efficient to perform 50 separate FISH studies to exclude all of these. Instead, start with a break-apart

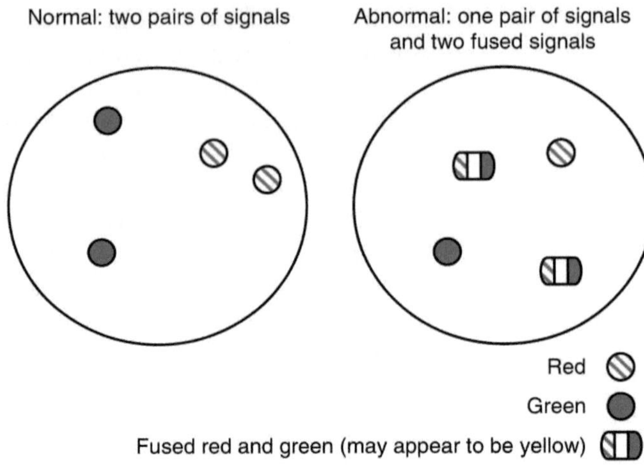

Fig. 2. Common signal patterns for the dual-fusion type of FISH probe.

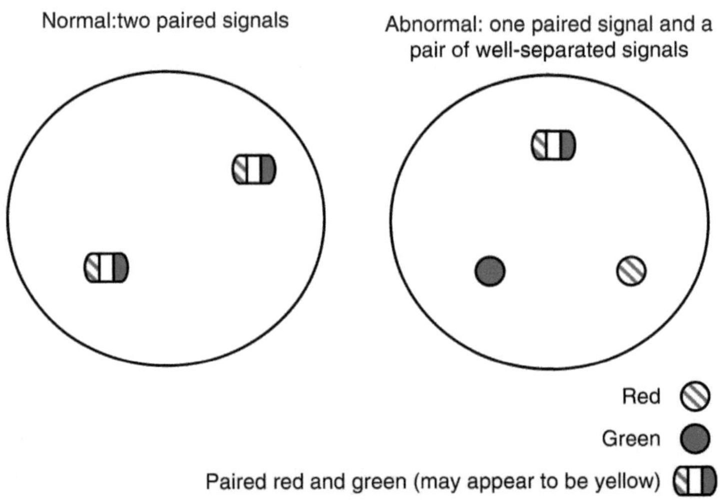

Fig. 3. Common signal patterns for the break-apart type of FISH probe.

probe that has a pair of red and green signals located on the gene of interest. The normal signal configuration is therefore two paired signals. If there is any translocation present involving this gene, then the two colors will be split apart: the signal pattern will be one paired (normal) signal, one red, and one green (see Fig. 3). The translocation partner is not identified in interphase cells, but looking at the location of signals in a metaphase cell should give a clue as to what the partner might be. Note that there is a small gap between the two colors in some probes, and in interphase nuclei, when the DNA is unwound, this gap may result in a clear space between the two colors. With careful focusing, it is sometimes possible to see a fine strand of DNA linking the two signals.

Therefore, scoring a positive split requires that only cells with widely separated signals are counted.

Probes are produced with the addition of Cot-1 human DNA. This is blocking or competitor DNA that hybridizes to the sequences that are common both to the probe and to other chromosomes, thereby preventing hybridization of the probe DNA to these sequences. It greatly improves signal clarity, as only sequences specific to the target remain available for probe hybridization. If this were not done, then the end result would be multiple signals occurring on many unwanted chromosome sites as well as on the site of interest. This blocking DNA also hybridizes to molecules in the nucleoplasm and cytoplasm that could also bind to the probe.

However, the blocking DNA may also prevent hybridization of probe DNA to closely adjacent DNA sequences by stearic hindrance. The method described here maximizes the rapid pairing of the sequences that are common to the probe and to the blocking DNA.

1.2. The Technique

For simplicity, it will be assumed that the material to be studied is a fixed cell suspension, as this is what is usually available in a cytogenetics laboratory. It is possible to study air-dried bone marrow or blood films, fresh tumor touch prints, and wax-embedded sections of solid tumors. The techniques are very similar to those described here (3, 4). It is also possible to perform FISH studies on slide preparations that have already been banded and analyzed as part of a metaphase karyotype study. However, the simplest and most reliable methods use freshly spread cells, and this is what will be described here.

A brief overview of the whole procedure is as follows:

(a) The cells to be studied are harvested, fixed, and spread, as for metaphase karyotype studies.

(b) The probe DNA is prepared and added to the slide.

(c) The target DNA and probe DNA are co-denatured at high temperature in the presence of formamide; this causes the DNA double strands to separate into single strands.

(d) The slide is incubated to allow the probe DNA to attach to the target DNA.

(e) The slide is washed to remove any remaining, unhybridized probe.

(f) A counterstain, e.g., DAPI, is added and a coverslip is placed on the slide.

(g) The slide is ready for assessment with a microscope that uses ultraviolet light. Fluorescence occurs when the electrons of a molecule of fluorochrome are excited by light of one wavelength and return to the unexcited state by emitting light of a longer wavelength.

2. Materials and Equipment

Note that many of the reagents used are potentially harmful by contact, inhalation, and/or ingestion. It is important that principles of good laboratory practice are followed, and that appropriate health and safety precautions are taken. Material Safety Data Sheets can be obtained from the manufacturers, either supplied with the reagent or available online. Most of these reagents can be obtained from any good supplier such as Sigma or Gibco. Other suppliers are indicated where necessary.

2.1. Solutions

1. 20× SSC (SSC = sodium chloride and sodium citrate solution): dissolve 175.3 g of sodium chloride and 88.2 g of sodium citrate in 900 mL of distilled water (dH_2O). Adjust the pH to 7.0 using sodium hydroxide or hydrochloric acid, then make up to 1 L with more dH_2O. This can be stored at room temperature for up to 6 months. For other concentrations, either dilute this stock or else modify these amounts accordingly; for example, for 4× SSC, use 34.1 g sodium chloride and 17.6 g sodium citrate per liter of solution.
2. 2× SSC.
3. 0.4× SSC.
4. Hybridization buffer (supplied by the probe manufacturer).
5. Pepsin (SIGMA).
6. NP40 (Vysis) or Igepal (SIGMA) or Tween 20.
7. 2× SSC/0.1% NP40/Igepal: Add 100 mL of 20× SSC to a 1,000-mL cylinder. Add 1 mL of NP40/Igepal with a clean disposable plastic pipette. Add dH_2O to the 1,000-mL mark. Mix well. Adjust pH if necessary to 6.8–7.5. Label and date the bottle. Store the stock solution in a fridge; the in-use solution can be kept on the bench at room temperature.
8. 0.4× SSC/0.3% NP40/Igepal: Add 20 mL of 20× SSC to a 1,000-mL cylinder. Add 3 mL of NP40/Igepal with a clean disposable plastic pipette. Add dH_2O to 1,000 mL mark. Transfer to a 1 L clean bottle. Mix well. If necessary, adjust pH to 6.8–7.5. Label and date the bottle. Store the stock solution in a fridge; the in-use solution can be kept on the bench at room temperature.
9. Counterstain: The choice of counterstain is affected by the choice of fluorochrome being used. DAPI (4', 6-Diamidino-2-phenylindole dihydrochloride) is better for red or green fluorochromes, such as SpectrumOrange and SpectrumGreen. An alternative to DAPI is propidium iodide, and this is better for yellow fluorochromes, such as fluorescein. The counterstain solutions and mountant can be bought separately and

diluted as needed to final concentrations of 0.3 μg/mL of PI or 0.1 μg/mL of DAPI. DAPI is used as a counterstain to highlight the nuclei and to contrast the fluorescence signals. It also produces faint G-banding of metaphase chromosomes, so that most of them can be identified. Suppliers include Vector Laboratories and Abbott (Vysis). This stain contains agents to reduce quenching (fading) of the fluorochromes, caused by oxidants and/or free radicals. DAPI stock is kept at 4°C in the fridge, ready to use.

10. Ethanol, used to dehydrate slides; a series of dilutions with water is prepared, 70, 85, and 100%, in containers such as Coplin jars.

2.2. Equipment

1. Pipettes: very small volumes of liquids need to be accurately dispensed; a Gilson pipette is suitable, to which sterile disposable pipette tips can be attached.
2. Microfuge tubes (e.g., Eppendorf) 1.5 mL.
3. Water bath, capable of maintaining a temperature of 75°C.
4. Vortex mixer.
5. Microfuge/microcentrifuge: This is used to spin down small amounts of DNA probe, hybridization buffer, and small quantities of fixed sample.
6. Incubator or oven set at 37°C.
7. pH Meter: this may need a special probe for testing the SSC wash solutions. These solutions have to be made fresh for every experiment and it is essential that the pH is between 6.8 and 7.5.
8. Rubber solution, sometimes called rubber cement, for sealing the edges of coverslips. If it cannot be obtained from your usual supplier, it can often be obtained from shops that sell or repair bicycles. It is convenient to have this in tubes with a fine opening, so that a fine line can be squeezed out. Alternatively, it can be drawn into a 5-mL syringe when needed and squeezed out through a blunt needle.
9. Coverslips: 22 mm × 22 mm (any thickness).
10. Coverslips: 22 mm × 50 mm, thickness grade 0.
11. Epifluorescence microscope: This microscope has an ultraviolet light source and a filter wheel; one that is electronically controlled is preferable.
12. Fluorescence filters (see Note 1).
13. A phase-contrast microscope is also very useful, for examining unstained slides to assess the quality of the spreading.
14. A programmable hotplate, such as a Hybrite™ or a Thermobrite™. This is not an essential item of equipment,

but is recommended as it helps to semi-automate the hybridization part of the FISH process. It can accommodate up to 12 slides. If this is not available, then the following two items will be needed:

15. Humidified container: This can simply be a plastic box containing a slide rack and damp towels or tissues, and which has a close-fitting lid.
16. Hotplate that can maintain a temperature of 75°C.

3. The FISH Method

Always read the probe manufacturer's protocols carefully. The method described here is robust and generally reliable, but some probes may need alterations to the procedure (see Note 2).

3.1. Slide Preparation

For FISH studies of metaphases, standard culturing, harvesting, and fixation procedures are used, and these are described in detail in other chapters in this book. If a rapid result is needed and can be obtained from interphase cells, it is not necessary to wait to collect cells in division: simply suspend the cells in hypotonic KCl for 10 min, fix using 3:1 absolute methanol:glacial acetic acid, and then change the fixative at least three times.

Metaphase preparations can be made from both freshly fixed and archived fixed samples. In the authors' laboratory, a large collection of fixed cytogenetic material going back over 15 years has been stored. Although there will be a degree of DNA degradation during this time, its quality is usually adequate for retrospective FISH studies. Resuspend the fixed cells in fresh fixative for a few hours, centrifuge, remove the supernatant, and then resuspend in fresh fixative for a few more minutes before spreading.

Slides for FISH studies are spread in the same way as those for karyotype studies. For high-quality FISH preparations, it is essential not to spread cells too densely on the slides, as this can increase background signal levels. Adequate fixative changes are also necessary to reduce cell debris, which can adversely interfere with a FISH analysis. The presence of large numbers of dead cells can also lead to difficulties; this is most common in samples from high-count diagnostic studies, in samples that have been delayed reaching the laboratory, and also sometimes in samples collected in EDTA instead of heparin.

As the quality of the slides affects the formation of the metaphase spreads, it is necessary to use thoroughly clean washed slides. Slides can be bought precleaned or else washed in ethanol and then kept in a freezer at –20°C before being used.

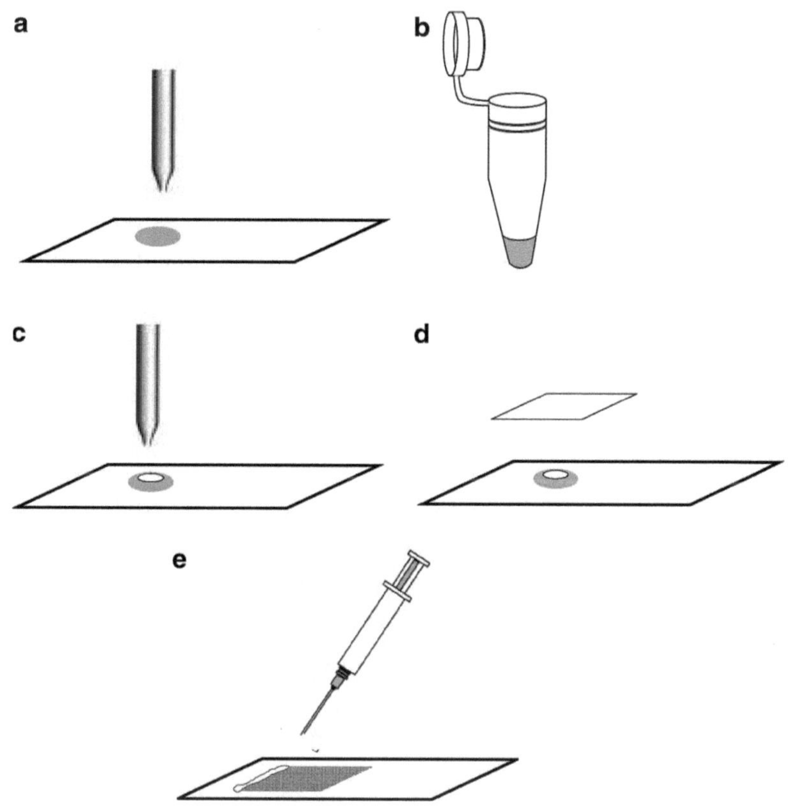

Fig. 4. (**a**) Dropping of fixed cell suspension on to a glass slide. (**b**) Probe stock stored frozen. (**c**) Probe applied to fixed cell suspension on glass slide. (**d**) Slide coverslipped. (**e**) Coverslip sealed with rubber solution.

1. Drop the fixed cell suspension (usually about 10–20 μL) on to a clean slide (see Fig. 4a).
2. Add 3–4 drops of fresh fixative on to the spread region and then leave to air-dry. It is helpful later if the area to be hybridized (usually about 20 mm diameter) is defined by scoring underneath the slide with a diamond marker.
3. Check the slide using a phase-contrast microscope. If a study of metaphases is planned, ensure that the chromosomes are well spread, with good contrast, and that there is little cytoplasm. The chromosomes should appear dark gray, not black and shiny, or pale. Some cytoplasm can be cleared, if necessary, by using a pretreatment with RNase prior to hybridization, as described below (see Note 3).
4. Slides should be left to age overnight before being treated for FISH. However, satisfactory results can be obtained from slides that are aged for just an hour in an incubator at 37°C.

If slides are being prepared for FISH studies more than a few days later, they should be stored at −20°C.

3.2. Slide Pretreatment

Freshly prepared slides do not usually require any enzyme pretreatment. However, metaphase spreads can be treated to facilitate disruption of the cell membrane and to allow efficient hybridization of the probe mixture to the target DNA; this may help when using archived fixed material (see Note 3).

3.3. Probe Dilution

Perform this procedure in reduced light. Some manufacturers supply their probe in diluted form. If this is not the case, then use the following procedure to make 10 μL of probe solution for immediate use (see Fig. 4b):

1. Remove the stock probe and hybridization buffer from the freezer and leave it in the dark for about 1 h at room temperature.
2. Flick with a finger to mix, and then briefly vortex the probe and the hybridization buffer to ensure thorough mixing.
3. Briefly pulse spin in the microfuge.
4. Remove 7 μL hybridization buffer and add to a labeled microfuge tube.
5. Add 2 μL sterile water.
6. Remove 1 μL DNA probe and add to the microfuge tube.
7. Flick the microfuge tube with fingers and briefly vortex.
8. Pulse spin in the microfuge for 10 s at 140 *g*.

The procedure can be adapted to dilute all of the probe as soon as it is received in the laboratory, instead of diluting small quantities as needed (see Note 4).

3.4. Co-denaturation and Hybridization

The target DNA on the slides is denatured, i.e., the DNA is rendered single-stranded to allow hybridization with the fluorescently labeled probe. Ordinarily, DNA needs prolonged exposure to temperatures of greater than 90°C to denature. However, using formamide, an organic solvent that is contained in the hybridization buffer, allows denaturation to take place at lower temperatures.

It is usual to leave hybridization to take place overnight. If an urgent result is needed, then the hybridization time can sometimes be reduced to as little as 2 h.

Hybridization is most efficiently done automatically using a programmable hotplate, such as a Thermobrite™ or Hybrite™, and use of equipment such as this is recommended. Instrument-specific usage instructions will be provided by the supplier and only a brief description is given below (see Subheading 3.4.2).

If such equipment is not available, then hybridization can be done perfectly satisfactorily in a humidified chamber, as described in Subheading 3.4.3.

3.4.1. Adding the Probe

1. Warm the slide(s) at 37°C for an hour in an incubator.
2. Take the slides through the alcohol series, i.e., 70, 85, and 100% ethanol for 2 min each.
3. 10 μL of probe is usually enough to cover the cells spread on the slide. Pipette the probe slowly onto the slide (see Fig. 4c).
4. Immediately place a 22 mm × 22 mm coverslip gently over the area (see Fig. 4d). Be careful to avoid the formation of air bubbles, as the probe will not hybridize uniformly around a bubble.
5. Seal around the edge of the coverslip with rubber solution. (The rubber solution can be squeezed directly from the tube onto the slide, or aspirated into a 5-mL syringe and squeezed out through a blunt needle.) (see Fig. 4e).

3.4.2. Hybrite Method

At this stage the slides are placed in a Hybrite, if one is available. This machine typically has four hybridization programs available. For FISH studies of fixed cells from blood or bone marrow, it is usual to run the program that uses a melt temperature of 75°C for 2 min, whereas for FISH studies of a section, 85°C for 5 min is usually more appropriate. The machine then lowers the temperature to 37°C and the slides are left for as long as needed, usually overnight.

3.4.3. Hotplate Method

If a Hybrite is not available, heat up the hotplate and ensure that the temperature is 75°C. (Slightly higher is acceptable but do not exceed 80°C.) Warm the humidified chamber to 37°C in the incubator. Place the slide on the hotplate for 2 min, then transfer to the humidified chamber, and leave at 37°C for as long as needed, usually overnight.

3.5. Posthybridization Washes and Signal Detection

After hybridization is complete, unbound probe is removed by a series of washes. These washes are usually carried out in a slightly more stringent solution than the hybridization buffer, to denature and remove weakly bound probe (see Note 5). This should leave only the positively bound probe-target DNA. It is important that the slides are prevented from drying out. Please note that all the wash buffers should have pH between 6.8 and 7.5

1. Place a Coplin jar containing 0.4× SSC and 0.3% NP40/ Igepal in a water bath and bring the temperature to 75°C.
2. Prepare a Coplin jar containing 2× SSC and 0.1% NP40/ Igepal at room temperature.

3. Carefully remove glue from around the coverslip using fine forceps.
4. Gently tap the slide on the side of a plastic beaker until the coverslip falls off. The coverslip must be removed gently from the slide to avoid damage to the cells. If the coverslip does not come off readily, check for residual glue which may be still holding it in place. If no traces of glue are found, then soak the slides in 1× SSC for 1–2 min at room temperature to gently lift the coverslip off the slide.
5. Place the slide(s) (no more than two at a time) into the Coplin jar containing 0.4× SSC and 0.3% NP40/Igepal at 75°C for 2 min.
6. Agitate for a few seconds.
7. Remove the slide and place it in a Coplin jar containing 2× SSC and 0.1% NP40/Igepal at room temperature for at least 1 min.
8. Remove the slide and leave to air-dry in the dark.

3.6. Counterstaining

1. Add 10–20 μL DAPI stain to a 22 × 50-mm cover slip.
2. Place cover slip (stain side down) on top of the FISH slide.
3. Carefully blot the slide to remove excess mountant solution.
4. It can be helpful to place the slide in a fridge for about 20 min before viewing under the microscope.
5. The slides may be stored for up to 6 months if kept at 4°C in the dark.

3.7. Assessing the Result

It can take a while for eyes to adjust to see fluorescence; the light emitted by fluorochromes is often very low, and it will be easier to see if the microscope is in a dark room or is surrounded by dark curtains.

It is prudent to systematically examine several fields across the slide if the first field examined does not appear to have any signals. However, the FISH procedure is not infallible and sometimes it fails to give a result even when performed in an experienced laboratory. The most likely causes of failure are listed in Note 6.

The fluorescence lasts for a brief time, after which it fades. The amount of light produced by fluorochromes is often very limited, especially if the probe is very small; sometimes it cannot be seen clearly by the human eye. Therefore, it is often necessary to photograph the cells with a digital camera, so that a record can be made before the fluorescence fades. Also, the digitized image can be enhanced to make the colors brighter. This is usually done using a computer program specifically designed for cytogenetic studies.

Do not issue a report if there is any doubt about the validity of the result. If the hybridization is poor and scoring is difficult, then do not hesitate to repeat the study using fresh slides.

3.8. Screening and Analysis

Visualization and analysis of FISH signals from larger probes, such as chromosome paints, alpha satellite probes, and yeast artificial chromosomes or bacteria artificial chromosomes, can be successfully effected using a simple epifluorescence microscope with appropriate filter sets. Such a microscope has an ultraviolet (UV) light source as well as white light. The pathway for each kind of light is different, with white light passing through the slide and UV light being projected onto it. The detection of smaller signals (e.g., from phages or cosmids) may require the help of a computer-based image analysis system: a video camera or a low light charge-coupled device (CCD) camera is used to create a digital image, which is enhanced by the computer using image-analysis software. Some systems are also able to control the filters, the camera, the exposure times, and the microscope focusing. The combination of CCD cameras and image analysis makes it possible to process very faint signals and produce images with remarkable clarity. Two of the major manufacturers are Applied Imaging International, and Metasystems, GmbH, Germany.

The slide is screened at low power to locate suitable cells, i.e., those with low background and with clear signals. It is good practice to take several photographs of each study as a record. A single photograph is usually composed of several images: these are usually collected using a digital camera mounted on the microscope, each image taken through a different filter, with the computer controlling the exposure time for each filter. The computer analyzes these images, calculates the contribution made through each filter, and produces a pseudo-colored image based on the combined data.

If the probe is large, the signals strong, and the hybridization efficient, then it is possible to screen a slide by eye and score the number of signals in each nucleus or metaphase. If the signals are small, or faint, or if the hybridization has been poor, then it can be difficult to see the signals by eye, and it may be necessary to capture (photograph) each nucleus or metaphase and then use a computer to enhance the image. A repeat study with careful attention to the process may be better than struggling with a poor preparation.

The hybridization efficiency can vary across a slide, so choose an area with well-spaced cells, low background, and clear signals. Systematically work across the area and record the signal pattern for each nucleus. Enough cells should be scored to give a clear and unambiguous result. For example, if a patient with a possible diagnosis of chronic myeloid is being screened and the first 20 cells all show a *BCR–ABL1* fusion, then it is not necessary to do more to confirm the diagnosis. Conversely, if the same patient is being studied after treatment, when low levels of positivity would be expected, then several hundred cells may need to be screened.

No report should be issued that is based on the analysis of just one person. Every case should be scored by two people, at least one of whom is an experienced analyst. The second scorer should not know the result obtained by the first. Afterward, the scores are compared: if the difference could be clinically significant, then a third person should be called on to undertake a further analysis.

Records should be kept of the hybridization efficiency of all the probes kept in stock. A probe with a reduced hybridization efficiency may fail to provide signals for all the genes in a cell, giving an underestimate of the true incidence. A control study (see Note 2) will indicate the expected distribution of signals. Traditionally, the mean number of normal/abnormal control signals has been determined, and then a range of ±3 standard deviations has been calculated. However, this custom presupposes that loss follows a normal distribution, when in fact it is more likely to follow a Poisson distribution. Any test results should be outside this range before they can be accepted as being significant. This is particularly important when screening using a single-color probe. In reality, it can be impractical to determine a precise "normal" range for each stock of probe, scored on different tissues by different analysts.

Scoring for loss of a single-color signal is particularly imprecise. If a cell appears to have just one signal, then the cell should be checked at different levels of focus through a single-band pass filter to ensure that the other signal is really missing. Note that two signals that happen to be on top of each other will be indistinguishable. A cell may also appear to have only one signal if there has been a failure of hybridization. It can also happen that a cell can lose a chromosome through random nondisjunction. For all of these reasons, it can be impossible to detect low levels of clonal loss, and many laboratories have decided that the level of loss has to be at least 10% before it can be regarded as being significant. It is safer to score loss using a break-apart probe, or a dual-color probe with both being located on the same chromosome. It is common practice, for example, to screen for monosomy 7 and deletion of the long arms of a chromosome 7 (7q-) by using a pair of probes, one located at the 7 centromere and one in the area commonly deleted.

Scoring gain of signals is more reliable. Some cells in a bone marrow aspirate may be tetraploid, and so a cell with three signals might be a tetraploid cell in which two of the four expected

Fig. 5. (continued) to that in (**c**), having five red (*RUNX1*) signals. However, the metaphase shows that there is one normal chromosome 21, and an isochromosome 21 that has multiple copies of the *RUNX1* signal on each arm. This amplification of the region around the *RUNX1* gene has been associated with a poor prognosis; (**f**) same probe: this nucleus has an *ETV6–RUNX1* fusion signal consistent with the presence of a t(12;21)(p13;q22), two red (*RUNX1*) signals on the normal and abnormal chromosomes 21, and loss of the green *ETV6* signal from the normal 12.

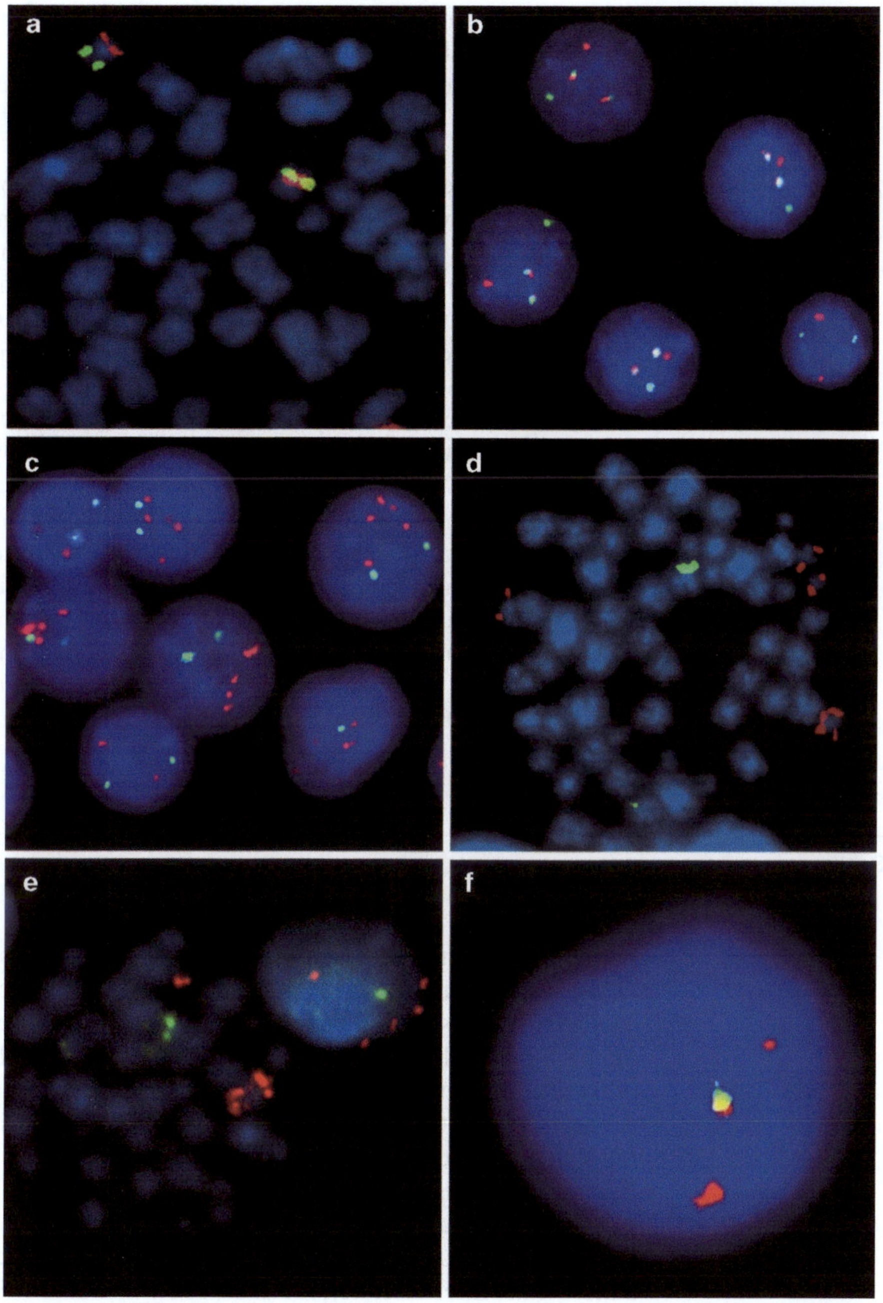

Fig. 5. Examples of FISH signal patterns (Vysis probes from Abbott). (**a**) Break-apart probe: in this illustration hybridizing to the *CBFB* gene locus at 16q22. The normal probe is composed of a red and green part, often appearing to have a yellow overlap. At the top of the metaphase is an inverted 16 with a split *CBFB* gene, showing the separation of the two parts of the probe; (**b**) dual-fusion probe, with a red signal on the *CCND1* gene at 11q13, and green on the *IGH@* gene at 14q32. The *lower right nucleus* has the normal signal pattern: two red and two green. The other cells all have one red, one green and two fusion signals, indicating a reciprocal translocation between the two genes; (**c**) dual-color probe, with the red signal located on the *RUNX1* gene on chromosome 21, and the green signal on the *ETV6* gene on chromosome 12. Three of these nuclei have the normal pattern, with two pairs of signals. The rest have up to three extra *RUNX1* signals; (**d**) same probe; same case as (**c**). This metaphase shows that there is a normal 21 (*on the left*), and two isochromosomes 21 (*on the right*), each with two *RUNX1* signals; (**e**) same probe: the pattern in the interphase nucleus is similar

signals are again superimposed. However, this coincidence is likely to occur at only a low frequency.

Some examples of different kinds of signal patterns are shown in Fig. 5.

3.9. Interpreting Results

It can be unsafe to report results in isolation; it is wise to be aware of the results of other laboratory tests. For example, a negative result can have two meanings: either the clonal cells do not have the abnormality being tested, or else there may be no clonal cells present in the sample. Knowledge of any immunophenotyping and/or morphological studies on the same sample can be particularly helpful.

It is not possible to identify all the different types of cells in a bone marrow aspirate using only the information given by DAPI stain. However, the shapes of the nuclei can give an indication of the quality of the sample being studied: if the sample is a follow-up from a patient with chronic myeloid leukemia, then the nuclei that are clearly from mature lymphocytes should be excluded from the score. These cells are likely to be an indication that the bone marrow sample was heavily contaminated with blood. Conversely, if the patient has a diagnosis of chronic lymphocytic leukemia, then the scoring of a bone marrow should exclude cells with polymorphic nuclei.

4. Notes

1. Filters: A filter is needed for each fluorochrome being used, e.g., DAPI, FITC, and Texas Red. A dual-band pass or triple-band pass filter block is also recommended so that two or three colors can be seen simultaneously. The bandwidth of a filter is the range of wavelengths at which at least 50% of the light is transmitted. A broad band filter allows more light to reach the specimen, so the signals appear brighter, but they tend to fade more quickly. Narrow band pass filters will produce fainter signals, but there will be less fading and less extraneous fluorescence.

   Most probes can be adequately identified using the broad-band Pinkel filter sets, i.e., for UV (405 nm), blue (490 nm), and yellow (570 nm). If more exotic fluorochromes are used, then the appropriate specific filter sets for the emission and excitation wavelengths will be needed.

   Although the advantage of using multiband pass filters is that more than one color can be visualized on the slide at one time, the signals emitted by the probes are less bright than that in single-band pass filters. This is because each filter reduces the amount of light falling onto the specimen.

   When scoring complex signal patterns, it will often be necessary to swap between single-band pass filters, when each color can be clearly viewed in isolation, and a dual or triple-band pass filter, when more than one color can be visualized simultaneously.

2. Quality control: It is essential to check that a new batch of probe is exactly what was ordered: a mistake can happen at any stage between an order being prepared in a lab and the probe being dispatched by the supplier. A way of checking the probe is described below.

   If several slides for FISH studies are being processed at the same time, and especially if different probes are being used, it is all too easy for an error to creep in. If the sample being studied contains no metaphases, or if the analyst is not a cytogeneticist, the mistake may not be discovered until it is too late, if at all. Meticulous record-keeping is essential: there should be an audit trail form on the lab bench, so that each slide is positively identified, and the batch number of each probe used is logged.

   Although commercially available probes are continually improving, and are produced to a high and consistent standard, the laboratory must always run its own controls before using the results from a new batch of probe. This will confirm that the correct probe has been supplied. At a minimum, the new probe should be tested on good-quality normal metaphase chromosomes to confirm that it locates to the right position. Ideally, it should be tested on metaphase chromosomes from an archive case that is already known to be positive for the abnormality being tested.

   In addition, batches of probe tend to vary in signal strength and in hybridization efficiency, and it may be necessary to establish new cut-off levels between background and positive signals. This is particularly important in the analysis of interphase nuclei, where it is necessary to distinguish between true and false positives and negatives. A probe with a low hybridization efficiency will produce fewer signals, giving an underestimate of the number of targets present. This can result in a false-negative result, i.e., a negative result from a sample that should have scored positive.

   As well as assessing variation between batches of probes, there should be an assessment of variation between individuals who are performing the scoring (5). Useful guidelines on assessing the cut-off values and determining the sensitivity of FISH analysis have been described (6).

3. Slide pretreatment: Three methods for slide pretreatment are given below; if the slides are over a month old or have been destained, then try method 1 or 2; if there is cytoplasm over the chromosomes, then try method 3.

   Method 1: Immerse slides in 2× SSC at room temperature for 30 min and rinse briefly in an ethanol series, then air-dry.

Method 2: Rinse through an acetic acid series diluted in water: 50, 70, and 100%. Rinse in an alcohol series (70, 85, and 100% absolute ethanol) and then air-dry.

Method 3: Enzyme digestion with RNase (RNase A, stock solution 10 mg/mL). Add 10 μL stock to 1 mL 2× SSC to give 100 μg/mL in 2× SSC. Place 100 μL on slide, add a coverslip, and incubate for 1 h in a humid chamber at 37°C. Rinse briskly in two jars of 2× SSC at room temperature for 3 min in each jar.

The response of the target cells to the above treatments varies according to the age of the slides. It may be necessary to experiment with different techniques and exposure times to obtain optimum results. Be warned that there is always a risk of losing the material from the slide, so run a test using a case with plenty of spare material.

4. Dilution of probes: An entire stock of probe may be mixed before the first use, so that it does not have to be freshly made up for each FISH procedure. Each batch of probe is supplied with a vial of hybridization buffer.
   (a) Add 150 μL of hybridization mixture to the probe.
   (b) Add 40 μL of sterile water.
   (c) Replace the cap on the probe container, ensuring that it is tight.
   (d) Vortex vigorously for a few seconds, or mix thoroughly by hand.
   (e) Label the probe container "mixed" and the date and the identity of the person making the dilution.
   (f) Store in a freezer at −20°C.

   Always check the supplier's instructions. However, if the probe has large, strong signals, it is usually possible to get good results using a dilution slightly greater than is recommended.
5. Background noise: The stringency of post-hybridization washes is a description of how severe the washing process is. It has to be a compromise between insufficient washing, which will leave a high level of background "noise" due to probe being attached to DNA outside the area of interest, and excessive washing, which will give cleaner but fainter signals. The stringency of washes is affected by temperature as well as by the composition of the mixture. In general, high stringency is associated with high temperature and low salt concentration, while low stringency is associated with low temperature and high salt concentration.
6. Troubleshooting: Commercial DNA probes for FISH studies are usually supplied with instructions about how to prepare them for use. The impression is sometimes given that a result

is guaranteed if the instructions are followed, and indeed this is often the case. However, problems do arise more often than the inexperienced might expect.

The most commonly encountered problems are high background and poor signal strength. There can be many reasons for this, from incorrect dilutions of the fluorochromes through to inadequate posthybridization washing procedures. Lack of any signal may be due to incorrect probe concentration through to labeling problems. An excellent guide to troubleshooting was contained in the Oncor FISH manual (Oncor, Gaithersburg), now out of print but which may still be available in some centers. In addition, most commercial probe kits usually contain troubleshooting guidelines.

The most common causes are:

(a) The sample is of poor quality; e.g., the cells may be dying due to age or because they were collected into EDTA instead of heparin.
(b) Faulty or incorrect probe was supplied;
(c) The probe has become contaminated;
(d) The probe has degraded (deteriorated) with age;
(e) Faults with the technical procedure, usually due to incorrect preparation of solutions, wrong pH, inaccurate temperature control, or omission of a stage during processing;
(f) Faults with the analysis system, due to incorrect filters, aging of the UV light source, incorrect alignment of bulb, and quenching of the fluorochrome due to very long exposure to light.

References

1. Tkachuk D., Westbrook C., Andreeff M., Donlon T., Cleary M., Suranarayan K., Homge M., Redner A., Gray J., and Pinkel D. (1990) Detection of BCR–ABL fusion in chronic myelogenous leukemia by two-color fluorescence in situ hybridization. *Science* **250**, 559–562.
2. Swansbury J. (ed) (2003) Cancer Cytogenetics: Methods and Protocols. Humana Press, New Jersey.
3. Dohner H. (1994) Detection of chimeric BCR–ABL genes on bone marrow samples and blood smears in chronic myeloid and acute lymphoblastic leukemia by in situ hybridization. *Blood* **83**, 1922–1928.
4. Muhlmann J., Thaler J., Hilbe W., Bechter O., Erdel M., Utermann G., and Duba H.-C. (1998) Fluorescence in situ hybridization (FISH) on peripheral blood smears for monitoring Philadelphia chromosome-positive chronic myeloid leukemia (CML) during interferon treatment: A new strategy for remission assessment. *Genes, Chromosomes and Cancer* **21**, 90–100.
5. Dewald G.W., Stallard R., Alsaadi A., Arnold A., Blough R., Ceperich T.M., Elejalde B.R., Fink J., Higgins J.V., Higgins R.R., Hoeltge G.A., Hsu W.-T., Johnson E.B., Kronberger D., McCorquodale D.J., Meisner L.F., Micale M.A., Oseth L.A., Payne J.S., Schwartz S., Sheldon S., Sophian A., Storto P., Van Tuinen P., Wenger G.D., Wiktor A., Willis L.A., Yung J.-F., and Zenger-Hain J. (2000) A multicenter investigation with D-FISH BCR/ABL1 probes. *Cancer Genetics and Cytogenetics* **116**, 97–104.
6. Drach J., Roka S., Ackermann J., Zojer N., Schuster R., and Fliegl M. (1997) Fluorescence in situ hybridization: Laboratory requirements and quality control. *Laboratory Medicine* **21**, 683–685.

Chapter 4

Chronic Myeloid Leukemia: Cytogenetic Methods and Applications for Diagnosis and Treatment

Christine M. Morris

Abstract

Chronic myeloid leukemia (CML) is a clonal myeloproliferative disease caused by recombination between the *BCR* gene on chromosome 22 and the *ABL1* gene on chromosome 9. This rearrangement generates the *BCR–ABL1* fusion gene that characterizes leukemic cells in all CML cases. In about 90% of cases, the *BCR–ABL1* rearrangement is manifest cytogenetically by the Philadelphia (Ph) chromosome, a derivative of the reciprocal translocation t(9;22)(q34;q11.2). For the remaining cases, recombination may be more complex, involving *BCR*, *ABL1*, and genomic sites on one or more other chromosomal regions, or it may occur cryptically within an apparently normal karyotype. Detection of the Ph and associated t(9;22) translocation is a recognized clinical hallmark for CML diagnosis. The disease has a natural multistep pathogenesis, and during chronic phase CML, the t(9;22) or complex variant is usually the sole abnormality. In 60–80% of cases, additional cytogenetic changes appear and often forecast progression to an accelerated disease phase or a terminal blast crisis. Because new frontline therapies such as imatinib specifically target the abnormal protein product of the *BCR–ABL1* fusion gene to eliminate *BCR–ABL1* positive cells, there is a new reliance on the cytogenetic evaluation of bone marrow cells at diagnosis, then at regular posttreatment intervals. Combined with other parameters, presence or absence of Ph-positive cells in the bone marrow is a powerful early indicator for clinical risk stratification. Cytogenetic changes detected at any stage during treatment, including in the *BCR–ABL1*-negative cells, may also provide useful prognostic information. Laboratory methods detailed here extend from initial collection of peripheral blood or bone marrow samples through cell culture with or without synchronization, metaphase or interphase harvest, hypotonic treatment and fixation, slide preparation for G-banding or fluorescent in situ hybridization (FISH), and final interpretation.

Key words: Chronic myeloid leukemia, Philadelphia chromosome, BCR–ABL1, t(9;22)(q34;q11.2), Karyotype evolution, Cell culture, Cytogenetics, GTG-banding, FISH, Minimal residual disease

1. Introduction

Chronic myeloid leukemia (CML) is a clonal myeloproliferative disease that arises following neoplastic transformation of a pluripotent bone marrow stem cell, and is characterized by excessive

Lynda J. Campbell (ed.), *Cancer Cytogenetics: Methods and Protocols*, Methods in Molecular Biology, vol. 730, DOI 10.1007/978-1-61779-074-4_4, © Springer Science+Business Media, LLC 2011

growth of myeloid cells and their progenitors (1). In Western countries, CML accounts for 15–20% of all leukemias, primarily affects older adults, and has an age-adjusted incidence rate of 1.5/100,000 individuals per year (2). Classically, the disease has a multistep pathogenesis and progresses through three recognized phases: chronic phase, accelerated phase, and blast phase (3, 4). Most patients are diagnosed during chronic phase, often serendipitously, with clinical symptoms when they occur including malaise, weight loss, and splenomegaly. Laboratory findings at diagnosis include peripheral blood leukocytosis, with basophilia and eosinophilia. Platelet counts are usually moderately elevated and leukocyte alkaline phosphatase activity is reduced. Bone marrow cellularity is increased, with granulocytic and often megakaryocytic hyperplasia, eosinophilia, basophilia, and suppression of other cellular elements. Marrow myelofibrosis occurs in about one-third of cases (1, 3).

CML develops in the genomic context of the bone marrow stem cell nucleus when two genes, *ABL1* and *BCR*, recombine to form a *BCR–ABL1* fusion gene. This rearrangement characterizes all cases and is typically marked by presence of a Philadelphia chromosome (Ph) in the leukemic cells. The Ph has great historical interest because it was the first cytogenetic marker found to recurrently characterize a clinically delimited group of human neoplasms (5). With the introduction of chromosome banding methods in 1970 (6), the Ph was identifiable as one product of a reciprocal exchange between the long arm of chromosome 22 and the end of the long arm of chromosome 9, represented t(9;22)(q34;q11.2) (7). Cytogenetic derivatives of the standard t(9;22) include a shortened chromosome 22, der(22)t(9;22)(q34;q11.2) (22q- or Ph) and an elongated chromosome 9, der(9)t(9;22)(q34;q11.2) (9q+) (Fig. 1).

At a molecular level, DNA breakage within the *BCR* and *ABL1* genes, followed by exchange and rejoining (recombination) of the broken ends gives rise to two fusion products: a 5′*BCR*–3′*ABL1* fusion gene on the der(22) that is transcribed and translated in all cases of CML, and a 5′*ABL1*–3′*BCR* fusion product on the der(9) that is transcribed in about 70% of cases (8, 9) (Fig. 2). The Ph and underlying *BCR–ABL1* rearrangement are classically associated with CML, but can also occur in other leukemias, particularly acute lymphoblastic leukemia (ALL) (see Chapter 8). The fundamental leukemogenic influence of the *BCR–ABL1* fusion gene has been studied extensively and demonstrated through a variety of experimental cell line and animal models (10).

1.1. BCR, ABL1, and Fusion Genes

Both *BCR* and *ABL1* are large genes, at approximately 138 kilobase pairs (kb) and 174 kb, respectively (Fig. 3). The *ABL1* gene harbors 11 exons, with alternative first exons 1a or 1b spliced to

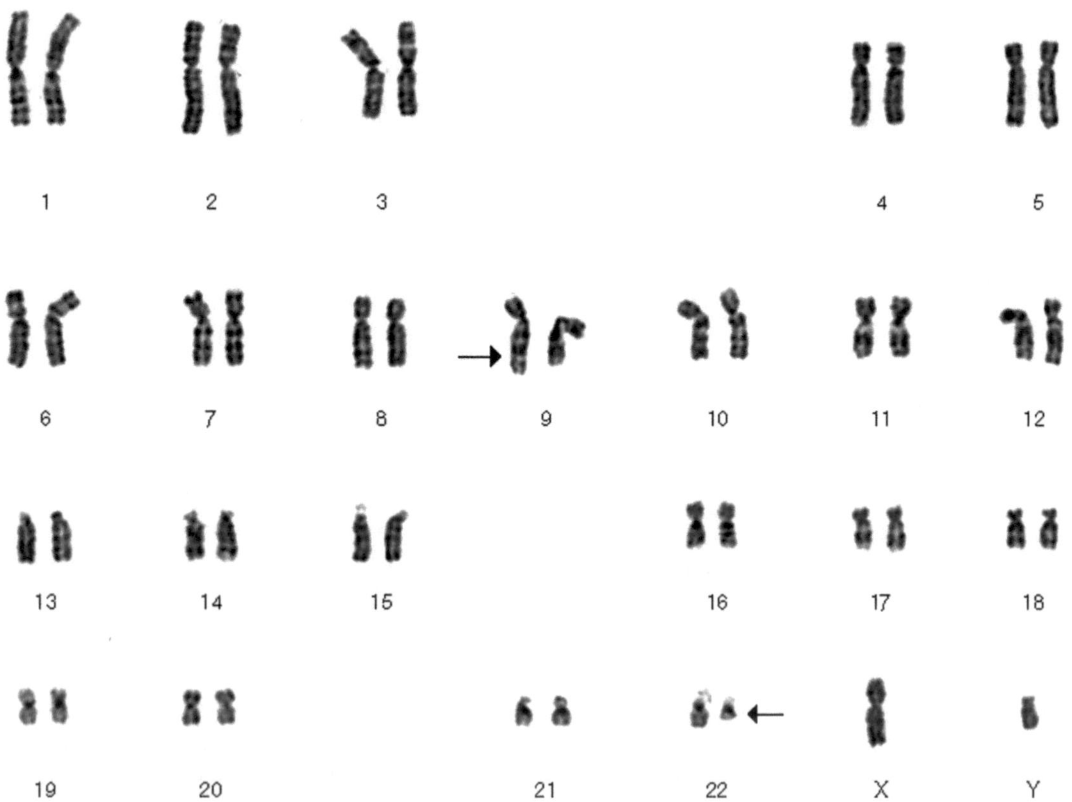

Fig. 1. Representative G-banded karyotype of a leukemic bone marrow metaphase cell showing t(9;22)(q34;q11.2). *Arrows* mark the shortened derivative chromosome 22 (Ph, 22q- or der(22)) and the elongated derivative chromosome 9 (9q+ or der(9)).

the common exons 2–11, and is expressed as either a 6- or 7-kb mRNA transcript. The *BCR* gene region spans 23 exons with alternative exons 1 and 2, and two transcripts of 4.5- and 7-kb are expressed. Both *BCR* and *ABL1* transcripts are expressed ubiquitously in most healthy human tissues and developmental stages (11). The ABL1 protein encodes a nonreceptor protein tyrosine kinase that localizes to both cytoplasm and nucleus, and has regulatory roles in DNA damage and cell stress response, cell differentiation, cell division, cell adhesion, and cell-cycle regulation (12). The normal BCR protein also localizes to both cytoplasm and nucleus, and complex functional domains implicate roles in intracellular phosphorylation and GTP-binding signaling pathways (13).

The significant molecular consequence of the Ph translocation is relocation of the 3′ part of *ABL1* from chromosome 9 to chromosome 22, adjacent to and telomeric of the truncated 5′ part of *BCR*, and in the same transcriptional orientation (Fig. 2). Preferential recombination sites occur in *ABL1*, but breakpoint locations overall within this gene are more variable than those in *BCR*, extending across a >200 kb region from 5′ of the entire

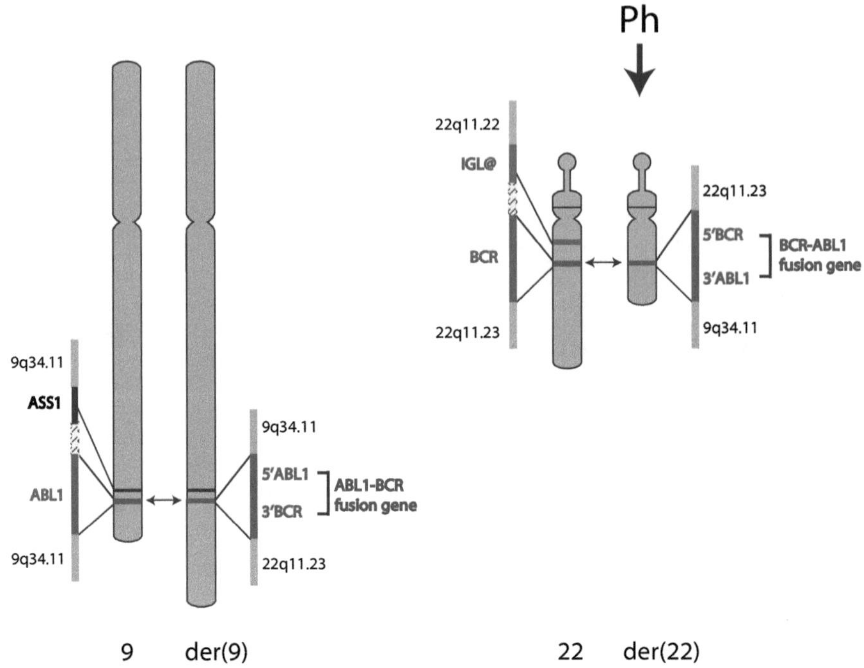

Fig. 2. Stylized representation of cytogenetic and molecular consequences when the *BCR* and *ABL1* genes break and recombine to form the t(9;22)(q34;q11.2). Normal chromosomes 9 and 22 are to the *left* and derivative chromosomes to the *right* of each chromosome pair. *Double-headed arrows* mark breakpoint locations within *ABL1* and *BCR*, respectively.

gene to exon 2 (14) (Fig. 3). By contrast, sites of breakage within *BCR* are tightly clustered, with most occurring within the 5 kb major breakpoint cluster region (M-Bcr) (15) (Fig. 3). The *BCR–ABL1* fusion gene is transcribed and spliced as a large 8 kb mRNA with *BCR* exon 13:*ABL1* exon 2 (e13a2) and/or *BCR* exon 14:*ABL1* exon 2 (e14a2) junctions, which is then translated to form a 210-kDa BCR–ABL1 fusion protein with acquired leukemogenic properties (8). In rare cases of CML, breaks occur more proximally in the minor breakpoint cluster region (m-Bcr) located between exons 1 and 2 of *BCR* (Fig. 3), to form an e1a2 *BCR–ABL1* transcript and 190-kDa protein. This variant is more commonly found in ALL, but was recently associated with poor outcome when found in CML (16). In a different subgroup of CML patients, breakage in *BCR* may occur downstream of exon 19 in a region designated μ-Bcr, to form the e19a2 mRNA transcript and a larger 230-kDa protein (17). The e19a2 *BCR–ABL1* variant is associated with neutrophilia, a lower white cell count than usual and with slower progression to blast crisis (18).

Because the standard t(9;22)(q34;q11.2) is a reciprocal exchange, a second fusion product is formed on the der(9)t(9;22) when the disrupted 5′*ABL1* gene joins with the translocated 3′

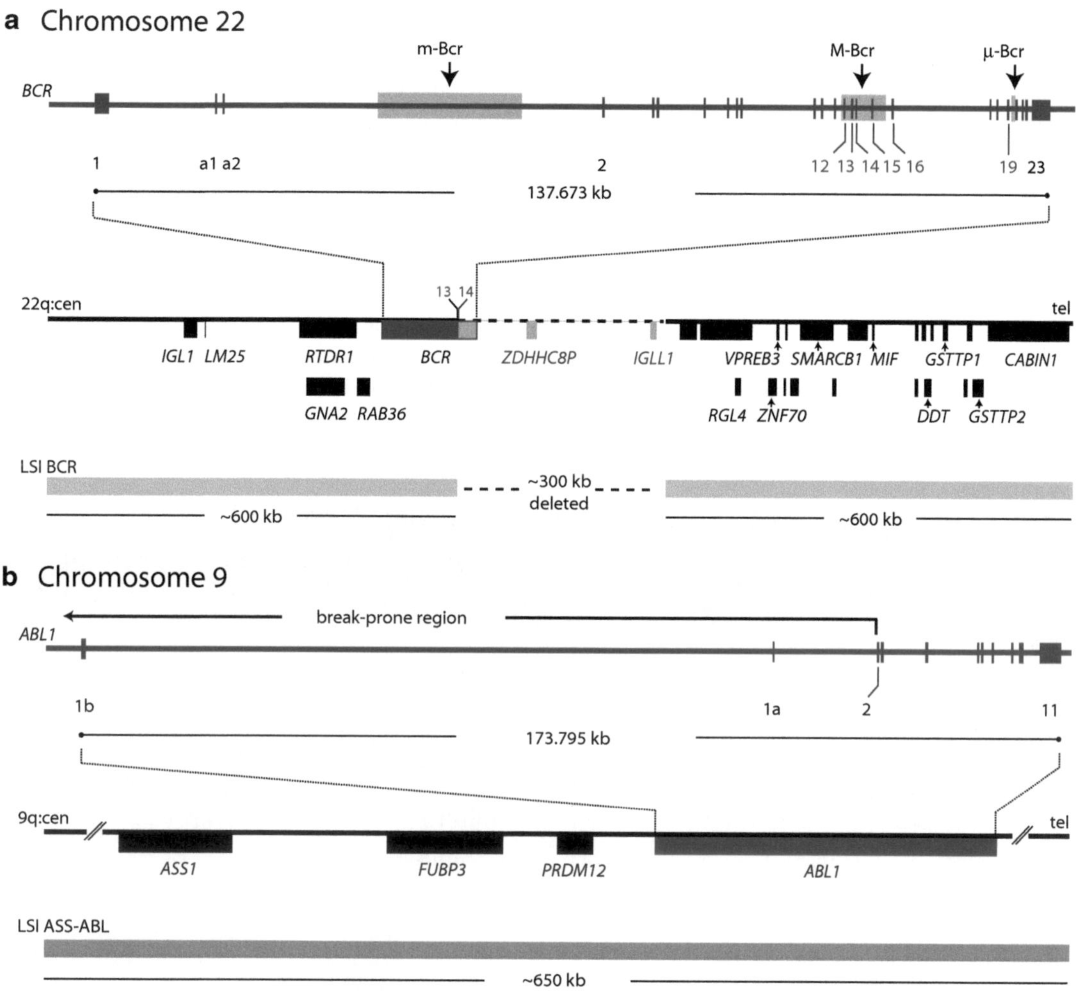

Fig. 3. Schematic representation of intron–exon structures of *BCR* (**a**) and *ABL1* (**b**), respectively, including breakpoint regions (*top*) together with genomic context on chromosomes 22q and 9q (*middle*) across regions spanned by the Vysis BCR/ABL FISH probe regions (*lower*). All RefSeq genes are marked and selectively named (68). LSI probe sizes and corresponding gene maps are to scale although exact 5′ and 3′ flanking endpoints of LSI BCR and LSI ASS-ABL are uncertain (http://international.abbottmolecular.com/; accessed 4 February 2010).

part of *BCR* (Fig. 2). This *ABL1–BCR* hybrid is transcribed in the leukaemic cells of about 70% of CML cases, but the transcript is not found in the remaining cases. This is because *ABL1–BCR* fusion is not possible for some variant translocations when 5′*ABL1* and 3′*BCR* are on different chromosomes, when large deletions extending from within *ABL1* or *BCR* occur, or for different conformational reasons ((8, 9, 19); see also below). Opinions are currently mixed as to whether or not the *ABL1–BCR* transcript, when present, is translated into a stable *ABL1–BCR* fusion protein (19, 20).

1.2. Variant Ph Translocations and Complex BCR–ABL1 Rearrangements

In about 10% of cases, *BCR–ABL1* fusion arises from more complex rearrangements than those that give rise to the standard t(9;22). These complex variants typically involve *BCR*, *ABL1*, and one or more additional chromosomal sites (21, 22). Cytogenetically, these variants may manifest as simple translocations involving chromosome 22 and a chromosome other than 9, or as complex translocations involving one or more chromosomes in addition to 9 and 22 (23). In some cases, the Ph chromosome may be masked by attachment of part of another chromosome (24). An apparently normal karyotype may also occur in leukemic cells of CML, when the *BCR–ABL1* fusion is detected using more sensitive molecular techniques, and is the result of complex submicroscopic genomic rearrangements (25–28) (Fig. 4). All chromosomes have

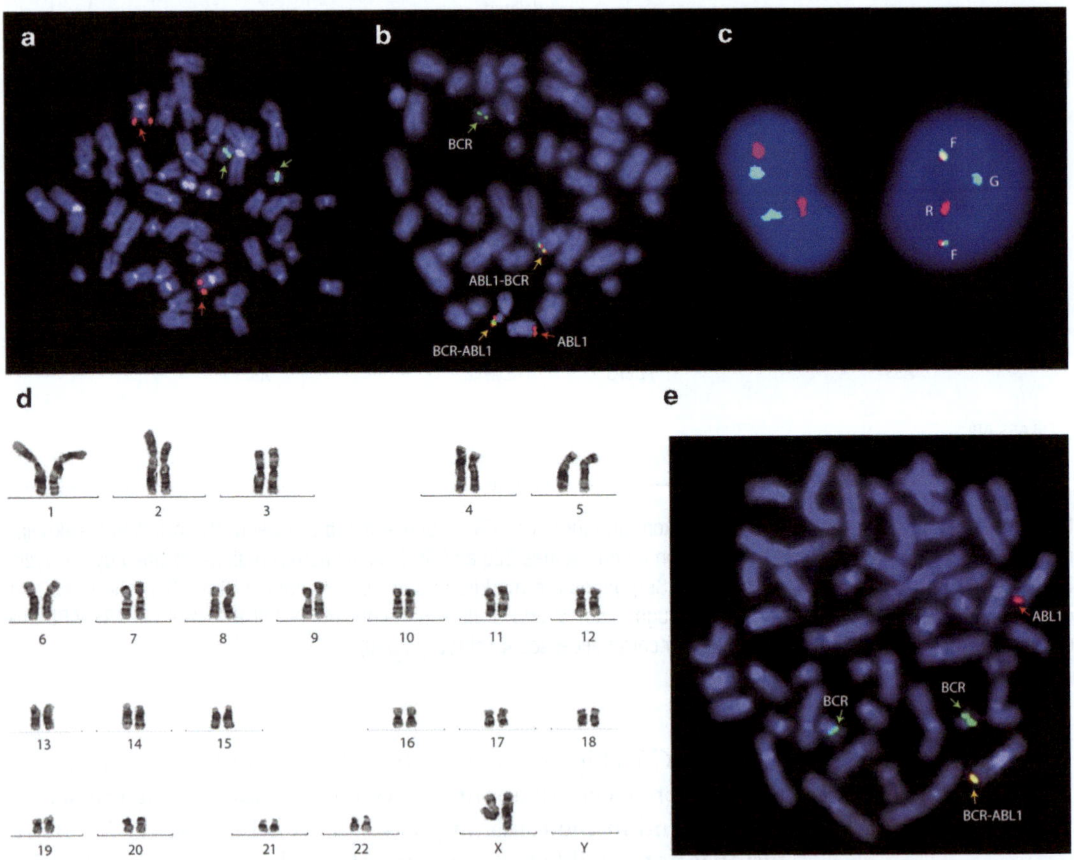

Fig. 4. Representative FISH images (**a**–**c**) showing signal patterns observed after FISH of the Abbott Vysis LSI BCR/ABL Dual-Color, Dual-Fusion Translocation Probe to DAPI counterstained normal 46,XY and *BCR–ABL1* rearrangement positive 46,XX,t(9;22)(q34;q11.2) metaphase cells (**a**, **b**), and interphase nuclei (**c**, *left* and *right*), respectively. *ABL1* or *BCR* signals on normal homologues are marked with *red* or *green arrows* on metaphase cells and with "R" or "G" on interphase nuclei, respectively. *BCR–ABL1* and *ABL1–BCR* fusion signals are as indicated on the t(9;22) metaphase and marked "F" on the interphase nucleus. An example of a cryptic *BCR–ABL1* rearrangement in an apparently normal 46,XX leukemic karyotype (**d**) is shown after FISH with the Vysis LSI BCR/ABL probe (**e**). In this case, 5′ *BCR* is presumed to have inserted adjacent to the disrupted 3′ part of *ABL1* within 9q34 to form the *BCR–ABL1* fusion gene. Green signals corresponding to residual 3′ *BCR* sequences are predicted to remain on the aberrant chromosome 22 recombinant.

been shown to participate in variant translocations, but there is a nonrandom pattern of involvement with breakpoints preferentially located in G-band light staining regions of the genome that are GC-rich and gene-rich (29). The prognostic impact of variant *BCR–ABL1* rearrangements remains controversial (30, 31).

1.3. Genomic Deletions

In ~10–15% of CML cases, deletion of large segments of genomic DNA may occur proximal to *ABL1* and/or distal to *BCR* as a consequence of *BCR–ABL1* rearrangement. These deletions were first identified fortuitously after the development of new fluorescent in situ hybridization (FISH) probe systems for detecting minimal residual disease (MRD) in interphase cells of CML patients (19, 32). The deletions are usually associated with the der(9) of the standard t(9;22) or with additional partner chromosomes in complex variant *BCR–ABL1* rearrangements. A recent study showed heterogeneous deletion sizes ranging from 230 kb to 12.9 Mb on chromosome 22 and from 260 kb to 41.8 Mb on chromosome 9. In the same study, regions spanning large domains (~2 Mb) on both chromosomes 9 and 22 were delineated, within which the majority of breakpoints proximal to *ABL1* (90%) and distal to *BCR* (88%) occurred (33). Similar to variant Ph translocations, impact on prognosis and survival outcomes for patients presenting with these deletions remains controversial (33).

1.4. Karyotype Evolution and Clinical Progression in CML

If untreated or when treatment is unsuccessful during chronic phase, CML will typically progress and transform after 3–5 years through accelerated phase to blast crisis, an aggressive short-term disease of 3–6 months duration that inevitably terminates fatally. CML at this stage resembles acute leukemia and is usually refractory to therapy (1, 3). Transformation from chronic to acute phase CML is evidenced, and often forecast, by evolution of the leukemic karyotype in up to 80% of cases (34). Diverse cytogenetic abnormalities are observed in addition to the t(9;22), both structural and numerical, either singly or in combination, with marked nonrandom involvement of certain chromosomes. The most common chromosomal abnormalities include +8 (34% of cases with additional changes), +Ph (30%), i(17q) (20%), +19 (13%), −Y (8% of males), +21 (7%), +17 (5%), and monosomy 7 (5%) (34). If not necessarily causal agents of transformation, these cytogenetic abnormalities are clear evidence for subliminal gene rearrangements that may drive transformation from chronic to acute phase (4).

Treatment interventions may also influence the spectrum of cytogenetic abnormalities observed in CML over time, with an increased frequency of unusual secondary changes following bone marrow transplantation or interferon therapies, and the identification of karyotype abnormalities in Ph-negative cells after treatment with imatinib (35).

1.5. Treatment of CML

As understanding of the molecular pathogenesis of CML has evolved, so too have therapeutic options for this disease become more rationally designed and administered (36–38). First tangible success came with the introduction of the alkylating agent busulphan (1950s) and later hydroxyurea (1970s) for CML treatment. Both drugs assisted to reduce leukocyte counts and produce hematological remission, although cytogenetic remission was not achieved and disease inevitably progressed from chronic phase to blast crisis (38). In contrast to these chemotherapies, interferon-α (IFN-α) treatments introduced in the 1980s were successful in extending survival times by a median of about 20 months (37). However, this benefit was restricted to chronic phase patients with favorable hematological features at presentation and minimal prior treatment. Importantly, IFN-α was found to induce cytogenetic responses in a significant number of patients, with up to 25% achieving a complete cytogenetic response (i.e., no Ph-positive metaphases) (39, 40).

The introduction of hematopoietic stem cell transplantation (SCT) proved another breakthrough in the management of CML, with unprecedented disease-free survival times extending beyond 20 years (41). Allogeneic SCT has been particularly effective, with the inherent graft-vs.-leukemia effect suggesting underlying immunological control of leukemic cell proliferation. Of note, both cytogenetic and molecular remissions are substantially more common following allogeneic SCT than after IFN-α (37). In fact, treatment of CML by SCT provided the first indication that it was possible to completely eradicate the Ph-positive leukemic clone (41). However, this treatment option is only available to a limited proportion of CML patients due to the advanced age at diagnosis, requirement to find suitable donors and need to consider the associated significant risks of early mortality and morbidity.

Once the molecular basis was known, including the critical role of constitutive BCR–ABL1 tyrosine kinase activity in CML pathogenesis, the development of new rationalized therapies designed to specifically target this activity, present only in the malignant cells, was quickly realized. The prototype drug imatinib was developed in the late 1990s, and has dramatically changed the prognosis of patients with newly diagnosed CML and their treatment algorithms. Latest outcomes of the IRIS trial [International Randomized Study of Interferon vs. STI571 (imatinib mesylate)], in which imatinib was first used for newly diagnosed CML patients, have shown 83% event-free survival at 6 years, and an estimated overall survival rate of 88% increasing to 95% when only CML-related deaths were considered (42). Because of this success, and with few exceptions, IFN-α is no longer used as initial therapy and SCT is mostly reserved for those who fail treatment with imatinib and related new agents (43).

Imatinib is now the recommended frontline therapy for CML, but not all patients respond well and resistance can occur. This resistance can arise during chronic phase, when it is usually driven by BCR–ABL1 independent mechanisms, or it may manifest with advancing disease as one or more acquired mutation(s) within the *BCR–ABL1* tyrosine kinase domain (44, 45). More than 100 different point mutations have been reported to date, but of these only a few represent 60–70% of all mutations. The T315I mutation, for example, has been described in 4–15% of patients resistant to imatinib, and is most frequently found in CML patients who progress to advanced stage disease with poor survival outcome (45, 46). Molecular screening, usually by direct sequencing, can detect tyrosine kinase mutations within the *BCR–ABL1* kinase domain providing the mutant clone comprises 10–20% of the sampled cell population (47). With raised awareness that imatinib resistance can be a problem, newer more potent tyrosine kinase inhibitors such as dasatonib or nilotinib have been developed to effectively induce remissions in many patients who fail imatinib therapy (48).

1.6. Diagnosis and Disease Monitoring of CML

The cytogenetic picture of CML provides unique information that is crucial to diagnosis, forecast of prognosis, and initial treatment of individual patients. Irrespective of therapeutic modalities, follow-up cytogenetic assessment on a regular basis also offers biologically rational verification of treatment response and likelihood of transition to acute phase. In this imatinib era, the achievement and maintenance of a complete cytogenetic response (i.e., no Ph-positive cells) is strong evidence of an optimal response to treatment (Table 1). In addition to examination of bone marrow metaphase cells (cytogenetic response), diverse clinical trial outcomes support complementary response definitions based on blood cell counts and differential (hematological response), assessment by real-time quantitative polymerase chain reaction (RT-QPCR) of the total *BCR–ABL1* transcript burden in peripheral blood buffy coat cells relative to an internal control gene, and *BCR–ABL1* kinase domain mutation status (molecular response) (43). In some situations, most particularly when a complete cytogenetic response is achieved, interphase FISH may substitute metaphase analysis and provide a more sensitive estimate of *BCR–ABL1*-positive cells at a given sampling time (49). However, like RT-QPCR, it is locus-specific and cannot substitute the global genomic information that karyotype analysis provides. Array technologies offer future promise for the assessment of genome-wide expression or high-resolution DNA copy-number characteristics in CML for more effective individualized medical intervention, but their place in the clinical arena has yet to be defined (50, 51).

In context of the above, and with awareness that current optimal treatment and prognostic risk stratification for CML relies on a

Table 1
Cytogenetic evaluation of CML: observations and clinical indications

Cytogenetic observation	Definition	Clinical indication
Diagnosis		
t(9;22)(q34;q11.2)	Ph+ve with standard translocation	CML diagnosis confirmed
t(9;22)(q34;q11.2) with additional abnormalities (e.g., +8, +Ph, i(17q))	Ph+ve with karyotype evolution	Suspect clonal progression and accelerated disease
der(9)t(9;V)(q34;V) and/or der(22)t(V;22)(V;q11.2)	Possible variant Ph translocation (simple, complex, or masked)	Request FISH to confirm *BCR–ABL1* status and resolve variant rearrangement
Normal karyotype or both cs 9 and cs 22 morphologically normal but with other cs abnormality	Possible cryptic complex *BCR–ABL1* rearrangement	Request FISH to confirm *BCR–ABL1* status and resolve complex rearrangement
Response to imatinib treatment (early chronic phase CML)[a]		
0% Ph+ve metaphases	Complete cytogenetic response	≤12 mos: optimal
1–35% Ph+ve metaphases	Partial cytogenetic response	3 mos: optimal 6 mos: optimal 12 mos: suboptimal 18 mos: probable failure
36–65% Ph+ve metaphases	Minor cytogenetic response	3 mos: optimal 6 mos: suboptimal 12 mos: probable failure
66–94% Ph+ve metaphases	Minimal cytogenetic response	6 mos: suboptimal 12 mos: probable failure
≥95% Ph+ve metaphases	No cytogenetic response	3 mos: suboptimal 6 mos: probable failure
Response to imatinib any time during treatment[b]		
Loss of complete cytogenetic response		Probable failure
Karyotype evolution of Ph+ve cells		Probable failure
Other cs abnormalities in Ph−ve (*BCR–ABL1* −ve) cells (e.g., +8, −7 or del(7))		Possible therapy-related myelodysplastic disease

cs chromosome, *+ve* positive, *−ve* negative, *mos* months after initiation of treatment, *V* one or more other participating chromosomes

[a]Adopted from ref. 43

[b]Adopted from refs. 35 and 43

combination of response definitions, the remainder of this chapter specifically addresses cytogenetic methods, including FISH, and their application to diagnosis and treatment monitoring.

1.7. Cytogenetic Monitoring of CML

Cytogenetic analysis of bone marrow or peripheral blood cells allows identification of the Ph translocation and other chromosome changes that are associated with the leukemic process. As the leukemic cells are naturally dividing, the addition of mitogens such as phytohemagglutinin is unnecessary, and in CML may stimulate nonleukemic populations of lymphocytes. The t(9;22)(q34;q11.2) is best visualized following cell culture and when leukemic cells are arrested at metaphase, the stage of cell division when chromosomal DNA is most contracted. Because of the varied sensitivity of leukemic cells to culture conditions, more than one protocol is adopted to ensure that the true cytogenetic status is reflected in the chromosome analysis. These different protocols are also directed at improving the mitotic index and chromosome morphology. Time in culture with supplementary growth factors combined with addition of cell synchronization agents such as methotrexate (MTX) or fluorodeoxyuridine (FdU) typically increases quality and yield of analyzable cells (52–55).

During metaphase, chromosomes are identifiable as morphologically discrete entities when fixed and spread on a microscope slide. After treating with trypsin and staining with Giemsa, identification of the t(9;22) and other abnormalities is achieved by karyotype analysis, when chromosomes are arranged according to international nomenclature systems (56) (Fig. 1). In the formal clinical laboratory setting, usually at least 20 G-banded metaphases are analyzed from cultured leukemic cell preparations, and for assessment of cytogenetic response to imatinib, the number of Ph-positive metaphase cells measured as a percentage of total cells analyzed (Table 1)

1.8. FISH Monitoring of CML

FISH is a natural extension of conventional cytogenetic procedures and enables microscopic assessment of the number, size, and location of specific DNA sequences in single cells (57, 58). The technique is applicable to a variety of cytological samples, including fixed preparations of interphase or metaphase chromosome spreads, interphase nuclei of tumor touch preparations, and bone marrow smears. Because FISH enables the study of interphase nuclei for genomic aberrations without the need for cell culture, some limitations associated with conventional cytogenetic techniques may be avoided. Target cell preparations are first pretreated with a protease such as pepsin to remove unwanted proteins that may obstruct probe access to chromosomal DNA target sites, then nuclear DNA is denatured and hybridized with fluorescently labeled and denatured DNA probes. Signal patterns in individual cell nuclei are subsequently visualized by fluorescent microscopy.

Requirement for cytogenetic and FISH assessment of *BCR–ABL1* status in CML varies according to disease stage and treatment (49, 59, 60). Both applications are important at diagnosis prior to initiation of treatment. At this stage, karyotype assessment of bone marrow will establish presence of the t(9;22)(q34;q11.2), complex Ph translocation variants, and any additional karyotype changes that may suggest advancing disease. FISH of the same cytogenetic preparation using appropriate *BCR* and *ABL1* probes will confirm *BCR–ABL1* fusion gene status and assist to identify unusual metaphase and interphase signal configurations consistent with complex *BCR–ABL1* variants or associated genomic deletions. Such prior knowledge of unusual signal conformations is useful when FISH is subsequently applied for MRD assessment.

Many different combinations of commercial or home-brew *BCR* and *ABL1* probes have been applied successfully by FISH to interphase or metaphase nuclei, but greatest sensitivity has come from the application of dual-color, double-fusion probes that detect both *BCR–ABL1* and *ABL1–BCR* fusion gene signals (32). Modern probes typically include a green-labeled *BCR* probe (G) and a red-labeled *ABL1* probe (R). Fusion of parts of the *BCR* and *ABL1* gene is represented in a yellow fusion (F) signal. Normal interphase nuclei will show a 2R2G pattern, whereas cells harboring the standard t(9;22)(q34;q11.2) show a 1R1G2F pattern (Fig. 4). Atypical FISH signal patterns indicative of submicroscopic der(9) deletions or complex variants include 1R2G1F, 2R1G1F, and 1R1G2F (e.g., see Fig. 4), and >2 F signals is consistent with increased *BCR–ABL1* copy number suggestive of clonal progression. Different locus-specific, centromeric or chromosome painting probes may be applied using FISH for the characterization of complex variants including der(9) deletions, to confirm clonal frequencies when additional karyotype changes are found or to clarify ambiguities.

2. Materials

2.1. Specimen Collection

1. Peripheral blood: 3 mL lithium-heparin vacutainer tubes (BD, NJ, USA).
2. Bone marrow transport medium: RPMI 1640 or F-10 nutrient mixture (Ham) (no HEPES or sodium pyruvate) (Invitrogen, Carlsbad, CA, USA), with L-glutamine and sodium bicarbonate, supplemented with 40 μg/mL preservative-free gentamicin sulfate and 10 U/mL preservative-free sodium heparin. Sterilize if necessary by membrane filtration (0.45 μM) and dispense 5 mL aliquots into 10 mL sterile

disposable round-bottom polystyrene centrifuge tubes. Store at 4°C for 6 months or according to medium expiry date.

2.2. Specimen Preparation and Estimation of White Blood Cell Count

1. Phosphate-buffered saline (PBS), Dulbecco A (Oxoid Ltd, Basingstoke, Hampshire, England) (see Note 1).
2. Turk's diluent: 3% (v/v) glacial acetic acid and 0.01% (v/v) gentian violet (see Note 2).
3. RPMI 1640 medium with L-glutamine (no HEPES or sodium pyruvate). If using powder, supplement with 2 g/L sodium bicarbonate.

2.3. 24 h Unstimulated Cell Culture ("A" Culture)

1. AmnioMAX C-100 complete medium (GIBCO, Invitrogen) supplemented to 10% (v/v) with fetal bovine serum (FBS) (GIBCO, Invitrogen).
2. Colcemid: 10 μg/mL (GIBCO, Invitrogen). *Caution* – refer MSDS (see Note 3).

2.4. Cell Synchronization ("B" and "C" Cultures)

1. RPMI 1640 medium (see Subheading 2.2, item 3) supplemented with 15% (v/v) FBS and 10% (v/v) 5637-conditioned medium (5637-CM) (see Note 4).
2. MTX: dissolve 25 mg methotrexate hydrate powder (Sigma–Aldrich, St. Louis, MO, USA) in 55 mL ddH_2O (1 mM stock solution). Dilute stock solution 1:100 with ddH_2O (10 μM working solution). Sterilize if necessary by membrane filtration (0.22 μm pore size) and dispense into aliquots. Store at −20°C for up to 1 year and working solution in use at 4°C for ≤1 month. *Caution* – refer MSDS.
3. 5-Fluoro-2′-Deoxyuridine (FdU): dissolve 25 mg FdU (Sigma–Aldrich) in 10 mL ddH_2O (100 mM). Dilute this solution 1:10 (v/v) with ddH_2O (10 mM stock solution), then dilute again 1:100 (v/v) (10 μM working solution). Filter sterilize (0.22 μm) and store aliquots at −20°C for up to 3 years. Once thawed, store working solution at 4°C for ≤1 month. *Caution* – refer MSDS.
4. Uridine: dissolve 100 mg uridine (Sigma–Aldrich) in 10 mL ddH_2O (40 mM stock solution), then dilute a further 1:100 (v/v) (400 μM working solution). Filter sterilize and store aliquots at −20°C for up to 2 years. Once thawed, store working solution at 4°C for ≤1 month.
5. Thymidine: dissolve 24.22 mg thymidine (Sigma–Aldrich) in 100 mL ddH_2O (1 mM working solution). Filter sterilize and store aliquots at −20°C for up to 1 year, working solution in use at 4°C for ≤1 month.
6. Colcemid: 10 μg/mL (GIBCO, Invitrogen). *Caution* – refer MSDS.

2.5. Cell Harvest

1. Hypotonic: 75 mM KCl in ddH_2O. Store working solution at 37°C several weeks.
2. Carnoy's fixative: 3:1 (v/v) AnalaR methanol/glacial acetic acid. Prepare fresh for each use. If capped tightly and stored at –20°C, fixative is stable for 2–3 h after preparation. *Caution* – refer MSDS.

2.6. Slide Preparation

1. Carnoy's fixative (see Subheading 2.5, item 2).
2. Microscope slides: soak precleaned glass microscope slides for a minimum of 12 h in a solution of 50 mM (0.05 N) HCl in methanol (standard grade). Immediately before use, rinse slides in SVM (clear methylated spirits) and wipe clean with lint-free cloth or tissue. *Caution* – refer MSDS.

2.7. Chromosome Banding (GTG-Banding)

1. 2× SSC (0.3 M NaCl/0.03 M Na Citrate), pH 7.0: prepare 1 L of 20× SSC stock solution (3.0 M NaCl/0.3 M Na Citrate) by mixing 175.5 g NaCl and 88.23 g Na Citrate to a volume of 900 mL with ddH_2O. Stir at room temperature until dissolved, adjust pH to 7.0 with 1 N HCl, and then add ddH_2O to 1 L. Stir to mix, sterilize by autoclaving, and store at room temperature.
2. Sorenson phosphate-buffered distilled H_2O (1.5 mM Na_2HPO_4/1.7 mM KH_2PO_4): mix 25 mL of 67 mM KH_2PO_4 and 25 mL of 60 mM Na_2HPO_4 to a final volume of 1 L in ddH_2O.
3. Trypsin solution: prepare 2% (w/v) stock solution by mixing 1 g trypsin powder (Invitrogen) with 50 mL ddH_2O until dissolved (*caution* – refer MSDS). Store 500 μL aliquots at –20°C. Prepare 0.02% trypsin working solution by thawing and diluting one aliquot to 50 mL with Sorenson phosphate-buffered distilled H_2O (see Note 5).
4. 75% (v/v) ethanol (AnalaR).
5. KaryoMAX Giemsa (Gurr's Improved R66, Invitrogen).
6. Gurr buffer, pH 6.8 prepared by dissolving one tablet (Invitrogen) in 1 L ddH_2O.

2.8. Fluorescent In Situ Hybridization

1. FISH probes: methods outlined below are for the Abbott Vysis LSI BCR/ABL Dual-Color, Dual-Fusion Translocation Probe (Abbott Laboratories, Abbott Park, IL, USA). The *BCR* component of this probe spans a tailored 1.5 Mb segment of chromosome 22 that begins within the variable segments of the immunoglobulin lambda light chain locus (*IGLV*) and extends to a region ~900 kb telomeric of *BCR*. An ~300 kb segment extending from exon 14 within M-Bcr and including the remaining 3′ part of the *BCR* gene is

deleted from this probe, which then extends a further ~600 kb towards the telomere. The *ABL1* component of LSI BCR/ABL is represented by a 650 kb segment of 9q34 that extends centromeric of the argininosuccinate synthetase gene (*ASS1*) to a region telomeric of the last *ABL1* exon. *BCR* and *ABL1* components of this probe are labeled with SpectrumGreen and SpectrumOrange, respectively (Fig. 3).

2. Pepsin: prepare 10% (w/v) pepsin stock solution in ddH_2O and freeze 1 mL aliquots at −20°C. Prepare working solution (100 μg/mL) by diluting stock solution 1:100 (v/v) with 0.01 N HCl. Working solution should be prepared immediately prior to use, allowing predigestion time of 1–2 h. Stock solution may be stored frozen for several years. *Caution* – refer MSDS.
3. 2× SSC (see Subheading 2.7, item 1).
4. Paraformaldehyde fixative 4% (w/v) in 1× PBS: prepare 50 mL volume by mixing 2 g paraformaldehyde with 22.5 mL ddH_2O, 2.5 mL 1 M $MgCl_2$, and ~3–5 drops of 1 N NaOH then standing to dissolve at 65°C for 1 h. Add 25 mL of 2× PBS, mix well then cool to room temperature. *Caution* – refer MSDS.
5. Ethanol gradient: 70, 80, and 95% ethanol (v/v) in ddH_2O. Prepare in coplin jars (50 mL) or larger volumes to fit slide rack container (250 mL). Solutions stored in clean, air-tight containers can be reused several times. *Caution* – refer MSDS.
6. Posthybridization wash I: 4× SSC/0.3% (v/v) *tert*-octylphenoxy poly(oxyethylene)ethanol (IGEPAL) (Sigma–Aldrich). *Caution* – refer MSDS.
7. Posthybridization wash II: 2× SSC/0.1% (v/v) IGEPAL. *Caution* – refer MSDS.
8. Antifade mounting medium: prepare 10 mL of 0.1% (v/v) *p*-phenylenediamine (PPD, Sigma–Aldrich) in a 9:1 mix of glycerol:1× PBS and stir on a magnetic stirrer for 1–2 h to dissolve. Adjust pH to 8.0 with carbonate–bicarbonate buffer, testing with pH paper after addition of first 12 drops (see Note 6). Antifade is also available commercially. *Caution* – refer MSDS.
9. DAPI counterstain (Sigma–Aldrich): prepare 1 mg/mL 4′,6-diamidino-2-phenylindole (DAPI) stock solution in ddH_2O and store 100 μL aliquots at −20°C. To prepare working solution (250 ng/mL), add 7.5 μL stock solution to 30 mL antifade, mix well, and store in 1 mL aliquots protected from light at −20°C. *Caution* – refer MSDS.

3. Methods

Preparation for cytogenetic or FISH interrogation of leukemic cells collected from CML patients involves a number of routinely applied steps. These extend from the collection of peripheral blood or bone marrow samples, through cell preprocessing and counting, culture with or without synchronization, harvest including cell-cycle arrest with colcemid, hypotonic treatment and fixation, slide preparation for G-banding or FISH analysis, and final interpretation.

3.1. Specimen Collection

1. Cytogenetic detection of the t(9;22)(q34;q11.2) and variants that are characteristic of *BCR–ABL1* rearrangement positive CML at diagnosis typically requires aseptically collected bone marrow. Approximately 0.2–1 mL of bone marrow is transferred immediately at the time of aspiration into 5 mL of bone marrow transport medium and then delivered to the laboratory on the same day. Approximately 5–10 mL of peripheral blood, collected into lithium-heparin tubes, may be used for FISH monitoring of MRD and in some circumstances for cytogenetic evaluation and/or detection of the *BCR–ABL1* fusion gene at diagnosis when bone marrow is not available (60, 61).
2. Because cytogenetic analysis involves culturing live cells, correct handling and transportation of samples is essential. Rapid same-day or overnight delivery at room temperature is optimal for these sample types.

3.2. Bone Marrow Sample Preparation and Estimation of White Blood Cell Count

Bone marrow or peripheral blood leukocytes are ideally cultured at a concentration of 1×10^6 cells/mL. For this reason, total white blood cell counts are determined using a hemocytometer.

1. Using a micropipettor fitted with a sterile tip, take 0.1 mL of bone marrow aspirate suspended in transport medium and place in a clean tube containing 0.9 mL of Turk's diluent ($\times 10^5$ dilution). Mix carefully, then transfer 0.1 mL of this first dilution to a second 0.9 mL volume of Turk's diluent ($\times 10^6$ dilution) and mix again.
2. Using a clean glass capillary tube, draw up an aliquot of the $\times 10^5$ dilution. Carefully fill one chamber of a hemocytometer by gently resting the end of the capillary at the edge of the chamber. Allow the sample to be drawn out of the pipette by capillary action; the fluid should run to the edges of the grooves only. Take care not to overfill the chamber. Repeat this process to fill a second hemocytometer chamber with an aliquot of the $\times 10^6$ dilution.

3. Focus on the grid lines of the hemocytometer using the 10× objective of a microscope under brightfield and with the condenser down, then count cells in all four squares of the first chamber ($\times 10^5$ dilution). Repeat for the second chamber ($\times 10^6$ dilution).
4. For each dilution, the final cell count/milliliter of the original bone marrow or peripheral blood buffy coat sample is represented in the following formula: [Total cells counted/ 4×10^5 cells/mL (first dilution)] or [Total cells counted/ 4×10^6 cells/mL (second dilution)]. If the cell count is high, a further dilution in Turk's diluent may be necessary ($\times 10^7$ dilution).
5. Discard Turk's dilutions and then mark the level of remaining bone marrow sample on tube to estimate the total volume. Multiply cells/mL and total volume to determine total number of cells in the sample.
6. Wash cell suspension by adding sterile PBS, prewarmed to 37°C, to a final volume of 10 mL, mix gently, and then centrifuge 5 min at 250–450 *g* in a clinical benchtop centrifuge (see Note 7).
7. Remove supernatant and repeat this wash step.
8. Resuspend cell pellet in RPMI 1640 to the same estimated total volume as the original sample after cell counting or so as to allow the removal of an accurate number of cells for culture (see Subheading 3.2, step 5).
9. Proceed to set up cultures (see Subheading 3.4).

3.3. Peripheral Blood Sample Buffy Coat Preparation and Estimation of White Blood Cell Count

When anticoagulated blood is centrifuged, a buff-colored coat ("buffy coat") containing most of the white blood cells and platelets forms a thin layer at the interface between a lower red blood cell layer and an upper straw-colored plasma fluid layer. Purification of buffy coat cells simplifies set-up and also reduces competitive interference from red blood cells through lysis or stimulation of cytokine/growth factor release that may not favor leukemic cell populations.

1. Mix heparanized blood sample and transfer to a sterile 15 mL polypropylene conical tube.
2. Replace cap and centrifuge sample for 5 min.
3. Discard plasma, taking care not to disturb the buffy coat interface. Remove buffy coat using a Pasteur pipette, and resuspend in 3–4 mL PBS in a 15 mL culture tube. Mark volume.
4. Estimate total available cell count, then wash cells in PBS as for bone marrow (see Subheading 3.2).

5. Resuspend cells in RPMI to the same total volume estimated after cell counting or so as to allow the removal of an accurate number of cells for culture (see Subheading 3.3, step 3).
6. Prepare to set up cultures (see Subheading 3.4).

3.4. 24 h Unstimulated Cell Culture ("A" Culture)

Bone marrow or peripheral blood samples drawn from patients diagnosed with CML in chronic phase typically have an increased number of early proliferating myeloid cells, including blast cells (1). For this reason, culture medium is tailored to suit growth characteristics of these cell populations and to attain a high mitotic index within a 24-h time frame without inducing extraneous chromosomal aberrations.

1. Working in a sterile and certified biological safety cabinet, label three 25 mL tissue culture flasks with the following information: laboratory identification number, patient name or other reference identifier as appropriate, date, set-up time, and A, B, or C to denote the culture type as 24-h unstimulated and MTX or FdU synchronized, respectively.
2. Take 10×10^6 freshly washed bone marrow or peripheral blood leukocytes and add to AmnioMAX C-100 complete medium supplemented with 10% v/v added FBS for a final culture volume of 10 mL in culture flask "A".
3. Incubate at 37°C in humidified 5% CO_2 incubator for 24 h, with the cap loosened to ensure good CO_2 exchange (see Note 8).
4. Add 50 μL of 10 μg/mL Colcemid (0.05 μg/mL final concentration) and incubate a further 30–60 min prior to cell harvest (see Subheading 3.6).

3.5. Cell Synchronization ("B" and "C" Cultures)

Synchronization of the cell-cycle of leukemic cells in culture often significantly improves both metaphase yield and quality to allow detection of subtle chromosomal rearrangements otherwise missed. Methods described below are based on those initially reported by Yunis (52) (MTX) and Webber and Garson (53) (FdU).

1. To each of the 25 mL culture flasks labeled "B" and "C", add 10×10^6 freshly washed bone marrow or peripheral blood leukocytes to a final volume of 10 mL in RPMI 1640 medium supplemented with 15% (v/v) FBS and 10% (v/v) 5637-condition medium. Incubate at 37°C for at least 24 h and up to 72 h with cap loosened.
2. Add DNA replication blocking reagents on the afternoon of the day before harvest (e.g., 3:30 p.m.). For MTX synchronization, add 100 μL of 10 μM MTX (final concentration 10^{-7} M) to flask "B". For FdU synchronization, add 100 μL of each of 10 μM FdU (final 10^{-7} M) and 400 μM uridine (final 4 μM) to flask "C". Return both flasks immediately to the 37°C incubator for 17 h (see Note 9).

3. The following morning (e.g., 8:30 a.m.), release cells from the DNA replication blocking agent with the addition of 100 µL of 1 mM thymidine to each of flasks "B" and "C". Incubate at 37°C in CO_2 atmosphere for 7 h 20 min (see Note 10).
4. Add 50 µL of 10 µg/mL colcemid (0.05 µg/mL final concentration) and incubate a further 10 min prior to harvest.

3.6. Cell Harvest

Although colcemid exposure times for metaphase arrest differ between the 24-h unstimulated and the synchronized cultures, harvest protocols thereafter are essentially the same.

1. Following Colcemid treatment, transfer cultures to 15 mL conical polypropylene centrifuge tubes prelabeled with laboratory sample number and patient identifiers when relevant.
2. Centrifuge 5 min, remove supernatant to 0.5 mL, and discard into a biohazard waste container. Loosen cell pellet into suspension by careful pipetting.
3. Add 5 mL of hypotonic solution and mix again by pipetting (see Note 11). Cap tubes and incubate for 15 min in a 37°C waterbath. Meanwhile prepare 20 mL of Carnoy's fixative for each culture in harvest and transfer to −20°C freezer.
4. At completion of hypotonic treatment, add 5 mL of freshly prepared chilled fixative to each sample and mix thoroughly. To avoid cell damage, do not vortex or pipette too vigorously.
5. Centrifuge 5 min. Remove supernatant to cell pellet then add 1 mL fixative and mix carefully by pipetting up and down to ensure that any clumped cells are dispersed. Add a further 4 mL fixative, mix again, and let stand for 10 min at room temperature.
6. Centrifuge 5 min. Remove supernatant to cell pellet then add 5 mL fixative, mix thoroughly by pipetting, or inversion of capped tube. Let stand for 10 min at room temperature.
7. Repeat the above step. At this stage, synchronized cultures are left to stand overnight at 4°C to facilitate the spread of prophase and prometaphase cells. Slides are prepared the following day after two or three more changes of fresh fixative as step 6. Slides for 24 h unstimulated cultures (culture "A", see Subheading 3.4) may be made on the day of harvesting.

3.7. Slide Preparation

1. Centrifuge cells after last fixation step for 5 min. Resuspend in 1–2 mL fresh fixative, more or less to achieve a milky cell suspension. Pipette carefully to avoid losing cells up the sides of the tube or the pipette.
2. Label precleaned glass microscope slides with sample identifier (see Subheading 3.4, step 1).

3. Breathe (exhale) onto slide until condensation occurs, then working quickly, after drawing a small volume of the fixed cell suspension into a Pasteur pipette, allow one drop to fall towards each end. For cultured preparations, the cell suspension is dropped onto the slide from a height of ~30 cm. To maximize spreading of synchronized preparations, fixed cell preparations are dropped from arms length, i.e., a height closer to 1 m.
4. Allow the slide to air dry, and inspect under 10× magnification using a brightfield microscope set for phase contrast or equivalent (e.g., condenser averted and field diaphragm setting low). If cells are evenly distributed, not overlapping and metaphase chromosomes are generally well separated and spread, proceed to G-banding (Subheading 3.8). If cell density is acceptable but spreading is poor, see Note 12.
5. If cells appear overcrowded on the slide, dilute suspension with more fixative. Conversely, if too thinly dispersed, add 2–3 mL fresh fixative to cell suspension, centrifuge 5 min, then resuspend in smaller volume and proceed with slide preparation as above. Monitor efficiency of spreading by phase microscopy as above.
6. Once cell spreading parameters have been optimized, prepare 6–8 slides for G-band analysis or FISH (see Note 13).
7. Harden slide preparations by baking 1 h at 65°C, e.g., on a heat-block (see Note 14). Store short term (1–2 weeks) at room temperature in a covered box (G-banding or FISH) or long term at –20°C in an air-tight box with a desiccant such as silica gel (FISH).

3.8. Chromosome Banding

Banding patterns are specific for each chromosome and arm, thus allowing precise identification of individual chromosomes, their homologues, and parts of chromosomes (56). One of the most commonly used chromosome banding techniques is the G-banding by trypsin using Giemsa (GTG-banding) method. This method requires pretreatment of metaphase slide preparations with the proteolytic enzyme trypsin, which selectively removes or modifies histone/nonhistone proteins so that after staining with Giemsa, light (GC-rich and gene-rich) and dark (AT-rich and gene poor) banding patterns appear along the length of each chromosome (Fig. 1).

1. Prepare a series of nine Coplin jars and, using forceps to grip the labeled end of one test slide, process through the following GTG-banding regime. All solutions are at room temperature unless otherwise stated.
   (a) Place slide in 2× SSC for 30 min at 60°C.
   (b) Remove slide and rinse by dipping up and down several times in Sorenson phosphate-buffered distilled H_2O.

(c) Drain slide briefly on paper towel and transfer to 0.02% (w/v) trypsin working solution for 10 s exactly (see Note 15).

(d) Rinse slide briefly in 75% (v/v) ethanol to deactivate trypsin.

(e) Rinse slide briefly in Sorenson phosphate-buffered distilled H_2O.

(f) Drain slide briefly and transfer to Giemsa stain diluted to 4% (v/v) in Gurr's buffer for 6–7 min (see Note 16).

(g) Transfer slide to Sorenson phosphate-buffered distilled H_2O and rinse briefly but thoroughly.

(h) Transfer slide to fresh Sorenson phosphate-buffered distilled H_2O and rinse again.

(i) Transfer slide to distilled H_2O for final rinse.

(j) Drain slide briefly, then leave to air dry at room temperature.

(k) Examine slide under brightfield microscope using 100× oil objective to assess banding quality. Repeat process for additional slides adjusting trypsin and staining times as necessary to achieve required band resolution (see Note 17).

3.9. G-Band Interpretation

Once slides have been successfully G-banded, it is necessary to analyze them. This requires careful observation using an optically well-configured brightfield microscope equipped with 10× (dry), 40× (dry), and 100× (oil) objectives. Slides are scanned systematically under low magnification using the 10× objective to identify analyzable metaphase cells (well spread, minimal overlap, and interpretable banding), then assessed in further detail at higher magnification using the 100× oil objective. Selection of cells for analysis should be as objective as possible, taking into account that mitoses with chromosomal abnormalities may have poorer morphology than those that are karyotypically normal. A typical assessment would include full G-band interpretation of five metaphase cells, and a chromosome count of a further 15 cells. Where possible, ten metaphase cells are analyzed from the unstimulated 24 h culture, and ten from the synchronized cultures. If clonal mosaicism is evident, a further ten cells should be counted. New clonal or nonclonal variant cells should be analyzed in full. Findings are recorded on an analysis sheet. If cytogenetic response to treatment is under assessment, a further 30 metaphases should be found, counted, and presence or absence of t(9;22) or variants recorded. A clonal cell line is confirmed when two or more cells show the same structural or numerical abnormality, or when three or more cells show the same missing chromosome (56).

3.10. Fluorescent In Situ Hybridization

1. All solutions are at room temperature unless otherwise specified.
2. Sample preparation: For CML diagnosis and MRD monitoring, FISH is most often performed using bone marrow or peripheral blood cells processed and spread on microscope slides according to cytogenetic methods outlined above. However, when samples are limited, FISH may be applied directly to bone marrow smears or to uncultured bone marrow or peripheral blood buffy coat cells. Bone marrow smears, if used, should be fixed in three changes, 5 min each, of freshly prepared Carnoy's fixative (see Subheading 2.5, step 2), then air-dried. For direct uncultured preparations, samples are washed and cell counts estimated (see Subheadings 3.2 and 3.3). Where possible, ~$10–20 \times 10^6$ cells are resuspended to a maximum volume of 0.5 mL in PBS. Cells are then directly exposed to hypotonic, fixed, and slides prepared as detailed above (see Subheadings 3.6, steps 3–6, and 3.7).
3. FISH probes are handled and prepared for hybridization according to the recommendations of the manufacturer. Probes should be protected from light both in storage (−20°C) and at all stages of the procedure to prevent the loss of signal intensity.
4. Pepsin pretreatment: Prepare pepsin working solution and incubate to predigest at 37°C for 1–2 h. Prepare ethanol gradient and paraformaldehyde fixative.
5. Incubate target slide preparations in prewarmed pepsin working solution for 10 min exactly (see Note 18).
6. Rinse slides in two changes of fresh 2× SSC, 3 min each.
7. Dehydrate through 70, 80, and 95% ethanol gradient, 2 min each.
8. Transfer air-dried slides to paraformaldehyde fixative for 10 min.
9. Rinse slides in two changes of 2× SSC, 3 min each.
10. Dehydrate through 70, 80, and 95% (v/v) ethanol gradient, 2 min each.
11. Air-dried slides may be stored long term after pepsin treatment at −20°C in plastic containers containing desiccant or short term (1–2 weeks) at room temperature prior to hybridization.
12. Hybridization: FISH probes prepared according to the manufacturer's recommendations are applied to a predetermined region on pretreated slides and covered with a clean glass coverslip. The final volume of probe mix may be scaled according to the coverslip size (e.g., 10 μL for 22 mm × 22 mm; 3 μL for 11 mm × 11 mm). Target chromosomal DNA and probe DNA are then codenatured by placing inside a humidified Dako Cytomation Hybridizer (Dako, Glostrup, Denmark), programmed to first heat the slides at 75°C for 5 min, then to cool to 37°C for 12–16 h to allow probe annealing (see Note 19).

13. After overnight hybridization, slides are lifted from the chamber and coverslips are removed, either directly with forceps or by soaking the slides in posthybridization wash II for 2–5 min at room temperature with gentle agitation. Slides are then washed once in posthybridization wash I for 2 min at 72°C and once in posthybridization wash II for 1 min at room temperature. Cell preparations are dehydrated through an ethanol gradient for 1 min each with gentle agitation, then air-dried in the dark before applying 10 μL antifade solution containing DAPI counterstain and overlaying with a coverslip. Slides are stored, protected from light, in air-tight containers at 4°C until required.
14. After FISH, slides are analyzed in a darkened room using a microscope equipped for fluorescence, fitted with 40× and 100× objectives, an appropriately sensitive digital camera and associated analysis software.
15. As for analysis of G-banded metaphases, accurate interpretation comes with experience, and many clinical and scientific organizations offer freely accessible guidelines for interpretation of FISH signal patterns (e.g., ref. 62). For locus-specific probes such as Vysis-Abbott LSI BCR/ABL Dual-Color Dual-Translocation Probe, at least ten metaphase cells should be scored to confirm or exclude an abnormality, and for interphase screening for malignant clones or clonal mosaicism, a minimum of 200 interphase cells should be scored (49). Increasing the numbers of cells analyzed will increase the sensitivity and accuracy of the final result.

4. Notes

1. Unless otherwise stated, all reagents are prepared using double distilled water (ddH_2O) or high-quality milli-Q purified water.
2. Turk's solution contains acid to lyse red blood cells and gentian violet to stain nuclei of white blood cells.
3. Reference to MSDS denotes hazardous substance: consult Material Safety Data Sheet and observe recommended cautionary procedures when handling.
4. The human bladder carcinoma cell line 5637 (ATCC HTB-9) secretes myeloid growth factors when grown in serum-deficient medium (63). For this reason, supplementation of short-term bone marrow or peripheral blood cultures with 5637-conditioned medium typically stimulates myeloid cell growth, with improved mitotic index and metaphase quality (64, 65). The cell line is serially passaged by trypsinization in the presence of EDTA and grows rapidly to form an adherent monolayer in

plastic tissue culture flasks. Routinely, 5637 cells are cultured in RPMI 1640 medium supplemented with 10% (v/v) FBS (see Subheading 2.2, item 3). After 48–72 h, following seeding of freshly thawed and washed cells, the conditioned medium is harvested, cell debris removed by centrifugation (10 min, 250–450 × *g*), and the resulting clear supernatant passed through a 0.2 μm filter and stored at −80°C until use. Further subculturing to loose confluency in 200 mL culture flasks followed by a change of medium and undisturbed incubation for 5 days will increase the yield.

5. Working trypsin solution should be prepared immediately prior to use, at least once daily and more frequently if deterioration of banding quality occurs. Trypsin is unstable and activity is pH and temperature dependent.
6. Carbonate–bicarbonate buffer is prepared by mixing 2 mL of a 200 mM solution of anhydrous sodium carbonate (2.12 g/100 mL) with 23 mL of a 200 mM solution of sodium bicarbonate (1.68 g/100 mL) and adding ddH_2O to a final volume of 200 mL when the pH is 9.2.
7. Unless otherwise specified, centrifugation steps are for 5 min at ~250–450 *g* in a clinical benchtop centrifuge at room temperature. Settings should be constant and will be determined by the type of centrifuge and available tubes.
8. All subsequent cell cultures are incubated in a humidified 37°C, 5% CO_2 incubator unless otherwise specified.
9. Because MTX is toxic to cells, the culture should be blocked no longer than 18 h.
10. The exact interval between release of blocking agent and harvest is critical, and determined by cell population kinetics and particularly cycling time for progression through G2 (typically about 4–6 h), to metaphase (typically ≤1 h) (53, 66, 67).
11. Hypotonic treatment times may be increased or decreased to achieve optimum metaphase spreading.
12. Laboratories vary in temperature and humidity, and it may be necessary to try several different methods to achieve optimal metaphase spreads. If metaphases are too tightly spread, options to try include (a) dropping from a greater height; (b) placing fixed cell suspension on ice or prechilling the slides on ice; (c) holding slide over steam (e.g., preheated waterbath) while dropping cell suspension onto the slide; (d) placing slides on prewet paper towels prior to dropping cell suspension.
13. Residual cells may be stored, suspended, in 1–2 mL fixative at 4°C for several months. If preparing new slides after this time, cells should be washed first in two to three changes of fresh fixative as described in Subheading 3.6, step 6.

14. The length of time and the temperature at which slide preparations are hardened will influence the quality of both G-banding and FISH.
15. Trypsin exposure times for optimal G-banding are affected by many factors including how the slides were prepared and hardened, the age of the slides, whether synchronized or not, the humidity and temperature at which the slides have been stored, and specifications of the trypsin reagent itself.
16. Time in Giemsa may vary between batches of stain. Test one or two control slides first to optimize.
17. Bands that appear solid and darkly stained are most likely undertrypsinized and can be improved by increasing the trypsin time. Bands that appear hollow and ghost-like are most likely overtrypsinized and can be improved by reducing the trypsin time.
18. Pepsin times may require adjustment according to batch. Over- or under-exposure may reduce signal strength, and titration of new pepsin stock activity using control slides is recommended.
19. The protocol detailed assumes the use of a programmable hybridization chamber. Alternatively, slide preparations may be denatured separately prior to application of the probe mix by immersing in 70% formamide/2× SSC (v/v), pH 7.0 at 72°C for 2 min exactly, followed by immediate transfer to ice-cold 70% ethanol for 1 min. Agitate to rinse off denaturing solution, then continue dehydration in ice-cold 80, 90, and 100% ethanol gradient, 1 min each, air dry, and store with desiccant at –20°C until required for FISH. Denatured probe is then applied under coverslip to denatured slides that have been prewarmed for 5–10 min on a heat-block set to 38°C, and the preparation incubated overnight at 37°C in a 2× SSC humidified chamber. If the probe mix is placed on a room temperature slide after denaturation, it may take a substantial time to reach the 37°C hybridization temperature, and irreversible binding to secondary target sites may occur. *Caution* – refer to MSDS before handling formamide.

Acknowledgments

The author acknowledges generous sponsorship of the Cancer Society of New Zealand that has made this work possible; past and present staff and students of the Cancer Genetics Research Group for their support and inspiration; Margaret McDonald and other members of the Cytogenetic and Molecular Oncology Sections, Canterbury Health Laboratories for devoted technical assistance and contributions to Figs. 1 and 4.

References

1. Vardiman, J. W., Meloni, G., Baccarani, M., and Thiele, J. (2008) Chronic myelogenous leukaemia, BCR–ABL1 positive, in *World Health Organisation Classification of Tumours of Haematopoietic and Lymphoid Tissues* (Swerdlow, S. H., Campo, E., Harris, N. L., Jaffe, E. S., Pileri, S. A., Stein, H., Thiele, J., and Vardiman, J. W. (eds.)), International Agency for Research on Cancer (IARC), Lyon.
2. SEER Cancer Statistics Review, 1975–2006, URL: http://seer.cancer.gov/statfacts/html/cmyl.html (Last updated November 2008), (Horner, M. J., Ries, L. A. G., Krapcho, M., Neyman, N., Aminou, R., Howlader, N., Altekruse, S. F., Feuer, E. J., Huang, L., Mariotto, A., Miller, B. A., Lewis, D. R., Eisner, M. P., Stinchcomb, D. G., Edwards, B. K. (eds.)), Accessed 16 February 2010.
3. Quintas-Cardama, A., and Cortes, J. E. (2006) Chronic myeloid leukemia: diagnosis and treatment. *Mayo Clin Proc* **81**, 973–88.
4. Melo, J. V., and Barnes, D. J. (2007) Chronic myeloid leukaemia as a model of disease evolution in human cancer. *Nat Rev Cancer* 7, 441–53.
5. Nowell, P. C., and Hungerford, D. A. (1960) A minute chromosome in human chronic granulocytic leukemia [abstract]. *Science* **132**, 1497.
6. Caspersson, T., Zech, L., Johansson, C., and Modest, E. J. (1970) Identification of human chromosomes by DNA-binding fluorescent agents. *Chromosoma* **30**, 215–27.
7. Rowley, J. D. (1973) Letter: A new consistent chromosomal abnormality in chronic myelogenous leukaemia identified by quinacrine fluorescence and Giemsa staining. *Nature* **243**, 290–3.
8. Deininger, M. W., Goldman, J. M., and Melo, J. V. (2000) The molecular biology of chronic myeloid leukemia. *Blood* **96**, 3343–56.
9. Goldman, J. M., and Melo, J. V. (2003) Chronic myeloid leukemia – advances in biology and new approaches to treatment. *N Engl J Med* **349**, 1451–64.
10. Wong, S., and Witte, O. N. (2001) Modeling Philadelphia chromosome positive leukemias. *Oncogene* **20**, 5644–59.
11. NCBI UniGene EST Profile Viewer (BCR UniGene Identifier Hs.517461; ABL1 UniGene Identifer Hs.431048), URL: http://www.ncbi.nlm.nih.gov/sites/entrez?db=unigene, Accessed 11 February 2010.
12. Silberman, I., Sionov, R. V., Zuckerman, V., Haupt, S., Goldberg, Z., Strasser, A., Ben-Sasson, Z. S., Baniyash, M., Koleske, A. J., and Haupt, Y. (2008) T cell survival and function requires the c-Abl tyrosine kinase. *Cell Cycle* 7, 3847–57.
13. Laurent, E., Talpaz, M., Kantarjian, H., and Kurzrock, R. (2001) The BCR gene and philadelphia chromosome-positive leukemogenesis. *Cancer Res* **61**, 2343–55.
14. Jeffs, A. R., Wells, E., and Morris, C. M. (2001) Nonrandom distribution of interspersed repeat elements in the BCR and ABL1 genes and its relation to breakpoint cluster regions. *Genes Chromosomes Cancer* **32**, 144–54.
15. Groffen, J., Stephenson, J. R., Heisterkamp, N., de Klein, A., Bartram, C. R., and Grosveld, G. (1984) Philadelphia chromosomal breakpoints are clustered within a limited region, bcr, on chromosome 22. *Cell* **36**, 93–9.
16. Verma, D., Kantarjian, H. M., Jones, D., Luthra, R., Borthakur, G., Verstovsek, S., Rios, M. B., and Cortes, J. (2009) Chronic myeloid leukemia (CML) with P190 BCR-ABL: analysis of characteristics, outcomes, and prognostic significance. *Blood* **114**, 2232–5.
17. Pane, F., Frigeri, F., Sindona, M., Luciano, L., Ferrara, F., Cimino, R., Meloni, G., Saglio, G., Salvatore, F., and Rotoli, B. (1996) Neutrophilic-chronic myeloid leukemia: a distinct disease with a specific molecular marker (BCR/ABL with C3/A2 junction). *Blood* **88**, 2410–4.
18. Verstovsek, S., Lin, H., Kantarjian, H., Saglio, G., De Micheli, D., Pane, F., Garcia-Manero, G., Intrieri, M., Rotoli, B., Salvatore, F., Guo, J. Q., Talpaz, M., Specchia, G., Pizzolo, G., Liberati, A. M., Cortes, J., Quackenbush, R. C., and Arlinghaus, R. B. (2002) Neutrophilic-chronic myeloid leukemia: low levels of p230 BCR/ABL mRNA and undetectable BCR/ABL protein may predict an indolent course. *Cancer* **94**, 2416–25.
19. Huntly, B. J., Bench, A., and Green, A. R. (2003) Double jeopardy from a single translocation: deletions of the derivative chromosome 9 in chronic myeloid leukemia. *Blood* **102**, 1160–8.
20. Zheng, X., Oancea, C., Henschler, R., Moore, M. A., and Ruthardt, M. (2009) Reciprocal t(9;22) ABL/BCR fusion proteins: leukemogenic potential and effects on B cell commitment. *PLoS ONE* **4**, e7661.
21. Fitzgerald, P. H., and Morris, C. M. (1991) Complex chromosomal translocations in the Philadelphia chromosome leukemias. Serial translocations or a concerted genomic rearrangement? *Cancer Genet Cytogenet* **57**, 143–51.

22. Morel, F., Herry, A., Le Bris, M. J., Morice, P., Bouquard, P., Abgrall, J. F., Berthou, C., and De Braekeleer, M. (2003) Contribution of fluorescence in situ hybridization analyses to the characterization of masked and complex Philadelphia chromosome translocations in chronic myelocytic leukemia. *Cancer Genet Cytogenet* **147**, 115–20.
23. Hagemeijer, A., Bartram, C. R., Smit, E. M., van Agthoven, A. J., and Bootsma, D. (1984) Is the chromosomal region 9q34 always involved in variants of the Ph1 translocation? *Cancer Genet Cytogenet* **13**, 1–16.
24. Hagemeijer, A., de Klein, A., Godde-Salz, E., Turc-Carel, C., Smit, E. M., van Agthoven, A. J., and Grosveld, G. C. (1985) Translocation of c-abl to "masked" Ph in chronic myeloid leukemia. *Cancer Genet Cytogenet* **18**, 95–104.
25. Morris, C. M., Reeve, A. E., Fitzgerald, P. H., Hollings, P. E., Beard, M. E., and Heaton, D. C. (1986) Genomic diversity correlates with clinical variation in Ph'-negative chronic myeloid leukaemia. *Nature* **320**, 281–3.
26. Hagemeijer, A., Buijs, A., Smit, E., Janssen, B., Creemers, G. J., Van der Plas, D., and Grosveld, G. (1993) Translocation of BCR to chromosome 9: a new cytogenetic variant detected by FISH in two Ph-negative, BCR-positive patients with chronic myeloid leukemia. *Genes Chromosomes Cancer* **8**, 237–45.
27. Jeffs, A. R., Benjes, S. M., Smith, T. L., Sowerby, S. J., and Morris, C. M. (1998) The BCR gene recombines preferentially with Alu elements in complex BCR-ABL translocations of chronic myeloid leukaemia. *Hum Mol Genet* **7**, 767–76.
28. Virgili, A., Brazma, D., Reid, A. G., Howard-Reeves, J., Valganon, M., Chanalaris, A., De Melo, V. A., Marin, D., Apperley, J. F., Grace, C., and Nacheva, E. P. (2008) FISH mapping of Philadelphia negative BCR/ABL1 positive CML. *Mol Cytogenet* **1**, 14.
29. Fisher, A. M., Strike, P., Scott, C., and Moorman, A. V. (2005) Breakpoints of variant 9;22 translocations in chronic myeloid leukemia locate preferentially in the CG-richest regions of the genome. *Genes Chromosomes Cancer* **43**, 383–9.
30. Gorusu, M., Benn, P., Li, Z., and Fang, M. (2007) On the genesis and prognosis of variant translocations in chronic myeloid leukemia. *Cancer Genet Cytogenet* **173**, 97–106.
31. Richebourg, S., Eclache, V., Perot, C., Portnoi, M. F., Van den Akker, J., Terre, C., Maareck, O., Soenen, V., Viguie, F., Lai, J. L., Andrieux, J., Corm, S., and Roche-Lestienne, C. (2008) Mechanisms of genesis of variant translocation in chronic myeloid leukemia are not correlated with ABL1 or BCR deletion status or response to imatinib therapy. *Cancer Genet Cytogenet* **182**, 95–102.
32. Dewald, G. W., Wyatt, W. A., Juneau, A. L., Carlson, R. O., Zinsmeister, A. R., Jalal, S. M., Spurbeck, J. L., and Silver, R. T. (1998) Highly sensitive fluorescence in situ hybridization method to detect double BCR/ABL fusion and monitor response to therapy in chronic myeloid leukemia. *Blood* **91**, 3357–65.
33. Albano, F., Anelli, L., Zagaria, A., Coccaro, N., D'Addabbo, P., Liso, V., Rocchi, M., and Specchia, G. (2010) Genomic segmental duplications on the basis of the t(9;22) rearrangement in chronic myeloid leukemia. *Oncogene* **29**, 2509–16.
34. Johansson, B., Fioretos, T., and Mitelman, F. (2002) Cytogenetic and molecular genetic evolution of chronic myeloid leukemia. *Acta Haematol* **107**, 76–94.
35. Loriaux, M., and Deininger, M. (2004) Clonal cytogenetic abnormalities in Philadelphia chromosome negative cells in chronic myeloid leukemia patients treated with imatinib. *Leuk Lymphoma* **45**, 2197–203.
36. Druker, B. J. (2008) Translation of the Philadelphia chromosome into therapy for CML. *Blood* **112**, 4808–17.
37. Silver, R. T., Woolf, S. H., Hehlmann, R., Appelbaum, F. R., Anderson, J., Bennett, C., Goldman, J. M., Guilhot, F., Kantarjian, H. M., Lichtin, A. E., Talpaz, M., and Tura, S. (1999) An evidence-based analysis of the effect of busulfan, hydroxyurea, interferon, and allogeneic bone marrow transplantation in treating the chronic phase of chronic myeloid leukemia: developed for the American Society of Hematology. *Blood* **94**, 1517–36.
38. Geary, C. G. (2000) The story of chronic myeloid leukaemia. *Br J Haematol* **110**, 2–11.
39. Pavlovsky, C., Kantarjian, H., and Cortes, J. E. (2009) First-line therapy for chronic myeloid leukemia: Past, present, and future. *Am J Hematol* **84**, 287–93.
40. Kantarjian, H. M., O'Brien, S., Cortes, J. E., Shan, J., Giles, F. J., Rios, M. B., Faderl, S. H., Wierda, W. G., Ferrajoli, A., Verstovsek, S., Keating, M. J., Freireich, E. J., and Talpaz, M. (2003) Complete cytogenetic and molecular responses to interferon-alpha-based therapy for chronic myelogenous leukemia are associated with excellent long-term prognosis. *Cancer* **97**, 1033–41.
41. Gratwohl, A., and Heim, D. (2009) Current role of stem cell transplantation in chronic myeloid leukaemia. *Best Pract Res Clin Haematol* **22**, 431–43.

42. Hochhaus, A., O'Brien, S. G., Guilhot, F., Druker, B. J., Branford, S., Foroni, L., Goldman, J. M., Muller, M. C., Radich, J. P., Rudoltz, M., Mone, M., Gathmann, I., Hughes, T. P., and Larson, R. A. (2009) Six-year follow-up of patients receiving imatinib for the first-line treatment of chronic myeloid leukemia. *Leukemia* **23**, 1054–61.
43. Baccarani, M., Cortes, J., Pane, F., Niederwieser, D., Saglio, G., Apperley, J., Cervantes, F., Deininger, M., Gratwohl, A., Guilhot, F., Hochhaus, A., Horowitz, M., Hughes, T., Kantarjian, H., Larson, R., Radich, J., Simonsson, B., Silver, R. T., Goldman, J., and Hehlmann, R. (2009) Chronic myeloid leukemia: an update of concepts and management recommendations of European LeukemiaNet. *J Clin Oncol* **27**, 6041–51.
44. Shah, N. P., Nicoll, J. M., Nagar, B., Gorre, M. E., Paquette, R. L., Kuriyan, J., and Sawyers, C. L. (2002) Multiple BCR-ABL kinase domain mutations confer polyclonal resistance to the tyrosine kinase inhibitor imatinib (STI571) in chronic phase and blast crisis chronic myeloid leukemia. *Cancer Cell* **2**, 117–25.
45. Cooper, S., Giles, F. J., and Savona, M. R. (2009) Overcoming resistance in chronic myelogenous leukemia. *Leuk Lymphoma* **50**, 1785–93.
46. Quintas-Cardama, A., and Cortes, J. (2008) Therapeutic options against BCR–ABL1 T315I-positive chronic myelogenous leukemia. *Clin Cancer Res* **14**, 4392–9.
47. Radich, J. P. (2009) How I monitor residual disease in chronic myeloid leukemia. *Blood* **114**, 3376–81.
48. Shah, N. P. (2009) Fine-tuning targeted therapy of CML. *Blood* **114**, 4914–5.
49. Testoni, N., Marzocchi, G., Luatti, S., Amabile, M., Baldazzi, C., Stacchini, M., Nanni, M., Rege-Cambrin, G., Giugliano, E., Giussani, U., Abruzzese, E., Kerim, S., Grimoldi, M. G., Gozzetti, A., Crescenzi, B., Carcassi, C., Bernasconi, P., Cuneo, A., Albano, F., Fugazza, G., Zaccaria, A., Martinelli, G., Pane, F., Rosti, G., and Baccarani, M. (2009) Chronic myeloid leukemia: a prospective comparison of interphase fluorescence in situ hybridization and chromosome banding analysis for the definition of complete cytogenetic response: a study of the GIMEMA CML WP. *Blood* **114**, 4939–43.
50. Yong, A. S., and Melo, J. V. (2009) The impact of gene profiling in chronic myeloid leukaemia. *Best Pract Res Clin Haematol* **22**, 181–90.
51. Nowak, D., Ogawa, S., Muschen, M., Kato, M., Kawamata, N., Meixel, A., Nowak, V., Kim, H. S., Kang, S., Paquette, R., Chang, M. S., Thoennissen, N. H., Mossner, M., Hofmann, W. K., Kohlmann, A., Weiss, T., Haferlach, T., Haferlach, C., and Koeffler, H. P. (2010) SNP array analysis of tyrosine kinase inhibitor (TKI) resistant chronic myeloid leukemia (CML) identifies heterogeneous secondary genomic alterations. *Blood* **115**, 1049–53.
52. Yunis, J. J. (1982) Comparative analysis of high-resolution chromosome techniques for leukemic bone marrows. *Cancer Genet Cytogenet* 7, 43–50.
53. Webber, L. M., and Garson, O. M. (1983) Fluorodeoxyuridine synchronization of bone marrow cultures. *Cancer Genet Cytogenet* **8**, 123–32.
54. Morris, C. M., and Fitzgerald, P. H. (1985) An evaluation of high resolution chromosome banding of hematologic cells by methotrexate synchronization and thymidine release. *Cancer Genet Cytogenet* **14**, 275–84.
55. Castagne, C., Muhlematter, D., Martinet, D., and Jotterand, M. (1999) Effect of conditioned medium, nutritive elements and mitotic synchronization on the accuracy of the cytogenetic analysis in patients with chronic myeloid leukemia at diagnosis and during alpha-interferon therapy. *Cancer Genet Cytogenet* **109**, 166–71.
56. Shaffer, L. G., Slovak, M. L., and Campbell, L. J. (eds.) (2009) *ISCN: An International System for Human Cytogenetic Nomenclature*, S. Karger, Basel.
57. Pinkel, D., Straume, T., and Gray, J. W. (1986) Cytogenetic analysis using quantitative, high-sensitivity, fluorescence hybridization. *Proc Natl Acad Sci U S A* **83**, 2934–8.
58. Lichter, P., Tang, C. J., Call, K., Hermanson, G., Evans, G. A., Housman, D., and Ward, D. C. (1990) High-resolution mapping of human chromosome 11 by in situ hybridization with cosmid clones. *Science* **247**, 64–9.
59. Hughes, T. P., and Branford, S. (2009) Measuring minimal residual disease in chronic myeloid leukemia: fluorescence in situ hybridization and polymerase chain reaction. *Clin Lymphoma Myeloma* **9 Suppl 3**, S266–71.
60. Tefferi, A., Dewald, G. W., Litzow, M. L., Cortes, J., Mauro, M. J., Talpaz, M., and Kantarjian, H. M. (2005) Chronic myeloid leukemia: current application of cytogenetics and molecular testing for diagnosis and treatment. *Mayo Clin Proc* **80**, 390–402.

61. Landstrom, A. P., Ketterling, R. P., Knudson, R. A., and Tefferi, A. (2006) Utility of peripheral blood dual color, double fusion fluorescent in situ hybridization for BCR/ABL fusion to assess cytogenetic remission status in chronic myeloid leukemia. *Leuk Lymphoma* **47**, 2055–61.

62. Requirements for Cytogenetic Testing (2007), National Pathology Accreditation Advisory Council (NPAAC), 2nd ed., Australian Government Department of Health and Ageing.

63. Quentmeier, H., Zaborski, M., and Drexler, H. G. (1997) The human bladder carcinoma cell line 5637 constitutively secretes functional cytokines. *Leuk Res* **21**, 343–50.

64. Vaughan, B. R., Scott, M. A., Howard, J. D., and Nacheva, E. P. (2007) Application of growth factor stimulants improves cytogenetic analysis of chronic myeloproliferative disorder patients without alteration to cell lineage or clonality. *Cancer Genet Cytogenet* **175**, 98–106.

65. Michaeli, J., Lerer, I., Rachmilewitz, E. A., and Fibach, E. (1986) Stimulation of proliferation of human myeloid leukemia cells in culture: applications for cytogenetic analysis. *Blood* **68**, 790–3.

66. Gallo, J. H., Ordonez, J. V., Brown, G. E., and Testa, J. R. (1984) Synchronization of human leukemic cells: relevance for high-resolution chromosome banding. *Hum Genet* **66**, 220–4.

67. Rapoza, P. A., Testa, J. R., Egues, M. C., and Rowley, J. D. (1984) Evaluation of techniques for increasing mitotic yield and chromosome length in leukemic cells. *Acta Cytol* **28**, 740–6.

68. UCSC Genome Bioinformatics, Human (Homo sapiens) Genome Browser Gateway, URL: http://genome.ucsc.edu/, (February 2009 (hg19) assembly), Accessed 1 Feb 2010.

Chapter 5

Cytogenetic Analysis in Acute Myeloid Leukaemia

Lynda J. Campbell and Joanne S. White

Abstract

Cytogenetic analysis is an integral part of the diagnostic work-up of the patient with acute myeloid leukaemia. Conventional cytogenetic analysis relies on obtaining a good quality bone marrow specimen in a timely fashion and setting up at least two short-term cultures. A 15–24-h culture and a 48-h synchronised culture are routinely set up but as the cytogenetics result is often required urgently to determine the type of therapy to be administered, analysis is undertaken using the overnight culture in the first instance. Rapid and accurate analysis relies on obtaining high-quality G-banding. Knowledge of the conditions affecting banding is therefore essential.

Key words: Acute myeloid leukaemia, Diagnosis, Chromosome analysis, G-banding, Synchronised cultures

1. Introduction

Over the past 40 years, cytogenetic analysis has become an integral part of the diagnosis and management of patients with acute myeloid leukaemia (AML). According to the World Health Organization (WHO) classification of haematopoietic and lymphoid tumours, the diagnosis of AML must include cytogenetic analysis to identify a series of recurrent genetic markers by which AML is subcategorised. Morphology alone is no longer sufficient to make the diagnosis (1).

The management of AML is reliant on cytogenetic analysis as karyotype is a strong independent predictor of outcome. The Medical Research Council (MRC) (UK), and Southwest Oncology Group (SWOG) and Eastern Cooperative Oncology Group (ECOG) (USA) have published AML prognosis categories based upon karyotype but with significant differences between the two groups (2, 3). A recent panel of experts from European

Lynda J. Campbell (ed.), *Cancer Cytogenetics: Methods and Protocols*, Methods in Molecular Biology, vol. 730,
DOI 10.1007/978-1-61779-074-4_5, © Springer Science+Business Media, LLC 2011

Table 1
Prognostic categories in AML (4)

Prognosis group	Chromosome and genetic abnormalities[a]
Favourable	inv(16)(p13.1q22)/t(16;16); *CBFB-MYH11*;+/− other abnormalities t(8;21)(q22;q22); *RUNX1-RUNX1T1*+/− other abnormalities Normal karyotype + mutated *NPM1* (without *FLT3*-ITD) Normal karyotype + mutated *CEBPA*
Intermediate-I	Normal karyotype (except for those in favourable category) and including: Normal Karyotype: mutated *NPM1* + *FLT3*-ITD Normal Karyotype: wild-type *NPM1* + *FLT3*-ITD Normal Karyotype: wild-type *NPM1* + no *FLT3*-ITD
Intermediate-II	t(9;11)(p22;q23); *MLLT3-MLL* Cytogenetic abnormalities not classified as favourable or adverse[b]
Adverse	inv(3)(q21q26)/t(3;3); *RPN1-EVI1* t(6;9)(p23;q34); *DEK-NUP214* t(v;11)(v;q23); *MLL* rearranged (except t(9;11) – see above) del(5q)/-5, -7, Abnormality of 17p Complex karyotype (three or more chromosome abnormalities in the absence of recurrent inversions or translocations specified above)

[a]Acute promyelocytic leukaemia has been excluded from these prognostic categories
[b]Appropriate risk group assignment for rarer cytogenetic abnormalities remains difficult due to lack of adequate numbers of cases treated with comparable treatment regimes

Union countries, the USA and Japan has produced a series of recommendations combining elements of both the MRC and SWOG/ECOG prognosis groups and also incorporating molecular prognostic markers (see Table 1) (4).

Whilst specific genetic targets can be detected using molecular techniques such as reverse transcriptase polymerase chain reaction, cytogenetic analysis provides a general, albeit low-powered, overview of genetic rearrangements within the malignant cell. There are a number of different methods available for producing metaphase spreads suitable for cytogenetic analysis. The methods described below are in routine use in the Victorian Cancer Cytogenetics Service (VCCS). The number and type of cultures established depend on the diagnosis but good practice dictates the setting up, where possible, of at least two cultures for each sample. For most myeloid disorders, a culture using an agent to produce cell synchrony is preferred and so, if limited patient material enables the establishment of only one culture, a synchronised culture is recommended. With new acute leukaemias, the subtype may not have been identified by the time the cultures are established and so, at the VCCS, an overnight un-synchronised culture is established preferentially and, if sample permits, a synchronised culture is set up as second choice. Synchronisation requires a 48-h culture and acute lymphoblastic leukaemia cells are generally less capable than AML cells of surviving more than 24 h in culture.

Cell synchronisation refers to a method of increasing the number of cells that have reached the metaphase stage of the cell cycle when harvesting of the culture is initiated. An agent is added to the culture to block DNA synthesis; the following method uses 5′-fluorodeoxyuridine (FdU) with added uridine. By blocking the cell cycle, a large number of cells are collected that have all arrived at the same point of division. Release of the blockage then allows them all to proceed through division together. FdU acts as an antagonist to thymidylate synthetase. 5-Bromo-2′-deoxyuridine (BrdU), an analogue of thymidine, is then used to release the block. The harvest is timed for approximately 7 h after release so that the maximum number of cells has arrived at metaphase (5). Metaphase is the stage of mitosis at which chromosomes are most contracted prior to the chromatids separating to travel to opposite poles of the cell; they are most distinguishable one from another at this point.

Cells are harvested by the addition of Colcemid®, a synthetic analogue of colchicine, which blocks the formation of the cell spindle fibres and thus prevents the onset of anaphase. The cells are treated with a hypotonic saline solution to increase the cell volume and so allow the chromosomes to disentangle from each other. Finally, the cells are fixed in a mixture of methanol and glacial acetic acid, which hardens and prevents degradation of the cellular structures and, most particularly, the nuclear DNA. The fixed suspension is dropped onto glass slides, which are then air-dried and ready for either G-banding or fluorescence in situ hybridisation (FISH) studies. G-banding stands for Giemsa banding, named after the German chemist Gustav Giemsa, but the characteristic banding pattern of dark and light bands along each chromosome can be induced using a number of different stains. The method described here uses Leishman's stain. Prior to G-banding, the slides require "ageing." When banding methods were first introduced, it was discovered that G-bands could only be induced in chromosome preparations after the slides had been allowed to sit for several days or even weeks. The pressures on clinical cytogenetics laboratories today require rather speedier turn-around times than this leisurely practice would allow. It has therefore been necessary to accelerate the ageing process. The method below describes the heating of slides on a 100°C hotplate for a few minutes to age slides prior to immersing them in a weak trypsin solution and staining with Leishman's stain to produce G-banding. After air-drying, the slides are then cover-slipped and are ready for cytogenetic analysis (6).

The following methods outline (a) the establishment of both overnight and synchronised cultures of bone marrow cells from patients with AML; (b) the harvesting of cells from each culture after varying periods of incubation; and (c) slide making and the production of banded metaphase spreads for microscopic analysis.

2. Materials

The materials required for each step have been listed in separate sections according to the stage at which they are required.

2.1. Materials Required for Bone Marrow Culture and Harvest

1. Bone marrow aspirate, 0.5–1.0 mL, collected in a sterile syringe or tube containing approximately 100 international units of preservative-free heparin.
2. RPMI 1640 medium – with L-glutamine and HEPES (see Note 1).
3. Foetal calf (bovine) serum (FCS): stored at –20°C in aliquots (i.e. 40 mL aliquots for addition to 200 mL bottles of medium to produce an approximately 18% solution). This should be thawed and added to medium immediately before use.
4. L-Glutamine–penicillin–streptomycin (PSG): a 1 mL mixture containing 200 mM L-glutamine, 10,000 international units of penicillin, and 10 mg streptomycin per millilitre. PSG is stored at –20°C and is added to a 200-mL bottle of medium immediately prior to use.
5. Sterile 50 mL tissue culture flasks.
6. Sterile pipettes.
7. FdU (MW = 246.2) and uridine (MW = 244.2) working solution: 1 mL uridine solution (4×10^{-3} M) and 1 mL FdU solution (5×10^{-5} M) added to 8 mL sterile distilled H_2O (dH_2O) to give working concentrations of 5×10^{-6} M FdU and 4×10^{-4} M uridine (stored at 4°C).
8. BrdU (MW = 307.4): a 2×10^{-3} M working solution (stored at 4°C).
9. Colcemid®: 10 μg/mL solution.
10. Hypotonic solution (potassium chloride 0.075 mol/L): a 5.59 g/L solution of KCl (MW = 74.55).
11. Carnoy's solution: 3:1 (v/v) analar methanol/glacial acetic acid, made fresh just before use.
12. 15 mL plastic, non-sterile screw-top centrifuge tubes.
13. Microscope glass slides: 76 mm × 26 mm superfrost slides with ground edges.

2.2. Materials Required for Banding

1. *Trypsin*: desiccated tryptic enzyme rehydrated in 10 mL sterile dH_2O and aliquots stored at –20°C. The amount of desiccated trypsin provided by the manufacturer is based on

its enzyme activity rather than weight and so the solution is prepared according to the manufacturer's instructions to produce a 5% solution of trypsin (1:250). A working solution is made fresh daily by adding 0.25 mL trypsin solution to 70 mL trypsin diluent (see below) in a coplin jar.

2. *Trypsin diluent*: NaCl (8.0 g), KCl (0.4 g), Na_2HPO_4 (0.06 g), KH_2PO_4 (0.06 g), and $NaHCO_3$ (0.5 g) dissolved in 2 L dH_2O and stored at 4°C prior to use.
3. *Ca^{2+}/Mg^{2+}-free solution*: NaCl (8.0 g), KCl (0.2 g), Na_2HPO_4 (1.15 g), and KH_2PO_4 (0.2 g) dissolved in 1 L dH_2O and stored at 4°C prior to use.
4. *Leishman's solution* (eosin methylene blue compound): stored in powdered form at room temperature (see Note 2). Leishman's stain powder (2.8 g) is added to 1 L analytical grade methanol and mixed with a magnetic stirrer for 3 h. The resulting solution is incubated at 37°C for at least 1 week. To prepare a working solution, 100 mL is filtered through Whatman No. 1 filter paper. The remainder is stored in the dark at room temperature until required. The filtered Leishman's stain is then diluted one in ten in stain buffer (see below) immediately prior to use.
5. *Stain buffer*: one Gurr® buffer tablet (pH 6.8) is dissolved in 1 L dH_2O and stored at 4°C prior to use.
6. *DPX mounting solution*: stored at room temperature.
7. *Glass coverslips*: 24 mm × 50 mm in size.

3. Methods

3.1. Conventional Cytogenetics

3.1.1. Culture 1: Overnight Culture

1. Place 10 mL RPMI 1640 medium supplemented with PSG and 18% FCS (20 mL FCS added to a 200-mL bottle of medium) into a sterile 50-mL tissue culture flask.
2. Using a sterile pipette, inoculate medium with appropriate amount of bone marrow (see Note 3).
3. Lie flask flat and incubate at 37°C for 15–24 h (see Note 4).
4. Add 0.2 mL of 10 μg/mL Colcemid® to culture and incubate at 37°C for a further 30 min.
5. Transfer culture to harvesting tube and centrifuge for 10 min at approximately 200*g* in a sealed bucket centrifuge (see Note 5).
6. Discard supernatant.
7. Resuspend cell pellet in 8 mL of KCl and place in 37°C water bath for 20 min.
8. Centrifuge for 10 min at 200 *g*.
9. Discard supernatant and resuspend pellet. It is critical that the pellet be resuspended thoroughly at this point.

10. Gradually add 5 mL of fresh fixative (3:1 analar methanol: glacial acetic acid) initially drop by drop with thorough but gentle mixing to avoid cell clumping.
11. Repeat steps 8–10 at least twice. The fixative should be replaced until suspension appears clear without any trace of a brown tinge. If brown tinge persists, the cell pellet may be "water-washed" (see Note 6). Finally, the suspension is spun, the supernatant discarded, and the pellet diluted if necessary with fixative to produce a slightly cloudy appearance (see Note 7). The cell suspension should be stored at −20°C until slide making (see below).

3.1.2. Culture 2: 48-h Synchronised Culture

1. Place 10 mL RPMI 1640 medium supplemented with PSG and 18% FCS into a sterile 50 mL tissue culture flask.
2. Using a sterile pipette, inoculate medium with appropriate amount of bone marrow (see Note 3).
3. Lie flask flat and incubate at 37°C for approximately 24 h.
4. After incubation for 24 h, add 100 μL of combined FdU/uridine solution to the flask and incubate at 37°C overnight (see Note 8).
5. The following morning, add 100 μL of BrdU (see Note 9) and incubate for a further 7 h.
6. To harvest, follow the procedure outlined for an overnight culture from step 4 (see Note 10).

3.2. Slide Making and G-Banding

Slide preparation is important for optimal G-banding. Ideally, slides are made when the temperature is 22°C and the humidity is approximately 40% (see Note 11).

1. Place cell suspension on the bench and allow it to warm to room temperature. This usually takes approximately 15 min (see Note 12).
2. Clean slides by filling a coplin jar with 100% ethanol, dipping the slides into ethanol, wiping the slides clean with a lint-free tissue, and allowing them to air-dry.
3. Using a clean Pasteur pipette, drop three drops of suspension evenly along the slide. Allow the slide to air dry.
4. Assess slide quality by phase contrast microscopy. The chromosomes should appear medium grey in contrast and be well spread (see Note 13). A decision to water-wash the cell suspension may be taken at this point, depending on the appearance of metaphases under phase contrast microscopy (see Notes 6 and 14 and Fig. 1).
5. The slides should be aged prior to banding to reduce fuzziness and to produce clear crisp G-bands (see Fig. 2). There are a number of methods available. The following steps are

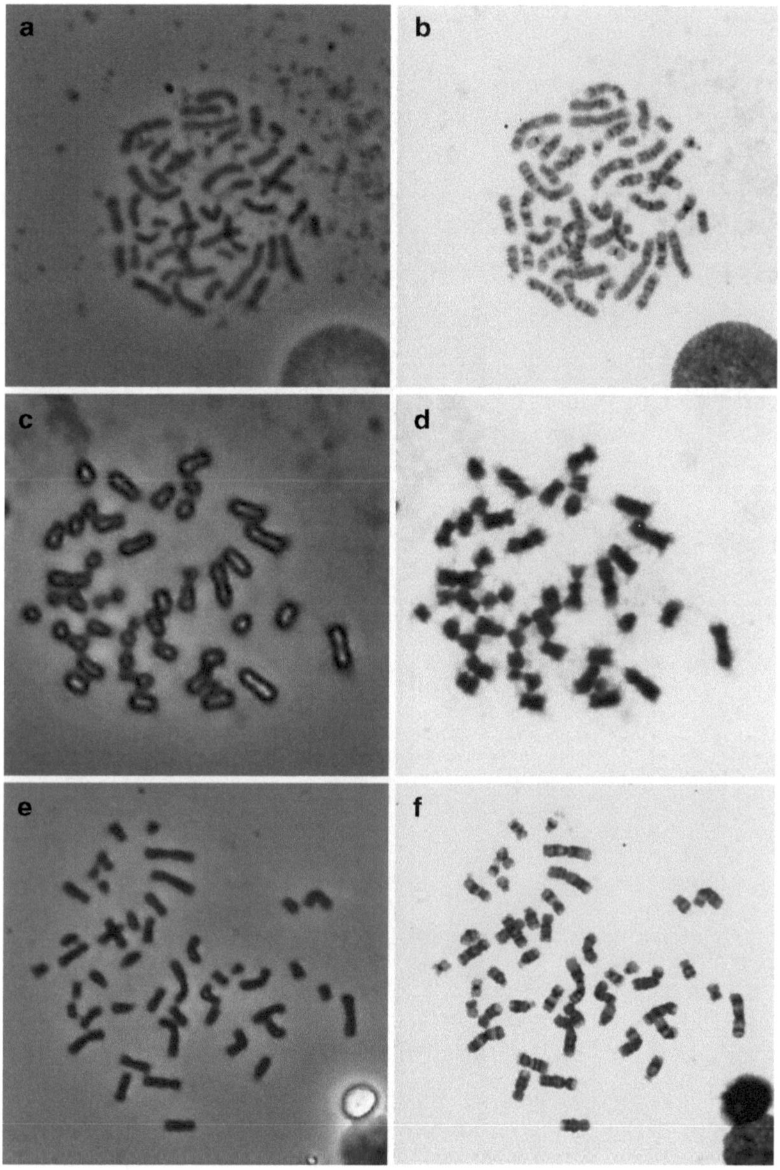

Fig. 1. A comparison of the effects of drying rates on banding quality as indicated by the appearance of metaphases viewed using a phase contrast microscope. All slides were aged in a 60°C oven overnight, placed on a 100°C hotplate for 10 min, exposed to trypsin for 8 s and flooded with Leishman's stain for 8 min. The phase contrast images are on the *left* and their corresponding G-banded images are on the *right*. (**a**, **b**) Metaphases that dry too quickly result in ill-defined, *pale grey*, poorly spread chromosomes that produce fuzzy, indistinct G-bands; (**c**, **d**) metaphases that have dried too slowly contain chromosomes that appear shiny, i.e. phase bright, which results in dark, poorly defined, and uneven banding; (**e**, **f**) metaphases that dry at an optimal rate have a matt, *dark grey*, or *black* appearance and produce well-defined and evenly contrasted banding.

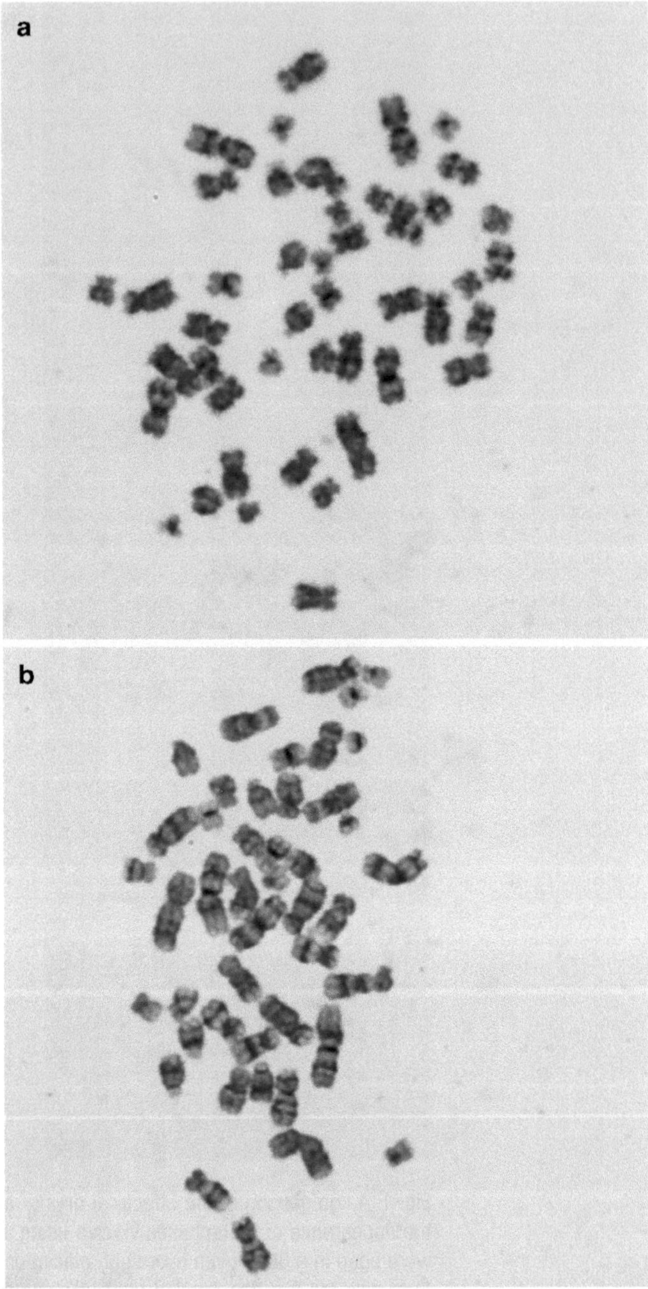

Fig. 2. A comparison of the effects of ageing on banding quality by comparing two metaphases from the same patient, banded following different ageing regimes: (**a**) an example of fresh banding produced on the same day as the slide was made following 12 min on a 100°C hotplate, 4 s in trypsin, and 8 min in Leishman's stain. Bands have a typical dark, hairy appearance indicating the freshness of the slide; (**b**) banding performed the next day on a slide that has been aged in a 60°C oven overnight, placed on a 100°C hotplate for 10 min, exposed to trypsin for 8 s, and stained with Leishman's stain for 8 min. The extra ageing has resulted in crisper, more well-defined bands. Note that the chromosomes generally become more resistant to trypsin with increasing ageing.

designed to produce successful G-banding on the day slides are made (see Note 15).

6. Allow the freshly made slide to air-dry at room temperature (see Note 16).
7. Prior to commencing banding, set up one coplin jar with a working solution of diluted trypsin and two coplin jars with Ca^{2+}/Mg^{2+}-free solution, and allow the jars to stand at room temperature.
8. Place slide on hotplate at 100°C for 8–10 min (see Note 17).
9. Without allowing the slide to cool, dip into the diluted trypsin and agitate for approximately 8–10 s (see Note 18).
10. Rinse in two changes of Ca^{2+}/Mg^{2+}-free solution.
11. Shake off excess moisture and place the slide on a staining rack. Pipette Leishman's solution onto the slide and let stand for 8–10 min or longer if necessary (see Note 19).
12. Rinse stain off under running tap water.
13. Allow slide to air dry.
14. Coverslip the slide by placing three small drops of DPX mounting medium at intervals along the coverslip. Gently place the air-dried slide face down onto the coverslip, invert, and place the slide onto a 37°C hotplate for sufficient time to allow the DPX to set.
15. The G-banded slide is now ready for microscopic analysis. All abnormalities are described according to the latest edition of the ISCN (see Note 20). An analysis may require more than one slide to be examined per culture, depending on the number and quality of metaphases available. Malignant cells tend to produce metaphase spreads with shorter chromosomes and poorer morphology than their normal counterparts (see Fig. 3). Therefore, abnormal clones may be overlooked if only well-banded metaphases of good morphology are analysed. The number of metaphases available for analysis on each slide varies greatly. Ideally, the 20–40 metaphases required for an adequate analysis will be found on one slide, but it may be necessary to band several slides to obtain sufficient metaphases (see Note 21).

4. Notes

1. RPMI 1640 medium, modified with HEPES buffer, is commercially available in liquid form. At the VCCS, the medium is made up from powder: 16.4 g RPMI 1640 powder is dissolved in 1 L dH_2O with 2 g sodium bicarbonate, adjusting

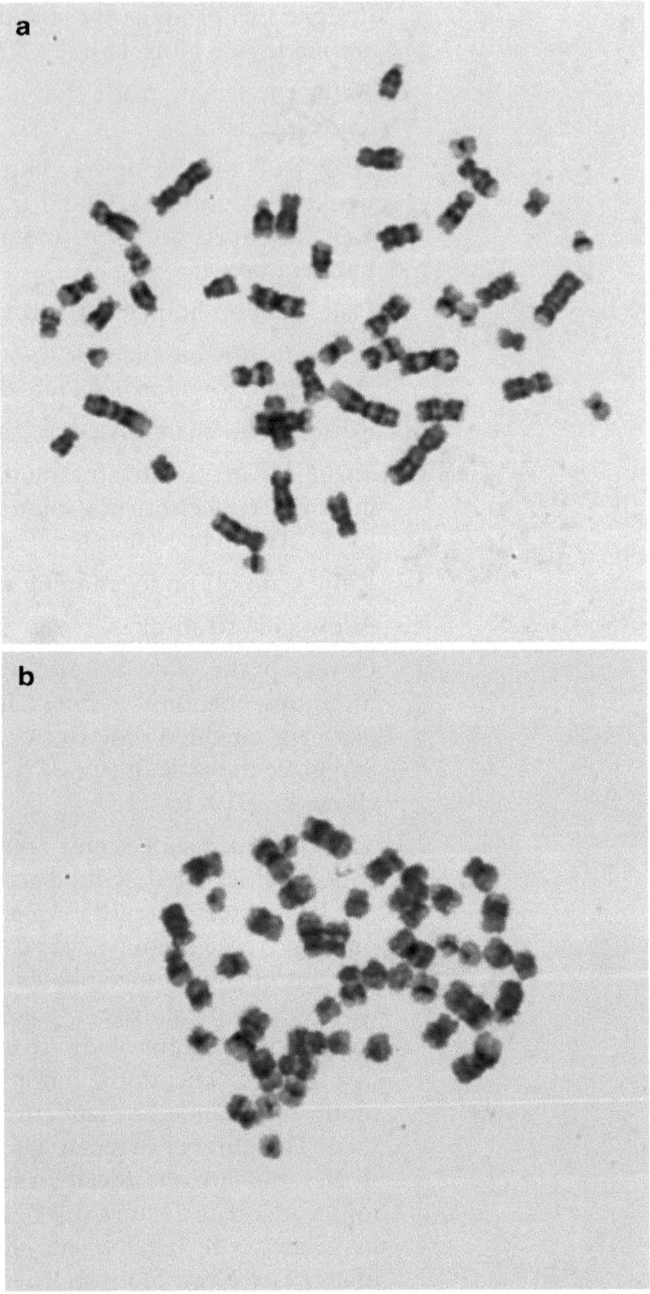

Fig. 3. The quality of chromosome morphology and banding varies between normal and abnormal metaphases in the same patient with a diagnosis of AML: (**a**) a normal metaphase (46,XX) showing reasonable chromosome morphology and banding. (**b**) An abnormal AML metaphase demonstrating the poor morphology and banding often exhibited by the malignant clone. This metaphase has the karyotype: 45,XX,-5,-12,-17,+mar1,+mar2.

the pH to 7.2–7.3 using 1 N HCl or 1 N NaOH and sterilising by filtration through a 0.20-μm filter. It is aseptically decanted into sterile bottles and stored at 4°C for 1–2 weeks. It should not be used if it is noted to become opaque, change colour, or acquire floating particles. This is certainly the most cost-effective method. However, if only small numbers of samples are being cultured, it may be simpler to purchase liquid medium. There are also "complete" media available that do not require the addition of FCS or other additives. Once a bottle of medium is opened and ready for use, it will only last a few days at 4°C in the dark. One way to avoid wasting expensive medium is to place 10 mL complete medium aliquots into tissue culture flasks and freeze at –20°C until required.

2. The G-banding method given here uses Leishman's solution as we have found this method to be the most reliable and least given to fading over time. However, other methods commonly used in cytogenetics laboratories involve the use of Giemsa stain or Wright's stain.

3. The amount of bone marrow aspirate inoculated into each culture is dependent upon the cellularity of the aspirate and the degree of blood dilution. If a cell count is performed on the aspirate, approximately 1×10^7 nucleated cells should be added to each 10 mL culture. Alternatively, an assessment of the viscosity of the marrow specimen may be made. Generally, if the marrow of a patient with AML appears to be quite thick and viscous, add three to five drops to each culture using a sterile pipette, if only slightly viscous, add six to eight drops and if quite thin and blood diluted, add eight to twelve drops of marrow. If the patient has a high white cell count, cultures may readily overgrow and so should be inoculated with only a few drops of marrow. In general, care should be taken not to exceed approximately 12–14 drops of marrow as an excess of red cells added to the culture may affect cell growth and the quality of the harvested cell suspension. If bone marrow is not available peripheral blood may be used, provided it contains sufficient cells capable of spontaneous division. Peripheral blood samples should be inoculated as for thin marrow, taking into account the white blood cell count (WBC) (i.e. fewer drops when high WBC). Note, however, that greater than eight drops tends to result in excessive red cell contamination and a poor suspension after fixing. When the WBC is not given, the specimen should be spun at 200 *g* for 10 min and the size of the buffy coat should be observed. If there is a small buffy coat layer, indicating a low WBC, set up from the buffy coat interface. If the buffy coat layer is large, remix the sample and use whole blood. Occasionally, bone marrow trephine samples may be induced to yield

analysable metaphases, but the success rate with trephines is generally low. Trephine specimens should be scraped or chopped-up under sterile conditions, to produce a single cell suspension that can then be used for inoculation.

4. Although the method was originally introduced as a 24-h culture, this time in culture has been shortened to approximately 15 h. The shorter culture time allows the harvest to be undertaken in the morning without any adverse consequences so that analysis of urgent specimens may be completed within 24 h.
5. All harvesting up to and including the point of initially adding fixative, should be performed wearing disposable gloves and in a class II biohazard cabinet. Likewise, all centrifuging prior to the first fix stage should be in a centrifuge with sealed, autoclavable buckets.
6. Water-washing is achieved by adding 5–10 mL dH_2O to the pellet; the cells are then resuspended in the water and centrifuged at 200 *g* for 5 min. The supernatant is discarded and the pellet is resuspended in 5–10 mL of fresh fix. At least one fix change is then required (7).
7. An over-crowded suspension indicates that the culture was over-inoculated and may result in few metaphases of poor morphology. If a culture yields more than 8 mL of cloudy cytogenetic suspension, a repeat culture should be attempted, if possible, using less marrow.
8. The blocking period should not be less than 14 h or greater than 17 h (8).
9. The incorporation of BrdU into the DNA renders the chromosomes susceptible to degradation on exposure to ultra-violet light; therefore, the culture should be shielded from light. As modern incubators are light-proof, no special precautions are required. However, with an incubator that allows light entry when the door is closed, the flasks should be placed in a light-proof box inside the incubator after the addition of BrdU.
10. Cells may be refrigerated overnight in the second change of fix if this is more convenient for the laboratory work flow and the remaining fix changes completed in the morning.
11. High humidity causes slides to dry too slowly and thus chromosomes to overspread. If humidity is above approximately 45%, the slide may be warmed briefly on a hotplate prior to dropping suspension onto the slide. Alternatively, low humidity causes slides to dry too fast and so chromosomes become clumped and under-spread. Below approximately 35% humidity, slides may be rested on a freezer block for a few seconds prior to dropping suspension onto the slide. An alternate method of slide making when the humidity is low is to use cell suspension that has just been removed from the freezer, rather than allowing the suspension to warm to room temperature.

12. Small amounts of suspension should not be left uncapped on the bench at room temperature for long periods of time. As soon as slides have been made, recap and store suspension in the freezer.

13. The rate at which slides dry greatly affects the spread and appearance of chromosomes. The drying rate is itself influenced by humidity and temperature; thus, the quality of slide preparations can be maximised by manipulating these variables. As a general rule, slides that dry slowly lead to over-spreading and to chromosomes that appear phase bright down the inverted (phase contrast) microscope, i.e. chromosomes have a black, shiny appearance with a halo effect (see Fig. 1). Phase bright chromosomes tend to be more resistant to trypsin and produce uneven, high contrast bands. Slides that dry too rapidly will be under-spread and can result in the retention of background cytoplasm. Chromosomes that have dried too quickly frequently appear pale grey and are more likely to exhibit chromatid separation. Subsequent banding tends to have a fuzzy, grey appearance, lacking contrast between bands. High humidity or low temperature will slow the rate of drying. Alternatively, low humidity or high temperature will increase the rate of drying. The following simple techniques can be applied to adjust these variables. If the ambient humidity is too low, (a) breathe on the slides to produce a film of moisture; this effect is enhanced if cold slides are used; (b) follow the drops of suspension on the slide with a drop of fix; (c) flood the slides with 60% acetic acid before dropping the suspension; this is most effective for cases involving persistent cytoplasm or for extremely tight metaphases, more typically encountered in acute lymphoblastic leukaemia than AML; (d) drop the slides with suspension taken straight from the freezer or use cold slides; slides can be stored in the fridge or freezer prior to dropping or rested on an ice tray; (e) dry the slides flat or at a reduced angle. If ambient humidity is too high, (a) pre-warm the slides on a hotplate, adjusting the temperature of the hotplate according to the degree of humidity or overspreading; in extreme cases of overspreading, the slide can be dried on the hotplate; (b) increase the airflow over the slide by blowing or waving; (c) blot off excess suspension immediately after dropping the slide; (d) increase the drying angle of the slide or dry upright. Other factors that can affect spreading and drying rates include the dilution of the suspension and the size of the drop. Traditionally, it has been claimed that the height from which the cell suspension is dropped onto the slide is critical for successful spreading. However, in truth, there is no advantage to dropping suspension from a great height although it may serve to distribute the spreads more evenly along the slide surface. The size of the drop is actually the more critical factor; a smaller drop will

dry quicker thereby reducing the extent of spreading and vice versa (9). Undivided nuclei can also interfere with spreading if the suspension is too concentrated.

14. Excessive red cell contamination of the culture may result in a sub-optimal cell suspension for slide-making. Uneven drying of fix leads to a mixture of under- and over-spread metaphases with "tide-marks" of fix on the slide. Chromosomes appear phase bright under phase contrast microscopy and under-spread metaphases often appear encased in a greyish patch of cytoplasm. Water-washing may assist in cleaning the suspension but usually results in fewer cells and metaphases, albeit of significantly better quality. It should therefore be used cautiously if there is only a small volume of suspension.
15. Slides may be aged by a variety of methods, including placing slides in a 60°C oven overnight or in a desiccator for 1–2 days (or as long as required) or a combination of these two methods.
16. Although hotplate ageing was introduced to allow banding of freshly made slides for cases where an urgent result was required, slides can continue to be aged in this manner for a number of hours after being made. Indeed, this method works well for slides made up to a week beforehand but the hotplate time should be reduced accordingly with increasing age of the slide.
17. Whilst most slides require between 8 and 10 min on the 100°C hotplate to age sufficiently, up to 20 min may be required in some cases when banding is being attempted on freshly prepared slides.
18. The time needed for each slide to be immersed in trypsin may vary depending on the quality of the chromosome morphology; e.g. less time may be required for metaphase chromosomes of poor quality, such as the metaphases typically seen in hyperdiploid acute lymphoblastic leukaemia. One slide should be tested at a time to estimate the optimum time for producing satisfactory G-bands. Note that the speed with which the slide is agitated in the trypsin will influence the trypsin time required: greater agitation, less time in trypsin, and vice versa. If the chromosomes appear dark with ill-defined bands, the time in trypsin may be extended. Fresh chromosomes tend to appear fuzzy and dark (see Fig. 2); the slides may need to be aged longer, either by increasing the time on the 100°C hotplate or by leaving the slide overnight in a 60°C oven. In contrast, fuzzy, pale chromosomes may indicate over-trypsinisation, and pale, under-banded chromosomes, often with prominent C-band regions, may have been left too long on the hotplate.

19. Hotplate-aged slides may stain paler than slides aged by alternative methods and so require a longer application of Leishman's solution.
20. Only clonal abnormalities can be included in the karyotype. Thus, structural abnormalities or gains of whole chromosomes must be observed in at least two metaphases and loss of a whole chromosome must be observed in at least three metaphases to establish the clonality of an abnormality (10).
21. A culture with few metaphase spreads available for analysis or a failed culture may result from a number of factors. The cause may be sample related, with few cells capable of division due to relative hypoplasia or blood dilution of the sample, prior therapy with chemotherapy agents, too long in transit, inappropriate storage in transit (either too hot or too cold conditions), inappropriate anticoagulation (EDTA), too much anticoagulant (haemolysing the sample), or too little anticoagulant (allowing the specimen to clot). Of the laboratory factors that may contribute to culture failure, the medium is critical; it must be maintained at the correct pH and a careful check of the colour of the medium provides an indication of pH change.

References

1. Swerdlow, S.H., Campo, E., Harris, N.L., Jaffe, E.S., Pileri, S.A., Stein, H., Thiele, J., Vardiman, J.W. (eds.) (2008). *WHO Classification of Tumours of Haematopoietic and Lymphoid Tissues.* IARC: Lyon.
2. Grimwade, D., Walker, H., Oliver, F., et al. (1998) The importance of diagnostic cytogenetics on outcome in AML: analysis of 1,612 patients entered into the MRC AML 10 trial. *Blood* **92**, 2322–33.
3. Slovak, M.L., Kopecky, K.J., Cassileth, P.A., et al. (2000) Karyotypic analysis predicts outcome of preremission and postremission therapy in adult acute myeloid leukemia: a Southwest Oncology Group/Eastern Cooperative Oncology Group study. *Blood* **96**, 4075–83.
4. Döhner, H., Estey, E.H., Amadori, S., et al. (2010) Diagnosis and management of acute myeloid leukemia in adults: recommendations from an international expert panel, on behalf of the European LeukemiaNet. *Blood* **115**, 453–74.
5. Webber, L.M. & Garson, O.M. (1983) Fluorodeoxyuridine synchronization of bone marrow cultures. *Cancer Genetics & Cytogenetics* **8**, 123–32.
6. Barch, M.J., Knutsen, T. & Spurbeck, J.L. (1997) *The AGT Cytogenetics Laboratory Manual*, 3rd ed. Philadelphia: Lippincott-Raven Publishers.
7. Swansbury, J. (ed.) (2003). *Cancer Cytogenetics Methods and Protocols.* Clifton: Humana Press, 83–91.
8. Rooney, D.E. (2001) *Human Cytogenetics: Malignancy and acquired abnormalities*, 3rd ed. London: Oxford University Press.
9. Henegariu, O., Heerema, N.A., Lowe Wright, L. et al. (2001) Improvements in cytogenetic slide preparation: controlled chromosome spreading, chemical aging and gradual denaturing. *Cytometry* **43**, 101–9.
10. ISCN (2009). *An International System for Human Cytogenetic Nomenclature.* L.G. Shaffer, M.L. Slovak, L.J. Campbell (eds.). Basel S.: Karger.

Chapter 6

Cytogenetics in Myelodysplastic Syndromes

Kazuma Ohyashiki, Atsushi Kodama, and Junko H. Ohyashiki

Abstract

Cytogenetic information in patients with myelodysplastic syndrome (MDS) is important in predicting prognosis and therapeutic direction. In MDS, the detection of numerical type abnormalities, either whole chromosome or partial chromosomal segments, is important. In general, conventional banding chromosome analysis is useful in detecting chromosome changes in MDS and is able to predict prognosis. More recently, uniparental disomy at various loci has been found in some MDS patients and target genes located within the deleted chromosome regions; these deletions are either cytogenetically detectable resulting in partial monosomy, or cryptic. Further therapeutic approaches for MDS patients may require more precise cytogenetic information in the near future.

Key words: Cytogenetic analysis, Myelodysplastic syndromes, Deletion, Monosomy, Trisomy, Prognosis, Uniparental disomy

1. Introduction

The myelodysplastic syndromes (MDS) are a heterogeneous category of hematopoietic stem cell malignancies characterized by various degrees of cytopenia in the peripheral blood (PB) and morphologic dysplasia in the bone marrow cells, suggesting ineffective blood cell production. Most MDS patients are elderly and the mean age is 75 years in the Caucasian population and approximately 65 years in the Asian population. About 30% of MDS patients, as defined by the French–American–British (FAB) classification, develop acute leukemia, and approximately 50% of MDS patients have clonal cytogenetic changes. In 2001, the WHO classification defined MDS patients as those who had less than 20% blasts in the marrow; thus, the category of refractory anemia with excess blast in transformation (RAEB-t) by the FAB

Lynda J. Campbell (ed.), *Cancer Cytogenetics: Methods and Protocols*, Methods in Molecular Biology, vol. 730,
DOI 10.1007/978-1-61779-074-4_6, © Springer Science+Business Media, LLC 2011

classification was deleted. Patients with chronic myelomonocytic leukemia are categorized as having a myelodysplastic/myeloproliferative disorder, and MDS patients with a prior history of exposure to carcinogenic/mutagenic agents are classified as having therapy-related MDS/acute myeloid leukemia (1). Moreover, the refractory anemia with excess blasts (RAEB) category has been divided according to the percentage of blasts in the marrow into RAEB-1 and RAEB-2. MDS patients with del(5q) are subcategorized as MDS associated with isolated del(5q) (2).

In MDS, translocations are rarely observed, except for the unbalanced der(1;7)(q10;10); therefore, the detection of numerical type abnormalities, either whole chromosome or partial chromosomal segments, is important (see Table 1) (3–7). Most deletion-type abnormalities are now known as interstitial deletions, e.g., del(5q) and del(20q) and these recurring anomalies in MDS patients have clinical significance (8, 9). Therefore, the fluorescence in situ hybridization (FISH) technique, using two different probes on the same chromosomal arm with one terminal portion (telomere region) might be useful to identify these abnormalities. In general, conventional banding chromosome analysis is useful in detecting chromosome changes in MDS and is able to predict

Table 1
Chromosome abnormalities in myelodysplastic syndromes

Cytogenetics	No of cases	Frequency (%)
Abnormalities	152	45.4
der(1;7)(q10;p10)	6	1.8
inv(3)(q21q26) or t(3;3)(q21;q26)	3	0.9
-5/del(5q)	36	10.7
-7/del(7q)	36	10.7
-5/del(5q) and -7/del(7q)	18	5.4
+8	34	10.1
11q23 anomalies	4	1.2
12p anomalies	9	2.7
del(13q)	5	1.5
del(20q)	15	4.5
3 or more anomalies	59	17.6
Normal karyotypes	183	54.6
Total	335	100

This research was originally published in *Blood* (19) © the American Society of Hematology

prognosis (see Note 1). In some special cases, FISH is useful to determine the effectiveness of certain therapeutic approaches. Deletion of the long arm of chromosome 5 in MDS patients was first described as "5q–syndrome," characterized by macrocytic anemia with normal or elevated platelets, marrow erythroid hypoplasia, hypolobulated megakaryocytes, and sole del(5q) (10, 11), as subcategorized by the WHO in 2001 (1). In the lenalidomide era, patients with del(5q) without particular morphologic changes in the marrow have been also included in this group and are designated as MDS associated with isolated del(5q) (2). Possible candidate tumor suppressor genes within the deleted chromosomal locus are the early growth response 1 (*EGR1*) and α-catenin (*CTNNA1*) genes (12, 13); thus, the deletion of these in cases with masked del(5q) could be also important to identify patients in whom lenalidomide might be effective (14). More recently, uniparental disomy at various loci has been found in some MDS patients, with some target genes located within the deleted chromosome region. These deletions may be either cytogenetically detectable, resulting in partial monosomy, or cryptic (15, 16). Thus, using the FISH technique to detect such genomic deletions might be a powerful tool.

2. Materials

2.1. Cells and Culture

1. Approximately 0.5–1 mL bone marrow aspirate is required, and is obtained by marrow aspiration using a heparinized syringe (see Note 2).
2. RPMI-1640 medium with 5% heparin (1,000 U/mL) and penicillin (100 U/mL).
3. RPMI-1640 medium with 15–20% fetal calf serum, penicillin (50 U/mL), and streptomycin (50 μg/mL).
4. Colcemid 0.2 μg/mL stock solution.
5. 15-mL Centrifuge tube.
6. 0.075 M KCl (hypotonic solution).
7. Fixative solution (1 volume of glacial acetic acid + 3 volumes of 100% methyl alcohol).
8. Pasteur pipettes.

2.2. Chromosome Slide Preparation

1. Phase contrast microscope.

2.3. Giemsa-Banding Methods

1. Trypsin solution: add 5 mL of 0.25% trypsin in 45 mL of Hanks' BSS (Ca^{++} and Mg^{++} free).
2. 50 mL of Hanks' BSS (Ca^{++} and Mg^{++} free).

3. 70% Ethyl alcohol.
4. Giemsa staining solution: 50 mL distilled water, 1.5 mL McIlvaine buffer (pH 6.8–7.0), 1.5 mL methanol, and 1.5 mL 2% Giemsa.
5. Light microscope.

2.4. Quinacrine-Banding Methods

1. McIlvaine buffer, pH 6.8–7.0.
2. Quinacrine mustard (QM) stock solution: 1 μg/mL QM in McIlvaine buffer, pH 6.8–7.0.
3. QM working solution: 3 mL QM stock solution in 50 mL McIlvaine buffer, pH 6.8–7.0.
4. Nail polish or rubber cement.
5. Fluorescence microscope.
6. 50% Ethanol.

3. Methods

3.1. Cell Culture

1. After the bone marrow aspirate is transported to the chromosome laboratory, aspirated marrow cells are transferred to 10 mL of RPMI-1640 medium with 5% heparin (1,000 U/mL) and penicillin (100 U/mL) in a centrifuge tube.
2. Culture in RPMI-1640 medium with 15–20% fetal calf serum, penicillin (50 U/mL), and streptomycin (50 μg/mL), and then incubate (37°C) for 1–2 days without any mitogens.

3.2. Chromosome Harvest

1. Add 0.015 μg/mL Colcemid (or roughly 2–3 drops of a 0.2-μg/mL stock solution) to 10 mL of the cell suspension in culture and incubate for 1–2 h.
2. Transfer the culture to a 15-mL centrifuge tube, and centrifuge for 5 min at approximately 200 *g*.
3. Decant the supernatant.
4. Slowly disturb the sediment by tapping your fingers on the centrifuge tube.
5. Add 4–5 drops of 0.075 M KCl (hypotonic solution) prewarmed at 37°C and disturb the pellet as given in step 4.
6. Add more 0.075 M KCl slowly to bring the volume up to 10 mL.
7. Mix with a Pasteur pipette, and incubate for 30 min at room temperature, while disturbing the cell suspension intermittently with a Pasteur pipette.
8. Add 4–5 drops of fixative solution (1 volume of glacial acetic acid + 3 volumes of 100% methyl alcohol).

9. Mix with a Pasteur pipette.
10. Centrifuge for 5 min at approximately 200 *g*.
11. Decant the supernatant.
12. Disturb the pellet as given in step 5.
13. Add 4–5 drops of fixative solution and disturb the pellet as given in step 5.
14. Add fixative solution slowly to make the volume up to 2 mL.
15. Mix with a Pasteur pipette.
16. Bring the volume to 10 mL with fixative.
17. Repeat step 15.
18. Leave the tube at room temperature for approximately 30 min, and then store at −15 to −20°C.

3.3. Chromosome Slide Preparation

1. Flame dry: This is useful when chromosomes are clumped and is suitable for Quinacrine banding. Slides are stored in 50% ethanol at 4°C and picked up with tweezers and gently wiped with Kimwipe. Place one drop of fixed cells on the slide and quickly heat the slide over a flame.
2. Air-dry in atmospheric conditions (best when chromosomes are widely spread and suitable for Giemsa banding). Slides are stored in 50% ethanol at 4°C and picked up with tweezers and wiped with Kimwipe. Place one or two drops of fixed cells on the slide and leave the slide to dry at room temperature.

Check the slide under a phase contrast microscope to select the best method for slide preparation.

3.4. Giemsa-Banding Methods

1. Leave the slides to air-dry in an incubator (overnight) at 60°C.
2. Arrange five Coplin jars with the following solutions.
   (a) Trypsin solution
   (b) 50 mL of Hanks' BSS (Ca^{++} and Mg^{++} free)
   (c) 70% ethyl alcohol twice
   (d) Giemsa staining solution
3. Immerse the slides in Coplin jar in (a) for about 3.5 min and rinse in (b) and then in (c) twice.
4. Air-dry and stain in Giemsa for 6–8 min, and air-dry.
5. Observe and take photographs by a light microscope (Fig. 1).

3.5. Quinacrine-Banding Methods

1. Immerse the slide in buffer (McIlvaine buffer, pH 6.8–7.0) for 1 min.
2. Stain in quinacrine mustard working solution for 10 min.
3. Wash in tap water.

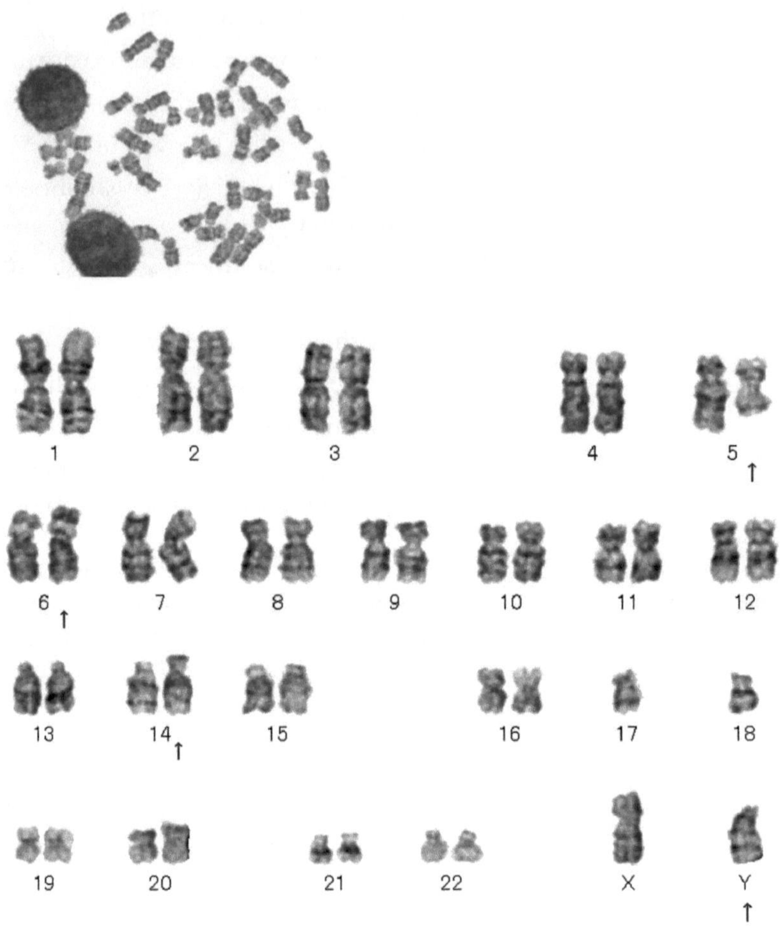

Fig. 1. A Giemsa-banding karyotype of a bone marrow cell obtained from a patient with refractory anemia with excess blasts-1 (*top panel*), and the karyotype showing multiple structural abnormalities of chromosomes 5, 6, and 14, and missing 17 and 18 chromosomes (*bottom panel*).

4. Rinse in three changes of buffer (pH, 6.8–7.0).
5. Mount in buffer and seal the coverslip with nail polish or rubber cement.
6. Observe and take photographs by fluorescence microscopy (see Note 3).

4. Notes

1. In 1993, a Japanese group and a French group independently proposed prognostic scoring systems, incorporating cytogenetic abnormalities, using a large number of MDS patients to

predict prognosis (17, 18). These studies promoted the establishment of the International Prognostic Scoring System for MDS (IPSS) in 1997 (19). The IPSS focused on three major points in predicting MDS prognosis: number of cytopenias, marrow blast percentage, and cytogenetic pattern (Table 2 and Fig. 2). Although the number of MDS patients classified with IPSS was still too small to establish prognostic value for patients with rare cytogenetic abnormalities, the IPSS is used worldwide and this scoring system contributes to estimating risk factors in introducing new agents for MDS. Most recently, the WHO-oriented prognostic scoring system has been proposed; this scoring system utilizes the cytogenetic scoring proposed by the IPSS (20).

Cytogenetic information in patients with MDS is important in predicting prognosis and therapeutic direction. For example, complex cytogenetic changes (involving three or more chromosomes) may indicate that the patient is a candidate for de-methylating agents. Recently, rare cytogenetic changes in MDS patients have been focused on as a means of obtaining information on individual patients to predict prognosis and choose an appropriate therapy. Further therapeutic approaches for MDS patients may require more precise cytogenetic information in the near future.

Table 2
International prognostic scoring system

Scoring					
	0	0.5	1	1.5	2
Marrow blasts	<5%	5–10%		11–20%	21–30%
Karyotypes	Good	Intermediate	Poor		
Cytopenia	0/1	2/3			

		Mean survivals			
Risk	**Score**	<60 y.o.	>60 y.o.	<70 y.o.	>70 y.o.
Low	0	11.8 years	4.8 years	9.0 years	3.9 years
INT-1	0.5–1.0	5.2 years	2.7 years	4.4 years	2.4 years
INT-2	1.5–2.0	1.8 years	1.1 years	1.3 years	1.2 years
High	>2.5	0.3 years	0.5 years	0.4 years	0.4 years

Cytopenia	Neutrophils < 1,800/mL Hb < 10 g/dL Thrombocytopenia < 100,000/mL
Karyotypes	Good: normal, del(20q), –Y, del(5q) Intermediate: others Poor: complex (3 or more), chromosome 7 abnormalities

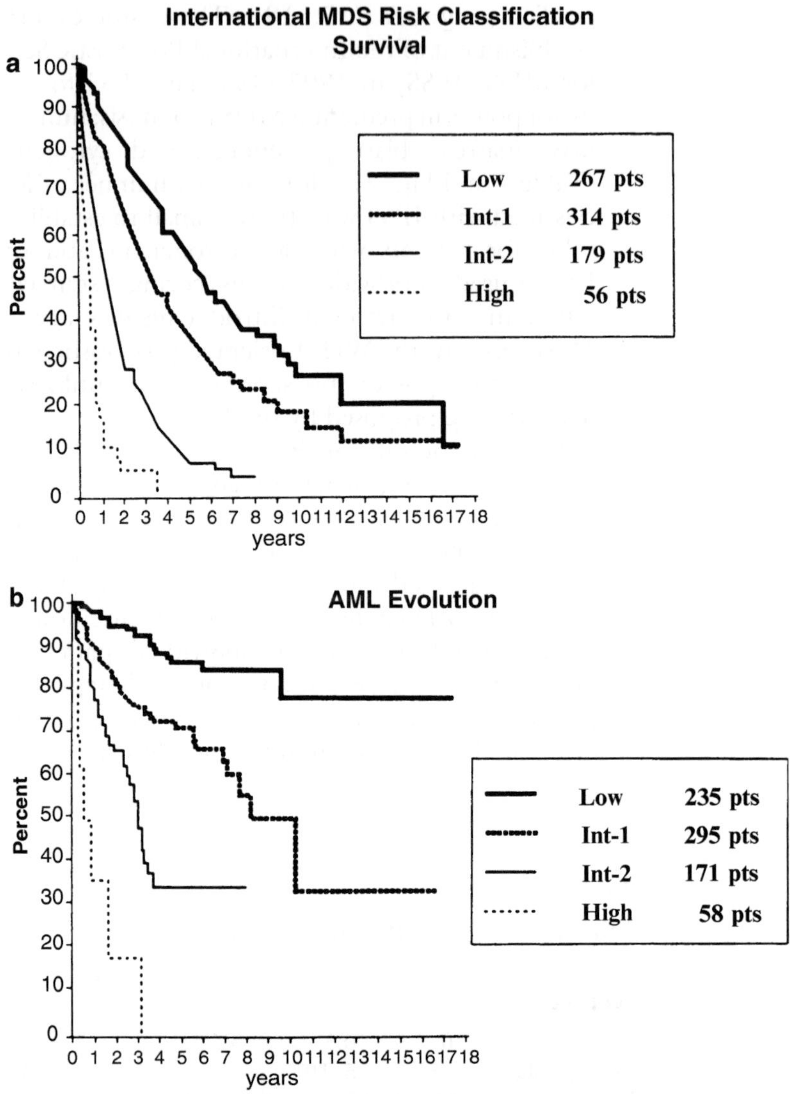

Fig. 2. Survival (**a**) and leukemic transformation (**b**) of MDS patients according to the International Prognostic Scoring System. This research was originally published in *Blood* (19) © the American Society of Hematology.

2. If the bone marrow examination results in a dry-tap, e.g., with marrow fibrosis, and immature cells are present in the peripheral blood, 10 mL of PB in a heparinized syringe is a suitable alternative to bone marrow.
3. A translocation-type abnormality should be considered as one anomaly. Chromosome abnormalities involving three or more chromosomes without three-way translocation should be counted as three or more. A missing Y chromosome is counted as one anomaly, but the significance of loss of Y as

representing any clonal change in MDS is doubtful. Karyotypically independent clones should be counted as two or more aberrations, depending on the numbers of independent clonal abnormalities. The interpretation for der(1;7)(q10;p10) is still controversial; i.e., whether or not this type of abnormality should be considered as a deletion of the long arm of chromosome 7 or a single translocation (21).

Acknowledgements

The authors are indebted to Prof. J. Patrick Barron of the Department of International Medical Communications of Tokyo Medical University for his review of this manuscript.

References

1. Brunning, R.D., Bennett, J.M., Flandrin, G., Matutes, E., Head, D., Vardiman, J.W., and Harris, N.L. (2001) Myelodysplastic syndromes. In: World Health Organization Classification of Tumours. Pathology and Genetics of Tumours of Haematopoietic and Lymphoid Tissues (Eds. Jaffe, E.S., Harris, N.L., Stein, H., and Vardiman, J.W.), pp 61–73, IARC Press: Lyon.
2. Brunning, R.D., Orazi, A., Germing, U., Le Beau, M.M., Porwit, A., Bauman, I., Vardiman, J.W., and Hellstrom-Lindberg, E. (2008) Myelodysplastic syndromes. In: WHO Classification of Tumours of Haematopoietic and Lymphoid Tissues (Eds. Swerdlow, S.H., Campo, E., Harris, N.L., Jaffe, E.S., Pileri, S.A., Stein, H., Thiele, J., and Vardiman, J.W.), pp 87–107, IARC Press: Lyon.
3. Olney, H.J. and Le Beau, M.M. (2007) Evaluation of recurring cytogenetic abnormalities in the treatment of myelodysplastic syndromes. *Leuk Res* **31**,427–34.
4. Haase, D. (2008) Cytogenetic features in myelodysplastic syndromes. *Ann Hematol* **87**, 515–26.
5. Bernasconi, P., Boni, M., Cavigliano, P.M., Calatroni, S., Giardini, I., Rocca, B., Zappatore, R., Dambruoso, I., and Caresana, M. (2006) Clinical relevance of cytogenetics in myelodysplastic syndromes. *Ann NY Acad Sci* **1089**, 395–410.
6. Panani, A.D. and Roussos, C. (2006) Cytogenetic aspects of adult primary myelodysplastic syndromes: clinical implications. *Cancer Lett* **235**, 177–90.
7. Steensma, D.P. and List, A.F. (2005) Genetic testing in the myelodysplastic syndromes: molecular insights into hematologic diversity. *Mayo Clin Proc* **80**, 681–98.
8. Nimer, S.D. (2006) Clinical management of myelodysplastic syndromes with interstitial deletion of chromosome 5q. *J Clin Oncol* **24**, 2576–82.
9. Campbell, L.J. and Garson, O.M. (1994) The prognostic significance of deletion of the long arm of chromosome 20 in myeloid disorders. *Leukemia* **8**, 67–71.
10. Van den Berghe, H., Cassiman, J.J., David, G., Fryns, J.P., Michaux, J.L., and Sokal, G (1974) Distinct haematological disorder with deletion of the long arm of no. 5 chromosome. *Nature* **251**, 437.
11. Boultwood, J., Lewis, S., and Wainscoat, J.S. (1994) The 5q–syndrome. *Blood* **84**, 3253–60.
12. Liu, T.X., Becker, M.W., Jelinek, J., Wu, W.S., Deng, M., Mikhalkevich, N., Hsu, K., Bloomfield, C.D., Stone, R.M., DeAngelo, D.J., Galinsky, I.A., Issa, J.P., Clarke, M.F., and Look, A.T. (2007) Chromosome 5q deletion and epigenetic suppression of the gene encoding alpha-catenin (CTNNA1) in myeloid cell transformation. *Nat Med* **13**, 78–83.
13. Joslin, J.M., Fernald, A.A., Tennant, T.R., Davis, E.M., Kogan, S.C., Anastasi, J., Crispino, J.D., and Le Beau, M.M. (2007) Haploinsufficiency of EGR1, a candidate gene in the del(5q), leads to the development of myeloid disorders. *Blood* **110**, 719–26.

14. Zou, Y.S., Fink, S.R., Stockero, K.J., Paternoster, S.F., Smoley, S.A., Tun, H.W., Reeder, C.B., Tefferi, A., and Dewald, D.W. (2007) Efficacy of conventional cytogenetics and FISH for EGR1 to detect deletion 5q in hematological disorders and to assess response to treatment with Lenalidomide. *Leuk Res* **31**, 1185–9
15. Gondek, L.P., Haddad, A.S., O'Keefe, C.L., Tiu, R., Wlodarski, M.W., Sekeres, M.A., Theil, K.S., and Maciejewski, J.P. (2007) Detection of cryptic chromosomal lesions including acquired segmental uniparental disomy in advanced and low-risk myelodysplastic syndromes. *Exp Hematol* **35**, 1728–38.
16. Gondek, L.P., Tiu, R., O'Keefe, C.L., Sekeres, M.A., Theil, K.S., and Maciejewski, J.P. (2008) Chromosomal lesions and uniparental disomy detected by SNP arrays in MDS, MDS/MPD, and MDS-derived AML. *Blood* **111**, 1534–42.
17. Toyama, K., Ohyashiki, K., Yoshida, Y., Abe, T., Asano, S., Hirai, H., Hirashima, K., Hotta, T., Kuramoto, A., Kuriya, S., et al. (1993) Clinical implications of chromosomal abnormalities in 401 patients with myelodysplastic syndromes: a multicentric study in Japan. *Leukemia* 7, 499–508.
18. Morel, P., Hebbar, M., Lai, J.L., Duhamel, A., Preudhomme, C., Wattel, E., Bauters, F., and Fenaux, P. (1993) Cytogenetic analysis has strong independent prognostic value in de novo myelodysplastic syndromes and can be incorporated in a new scoring system: a report on 408 cases. *Leukemia* 7, 1315–23.
19. Greenberg, P., Cox, C., LeBeau, M.M., Fenaux, P., Morel, P., Sanz, G., Sanz, M., Vallespi, T., Hamblin, T., Oscier, D., Ohyashiki, K., Toyama, K., Aul, C., Mufti, G., and Bennett, J. (1997) International scoring system for evaluating prognosis in myelodysplastic syndromes. *Blood* **89**, 2079–88.
20. Malcovati, L., Porta, M.G., Pascutto, C., Invernizzi, R., Boni, M., Travaglino, E., Passamonti, F., Arcaini, L., Maffioli, M., Bernasconi, P., Lazzarino, M., and Cazzola, M. (2005) Prognostic factors and life expectancy in myelodysplastic syndromes classified according to WHO criteria: a basis for clinical decision making. *J Clin Oncol* **23**, 7594–603.
21. Slovak, M.L. and Dewald, G.W. (2008) International Working Group on MDS cytogenetics: October 2007 meeting report. *Leuk Res* **32**, 1329–32.

Chapter 7

Cytogenetics of Myeloproliferative Neoplasms

Lynda J. Campbell

Abstract

The introduction of *JAK2* mutation testing has changed dramatically the diagnostic algorithms for myeloproliferative neoplasms (MPNs) but there is still a place for conventional cytogenetic analysis in the initial work-up of MPN cases, particularly as this group of myeloid disorders has been expanded to include chronic eosinophilic leukaemia and myeloid neoplasms with abnormalities of the *PDGFRA*, *PDGFRB*, and *FGFR1* genes. Mastocytosis is also included under the umbrella of MPN but the cytogenetic abnormalities observed usually reflect any associated clonal haematological non-mast cell lineage disease.

Key words: Myeloproliferative neoplasm, Polycythaemia vera, Essential thrombocythaemia, Primary myelofibrosis, Chronic neutrophilic leukaemia, *JAK2* mutation, *PDGFRA*, *PDGFRB*, *FGFR1*

1. Introduction

The myeloproliferative neoplasms (MPNs) are clonal haematopoietic stem cell disorders that are characterized by dysregulated proliferation of one or more cell lineages, resulting in over-production of granulocytes, red cells, megakaryocytes, or mast cells or a combination of these lineages. Chronic myeloid leukaemia (CML) has always been regarded as an MPN but separated from the other subtypes by its genetic marker, the Philadelphia chromosome. The cytogenetics of CML is covered in Chapter 4. Polycythaemia vera (PV), essential thrombocythaemia (ET), primary myelofibrosis (PMF), and CML were first grouped together as inter-related entities by William Dameshek in 1951. The 2008 WHO Classification of Tumours of Haematopoietic and Lymphoid tissues also includes chronic neutrophilic leukaemia (CNL), chronic eosinophilic leukaemia (CEL) (not otherwise specified), mastocytosis, and MPN unclassifiable, within the definition of MPN (1). The new entity of myeloid and lymphoid

Lynda J. Campbell (ed.), *Cancer Cytogenetics: Methods and Protocols*, Methods in Molecular Biology, vol. 730, DOI 10.1007/978-1-61779-074-4_7, © Springer Science+Business Media, LLC 2011

neoplasms with eosinophilia and abnormalities of *PDGFRA*, *PDGFRB*, and *FGFR1* also includes MPN.

A number of MPN and other myeloid disorders have been associated with mutations or rearrangements of tyrosine kinase genes (see Table 1). In particular, PV, ET, and PMF have all been shown to contain *JAK2* mutations (see Table 2) and so, in consequence, the criteria for diagnosing MPN have been modified markedly since the advent of *JAK2* testing. However, cytogenetic analysis is still recommended as part of the diagnostic work-up, particularly for essential thrombocythaemia and primary myelofibrosis (PMF) (2).

Most chromosome abnormalities that are observed in MPN are not necessarily specific for MPN or for a particular subtype but certain changes are more commonly observed in one subtype than another and so assist in confirming the diagnosis. The commonest

Table 1
Possible activating mutations in MPN and MDS/MPN

Disorder	Activating mutations
Mastocytosis	*KIT* D816V
Polycythaemia vera	*JAK2* V617F, *JAK2* exon 12
Essential thrombocythaemia (including ET with RARS)	*JAK2* V617F, *MPL* W151L/K
Primary myelofibrosis	*JAK2* V617F, *MPL* W151L/K
Chronic myeloid leukaemia	*BCR–ABL1*
Myeloid neoplasms with eosinophilia	*PDGFRA*, *PDGFRB*, *FGFR1*

Table 2
Incidence of *JAK2* mutations in myeloid neoplasms

Disorder	*JAK2* exon 12 (%)	*JAK2* V617F exon 14 (%)	*MPL* W515L/K (%)
PV	3	95	–
ET	–	50	3
PMF	–	58	5
AML	–	1	–
MDS	–	<5	–
RARS-T	–	50	<2

AML acute myeloid leukaemia, *MDS* myelodysplastic syndrome, *RARS-T* refractory anaemia with ring sideroblasts and thrombocytosis

Table 3
Common cytogenetic abnormalities in MPN

Chromosome	PMF	ET[a]	PV	CNL
1	Partial trisomy of 1q (1q21–1q32) der(6)t(1;6)(q23–25;p21–22) in <3% PMF		Partial trisomy of 1q[b]	t(1;20)
8		+8	+8	
9		+9	+9	
13	del(13q) – commonest abnormality in PMF (seen in 25% cases with abnormal karyotype)	del(13q)	del(13q)[b]	
20	del(20q) second most common abnormality in PMF	del(20q)	del(20q)	del(20q) most frequent finding
21				+21

[a]No chromosome abnormalities are common in ET
[b]Both trisomy of 1q and del(13q) have been correlated with transformation to myelofibrosis in PV (4)

Table 4
MPN diagnostic work-up

Diagnosis	Conventional cytogenetics	FISH	Mutation analysis
PV	+/–	–	*JAK2*V617F; If negative, *JAK2* exon 12 testing
ET	Yes, to exclude t(9;22)	–	*JAK2*V617F
PMF	Yes	–	*JAK2*V617F
CNL	Yes, especially to exclude t(9;22)	–	–
Myeloid neoplasms with eosinophilia	Yes	FIP1L1–CHIC2–PDGFRA probe; PDGFRB break apart probe if CC indicates 5q abnormality	–
Mastocytosis	Yes, to identify any abnormalities associated with SM-AHNMD	If associated eosinophilia, see above	*KIT* mutation

abnormalities found in the different subtypes are listed in Table 3. In some specific subtypes of MPN, fluorescence in situ hybridization (FISH) studies should also be incorporated into the diagnostic work-up (see Table 4).

1.1. Cytogenetics of Polycythaemia Vera

Almost all cases of PV contain a *JAK2* mutation and so a case could be made for dispensing with a bone marrow examination including cytogenetic analysis in patients with a clinical picture consistent with PV and a *JAK2* mutation identified in the peripheral blood.

For those cases that are shown to be *JAK2* mutation negative, however, further investigation is required and cytogenetic analysis is an important component of the process, together with mutation testing for *JAK2* exon 12 and possibly *MPL* mutations (3).

The incidence of chromosome abnormalities in PV is less than 20% (4). The most common abnormalities observed in PV are summarized in Table 3. The finding of trisomy 9 almost always suggests the diagnosis of PV. As *JAK2* is located on the short arm of chromosome 9, trisomy 9 increases the mutated gene dose and increased gene dosage of *JAK2* has been shown to affect the MPN phenotype. Thus, the tight connection between trisomy 9 and PV does not appear to be accidental (5).

1.2. Cytogenetics of Essential Thrombocythaemia

JAK2 mutation testing also plays a major role in the diagnosis of ET but only approximately 50% ET cases are positive and so the diagnostic work-up for patients with a consistently elevated platelet count must include a bone marrow examination looking for morphological evidence of dysplastic changes and cytogenetic abnormalities.

The incidence of chromosome abnormalities in ET is cited as approximately 5–10% but our experience suggests that the abnormality rate is lower. This makes the cytogenetic analysis of these cases largely unrewarding. However, in about 1 in every 200 cases, an unexpected Philadelphia 9;22 translocation is observed, justifying the tedium of the majority of ET cytogenetic analyses.

The other major differential diagnosis of ET when the platelet count is only mildly elevated and the patient is also anaemic is a myelodysplastic syndrome (MDS), particularly with a deletion of the long arm of chromosome 5. MDS with del(5q) should be readily differentiated from ET by morphology, especially the finding of monolobular megakaryocytes on bone marrow examination. However, there are cases with ambiguous morphology where the finding of an isolated del(5q) makes the diagnosis. Moreover, the identification of this subset of MDS is of significant importance as lenalidomide now appears to be an effective therapy in this disorder (6).

Gangat et al. looked at the cytogenetic results of 402 consecutive patients with ET seen at the Mayo Clinic between 1956 and 2005. Median follow-up for their cohort was 70 months. Only 28/402 (7%) had an abnormal karyotype at diagnosis. The most frequent abnormality was trisomy 9 (four patients). Other notable abnormalities were trisomy 8 (two patients) and del(20q) (two patients). Interestingly, two patients who met the diagnostic criteria for ET had a del(5q), an abnormality more commonly encountered in MDS. Ten patients had documented AML transformation and all ten had a normal karyotype at diagnosis but nine of the ten had a cytogenetic abnormality at AML transformation, mostly complex karyotypes. An additional ten patients developed post-ET myelofibrosis. Initially, all had a normal karyotype but one developed a del(20q) at transformation. Cytogenetic abnormalities were significantly associated with palpable splenomegaly, haemoglobin less than 100 g/L, venous thrombosis, and current smoking. Most importantly, detection of a cytogenetic abnormality at diagnosis had no impact on overall survival in this study (7).

1.3. Cytogenetics of Primary Myelofibrosis

In contrast to the paucity of cytogenetic abnormalities found in ET, the incidence of chromosome abnormalities in PMF has been described as between 32 and 48% (8) although the higher abnormality rates have not been observed in patients evaluated at their initial diagnosis. Studies from Lille, Sheffield, and the Mayo Clinic of 47, 106, and 165 patients, respectively (summarized in (9)) reported that the presence of a cytogenetic abnormality was generally associated with a reduced survival, although those patients with deletions of 13q or 20q tended to have a similar survival to patients with a normal karyotype. These findings were expanded by Tam et al. who studied 256 patients with PMF and determined that those patients with sole deletions of 13q or 20q, or those with trisomy 9 with or without one other abnormality had a similar survival (median 63 months) to those with a normal diploid karyotype (median 46 months), whereas patients with what they deemed "unfavourable" abnormalities (abnormalities of chromosomes 5 or 7 or 3 or more chromosome abnormalities) had a median survival of only 15 months, significantly less than the median survival of those with favourable abnormalities and a normal karyotype (8). The inclusion of patients with trisomy 9 with or without an extra abnormality in the favourable prognosis group requires confirmation by other groups.

1.4. Cytogenetics of Chronic Neutrophilic Leukaemia

CNL is a rare entity characterized by a proliferation of mostly mature neutrophils. No specific genetic abnormalities have been associated with CNL but an abnormal cytogenetics result is helpful to differentiate CNL from a reactive process secondary to infection or an underlying malignancy. It is also necessary to exclude CML as there have been case reports of CNL with a t(9;22) resulting in a 230-kDa BCR–ABL1 fusion protein (10).

An abnormal karyotype is observed in approximately one-third of CNL cases. Del(20q) is the commonest abnormality but trisomies of chromosomes 8 and 21 have also been reported. None of these changes is specific for CNL.

1.5. Cytogenetics of Myeloid/Lymphoid Neoplasms with PDGFRA, PDGFRB, and FGFR1 Rearrangements

A diagnosis of hypereosinophilic syndrome (HES) or CEL warrants a cytogenetic analysis. Some cases of persistent idiopathic eosinophilia have been shown to be caused by abnormalities of the alpha or beta subunits of the platelet derived growth factor receptor (*PDGFRA* and *PDGFRB*). Abnormalities of *PDGFRA* on chromosome 4 (4q12) generally result from a cryptic deletion of 4q12. The sub-microscopic deletion of approximately 800 kb disrupts both *FIP1L1* and *PDGFRA* genes, removing the intervening sequences between the two and causing the formation of a *FIP1L1–PDGFRA* fusion gene (11). A handful of HES cases has been described with balanced reciprocal translocations involving breakpoints either near to or within the *PDGFRA* and *KIT* tyrosine kinase genes; both genes are located close to one another on chromosome 4.

Translocations involving the *PDGFRB* gene are rare. The most common is the t(5;12)(q31;p13), involving the *ETV6* gene at 12p13. It is observed in chronic myelomonocytic leukaemia (classified as a myeloproliferative/myelodysplastic syndrome (MPN/MDS) rather than MPN), and also in rare cases of atypical CML (also classified as MPN/MDS), CEL, and MPN; all translocations involving the *PDGFRB* gene are usually accompanied by eosinophilia. The *PDGFRB* translocations reported to date are shown in Fig. 1 (12). It is advisable to confirm the involvement of the *PDGFRB* gene in a translocation as its involvement is critical for predicting the disease responsiveness to tyrosine kinase inhibitors such as imatinib. Thus, any chromosomal abnormality of

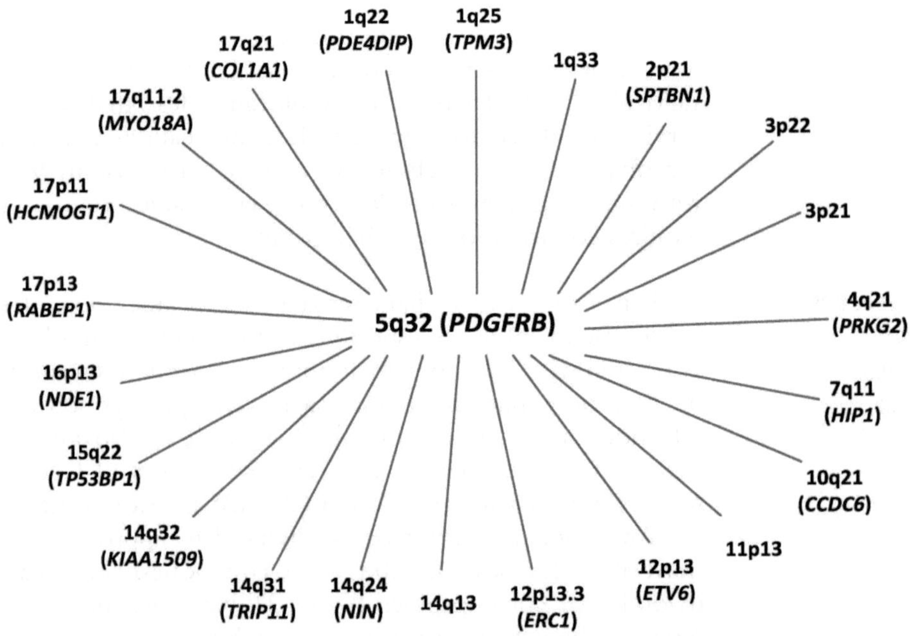

Fig. 1. Diagram of the reported *PDGFRB* translocation partner chromosome breakpoints and genes (if identified) (12).

5q31–33 in the setting of an MPN, with or without eosinophilia, should be investigated by FISH with a *PDGFRB* break apart probe. However, if there is no 5q or 12p13 abnormality observed, further molecular testing is not warranted as, to date, all reported cases have had a chromosome abnormality that was readily detected by conventional cytogenetics, albeit occasionally involving complex rearrangements.

The 8p11 myeloproliferative syndrome, now known as "myeloid and lymphoid neoplasm with *FGFR1* abnormality" is a curious entity as patients may present with one or more of the following disorders: T cell lymphoma, rarely B cell lymphoma, MPN (usually with eosinophilia), or even with acute myeloid leukaemia (AML). Patients may develop these disorders simultaneously or sequentially and often have a terminal transformation to AML. What unites all cases is the involvement of the *FGFR1* locus on 8p11. The most common translocation is the t(8;13)(p11;q12) causing the formation of the *ZNF198–FGFR1* fusion gene but there are a number of variant translocations reported (13). All are readily identified by conventional cytogenetics.

1.6. Cytogenetics of Chronic Eosinophilic Leukaemia (Not Otherwise Specified)

Once *PDGFRA*, *PDGFRB*, and *FGFR1* rearrangements or translocations have been eliminated, most cases of HES or CEL exhibit a normal karyotype by conventional cytogenetics. Occasional cases contain trisomy 8, a common finding in all myeloid malignancies. The diagnosis of CEL, not otherwise specified, is reliant on the exclusion of all genetic and other causes of a persistent eosinophilia.

1.7. Cytogenetics of Mastocytosis

An activating *KIT* mutation is the most frequent genetic abnormality observed in mastocytosis. Those patients with an associated clonal haematological non-mast cell lineage disease (SM-AHNMD) may carry genetic abnormalities that are specific to the particular haematological disease. In MPN with eosinophilia and mastocytosis, a 4q12 deletion resulting in the *FIP1L1–PDGFRA* fusion may be observed but these patients are usually classified as a myeloid neoplasm with *PDGFRA* abnormality.

1.8. Cultures Required for Cytogenetic Analysis of MPN

The conventional cytogenetics analysis of myeloproliferative disorders at the Victorian Cancer Cytogenetics Service requires the establishment of overnight and synchronized cultures from the bone marrow of patients with suspected MPN. These methods are described in Chapter 5.

In patients with significant marrow fibrosis, obtaining an adequate bone marrow specimen for culture may be problematic as the specimen may be heavily blood contaminated. This pertains particularly to PMF cases but some ET and PV cases may also provide relatively aparticulate aspirates due to increased marrow fibrosis. Our practice is to set up an additional culture when the volume of specimen allows or as an alternative to the synchronized culture, with Colcemid added for 15–24 h rather than the standard 30 min so that the chance of obtaining metaphase spreads for analysis is maximized. The following method outlines the establishment of an overnight Colcemid culture of bone marrow cells from patients with MPN (see Note 1). Slide-making and the production of banded metaphase spreads for microscopic analysis from this culture are covered in Chapter 5. FISH studies required in MPN are performed in accordance with standard methods, as outlined in Chapter 3.

2. Materials

Materials required for bone marrow culture and harvest:

1. Bone marrow aspirate, 0.5–1.0 mL, collected in a sterile syringe or tube containing approximately 100 IU of preservative-free heparin.
2. RPMI 1640 medium – with L-glutamine and HEPES.
3. Foetal calf (bovine) serum (FCS): stored at −20°C in aliquots (i.e. 40-mL aliquots for addition to 200-mL bottles of medium to produce an 18% solution). This should be thawed and added to medium immediately before use.
4. L-Glutamine–penicillin–streptomycin (PSG): a mixture containing 200 mM L-glutamine, 10,000 IU of penicillin, and 10 mg streptomycin per millilitre. PSG is stored at −20°C and 1 mL is added to a 200-mL bottle of medium immediately prior to use.
5. Sterile 50-mL tissue culture flasks.
6. Sterile pipettes.
7. Colcemid®: 10 μg/mL solution.
8. Hypotonic solution (potassium chloride 0.075 mol/L): a 5.59-g/L solution of KCl (MW = 74.55).
9. Fixative: 3:1 (v/v) analar methanol/glacial acetic acid, made fresh just before use.
10. 15-mL plastic, non-sterile screw-top centrifuge tubes.

3. Methods

3.1. Overnight Colcemid Culture

1. Place 10-mL RPMI 1640 medium supplemented with PSG and 18% FCS into a sterile 50-mL tissue culture flask.
2. Using a sterile pipette, inoculate medium with appropriate amount of bone marrow (see Note 2).

3. Add 0.2 mL of 10 μg/mL Colcemid® to culture.
4. Lie flask flat and incubate at 37°C for 15–24 h (see Note 3).
5. Transfer culture to harvesting tube and centrifuge for 10 min at approximately 200 *g* in a sealed bucket centrifuge.
6. Discard supernatant.
7. Resuspend cell pellet in 8 mL of KCl and place in 37°C water bath for 20 min.
8. Centrifuge for 10 min at 200 *g*.
9. Discard supernatant.
10. Resuspend in 5 mL of fresh fixative (3:1 analar methanol:glacial acetic acid) by adding the fix drop by drop initially with thorough mixing to avoid cell clumping.
11. Repeat steps 8–10 at least twice. The fixative should be replaced until the suspension appears clear without any trace of a brown tinge. Finally, the suspension is spun and diluted if necessary with fixative to produce a slightly cloudy appearance. The cell suspension should be stored at -20°C until slide-making (see Chapter 5).

4. Notes

1. The exposure of the cultured bone marrow cells to Colcemid for 15–24 h results in an improved mitotic yield but many of the resulting metaphase spreads may be difficult to analyse as the chromosomes may be extremely short.
2. The amount of bone marrow aspirate inoculated into the culture is dependent upon the cellularity of the aspirate and the degree of blood dilution. In general, it is recommended that approximately 1×10^7 nucleated cells be added to each 10-mL culture. If a cell count is not available, approximately 5–7 drops of thick, viscous marrow or 8–11 drops of thin, blood diluted marrow may be added. Bone marrow aspirates from patients with myelofibrosis are frequently considerably blood diluted and so if more than 2 mL of bone marrow aspirate is provided, the specimen should be spun and the buffy coat is used as the inoculum.
3. The amount of time spent in culture depends upon laboratory practice and does not appear to affect the result. Our laboratory practice is to harvest all un-synchronized cultures in the morning after setting up the culture, resulting in a culture time of approximately 15 h.

References

1. Swerdlow SH, Campo E, Harris NL, Jaffe ES, Pileri SA, Stein H, Thiele J, Vardiman JW (Eds.). WHO Classification of Tumours of Haematopoietic and Lymphoid Tissues. Lyon: IARC; 2008.
2. Vardiman JW, Thiele J, Arber DA, et al. The 2008 revision of the World Health Organization (WHO) classification of myeloid neoplasms and acute leukemia: rationale and important changes. Blood 2009;114:937–51.
3. Smith CA, Fan G. The saga of JAK2 mutations and translocations in hematologic disorders: pathogenesis, diagnostic and therapeutic prospects, and revised World Health Organization diagnostic criteria for myeloproliferative neoplasms. Human Pathology 2008;39:795–810.
4. Reilly JT. Pathogenetic insight and prognostic information from standard and molecular cytogenetic studies in the BCR–ABL-negative myeloproliferative neoplasms (MPNs). Leukemia 2008;22:1818–27.
5. Vannucchi AM, Antonioli E, Guglielmelli P, Pardanani A, Tefferi A. Clinical correlates of JAK2V617F presence or allele burden in myeloproliferative neoplasms: a critical reappraisal. Leukemia 2008;22:1299–307.
6. Kelaidi C, Eclache V, Fenaux P. The role of lenalidomide in the management of myelodysplasia with del 5q. British Journal of Haematology 2008;140:267–78.
7. Gangat N, Tefferi A, Thanarajasingam G, et al. Cytogenetic abnormalities in essential thrombocythemia: prevalence and prognostic significance. European Journal of Haematology 2009;83:17–21.
8. Tam CS, Abruzzo LV, Lin KI, et al. The role of cytogenetic abnormalities as a prognostic marker in primary myelofibrosis: applicability at the time of diagnosis and later during disease course. Blood 2009;113:4171–8.
9. Hussein K, Dyke DLV, Tefferi A. Conventional cytogenetics in myelofibrosis: literature review and discussion. European Journal of Haematology 2009;82:329–38.
10. Pane F, Frigeri F, Sindona M, et al. Neutrophilic-chronic myeloid leukemia: a distinct disease with a specific molecular marker (BCR/ABL with C3/A2 junction). Blood 1996;88:2410–4.
11. Gotlib J, Cools J, Malone JM III, Schrier SL, Gilliland DG, Coutre SE. The FIP1L1–PDGFR{alpha} fusion tyrosine kinase in hypereosinophilic syndrome and chronic eosinophilic leukemia: implications for diagnosis, classification, and management. Blood 2004;103:2879–91.
12. Walz C, Haferlach C, Hänel A, et al. Identification of a MYO18A-PDGFRB fusion gene in an eosinophilia-associated atypical myeloproliferative neoplasm with a t(5;17)(q33-34;q11.2). Genes, Chromosomes and Cancer 2009;48:179–83.
13. Soler G, Nusbaum S, Varet B, et al. LRRFIP1, a new FGFR1 partner gene associated with 8p11 myeloproliferative syndrome. Leukemia 2009;23:1359–61.

Chapter 8

Acute Lymphoblastic Leukaemia

Claire Schwab and Christine J. Harrison

Abstract

Cytogenetics plays an important role in the diagnosis of acute lymphoblastic leukaemia (ALL), particularly in relation to the association of specific chromosomal abnormalities with outcome. The karyotype at diagnosis is used in the risk stratification of patients for treatment within trial-based protocols. Chromosomal analysis of the leukaemic cells of patients with ALL is challenging as the mitotic index may be low and the chromosomal morphology is often poor. Therefore, the use of fluorescence in situ hybridisation (FISH) in parallel with cytogenetic analysis is important for the detection of those chromosomal abnormalities of prognostic significance. This chapter is dedicated to the preparation of ALL samples for cytogenetic and FISH analysis, with emphasis on the modifications to standard protocols which may be used to improve their quality. The specific difficulties encountered in the analysis of ALL metaphases and suggestions for overcoming them are provided. The chapter also includes an overview of the abnormalities that are expected to be found in this disease and how the results from both cytogenetics and FISH should be interpreted.

Key words: Acute lymphoblastic leukaemia, Chromosomal abnormalities, FISH interpretation, Prognostic significance

1. Introduction

It has been established for some time that specific chromosomal abnormalities found in acute lymphoblastic leukaemia (ALL) are important in the diagnosis of the disease, although of most importance is their association with outcome. Developments from these observations have been that the karyotype found at diagnosis is utilised for the risk stratification of patients for treatment, particularly within trial-based protocols. This applies especially to those abnormalities classified as poor risk, e.g. children with hypodiploidy (<44 chromosomes) and *BCR-ABL1* positive ALL, as well as those under the age of 2 years with rearrangements of the *MLL* gene.

Lynda J. Campbell (ed.), *Cancer Cytogenetics: Methods and Protocols*, Methods in Molecular Biology, vol. 730,
DOI 10.1007/978-1-61779-074-4_8, © Springer Science+Business Media, LLC 2011

Chromosomal analysis of the leukaemic cells of patients with ALL is challenging as the mitotic index is usually low and the chromosomal morphology is poor. Therefore, in these cases, the use of fluorescence in situ hybridisation (FISH) or other molecular procedures in parallel with cytogenetic analysis is particularly important to identify those chromosomal abnormalities of prognostic significance. Although the genetic picture of ALL is rapidly advancing, with the continual discovery of novel genetic changes using state-of-the-art technologies, many are currently research tools and the true prognostic significance of the novel changes in ALL remains to be determined.

This chapter is dedicated to the preparation of ALL samples for cytogenetic and FISH analysis, with emphasis on the modifications to standard protocols which may be used to improve the quality of the leukaemic cells. Alternative molecular procedures are mentioned but not fully described. The specific difficulties encountered in the analysis of ALL metaphases and suggestions for overcoming them are provided. The chapter also includes an overview of the abnormalities that are expected to be found in this disease and how the results from both cytogenetics and FISH should be interpreted.

2. Materials

The main reagents required for the different procedures are given below. Suggestions for suppliers are provided for the less common items. The techniques should be carried out in laboratories provided with standard equipment and good laboratory practise should be followed at all times.

2.1. Culturing and Slide Preparation

1. RPMI 1640 culture medium, supplemented with 25% foetal calf serum and 1% antibiotics (penicillin/streptomycin), stored at 4°C.
2. Lysis Buffer (Qiagen Kit, 100 mL bottles).
3. Colchicine solution (Colcemid, Sigma) stored at 4°C.
4. Bromodeoxyuridine (BrdU) (Sigma, 100 mg). Stock solution 10.0 mg per mL in sterile phosphate-buffered saline (PBS) stored at −20°C.
5. Dilute colchicine/BrdU (1.0 mL Colcemid plus 0.64 mL BrdU stock solution) in 8.36 mL sterile PBS. Store bottle wrapped in aluminium foil at 4°C.
6. Potassium Chloride (KCl) hypotonic solution 0.075 M (5.56 g/L) dissolved in distilled water (dH_2O) and stored at room temperature.
7. Virkon disinfectant (3% solution) stored at room temperature.

8. Carnoy's fixative: Analar methanol:Analar glacial acetic acid ratio 3:1, made fresh daily. Fixation should be carried out in a fume extraction hood.
9. 70% acetic acid, stored at room temperature.

2.2. Cytogenetic Analysis

1. Trypsin solution (one vial of Bacto-trpsin reconstituted with 10 mL dH_2O).
2. Buffer pH 6.8 prepared from buffer tablets.
3. Giemsa stain (2.5 mL of stock Giemsa solution) in 45 mL pH 7.0 buffer.
4. Saline (8.5 g NaCl in 1 L dH_2O).
5. Slide mountant (DPX or XAM).

2.3. Fluorescence In Situ Hybridisation

1. Directly labelled FISH probes.
2. Coverslips of various sizes.
3. Slide mounting medium: Vectorshield (Peterborough) containing 4′,6-diamidino-2-phenylindole (DAPI) (Vector Laboratories).
4. Rubber solution.
5. PBS, prepared according to the manufacturer's instructions, stored at room temperature.
6. Nail varnish for long-term slide storage.

2.4. Analysis

1. A bright-field microscope is required for the observation of G-banded metaphase chromosomes.
2. A microscope equipped for epifluorescence is required for FISH analysis. The minimum filter set required for standard three-colour FISH is DAPI (counterstain) and appropriate filters for the red and green fluorochromes.
3. Immersion oil for high-power examination of G-banded and FISH slides.

3. Methods

3.1. Setting Up Bone Marrow Cultures for Cytogenetic Analysis and FISH

Bone marrow samples may be received by post or courier. They should be processed as soon as possible to improve the chance of achieving a successful result. The sample should be handled in a class II safety cabinet.

1. Centrifuge the sample for 5 min at 200 *g* (see Note 1).
2. Remove the supernatant and discard into 3% Virkon solution (see Note 2). Virkon should be used each time that the cells are washed prior to fixation.

3. Re-suspend the cell pellet in a small volume of supplemented RPMI culture medium and count the cells (see Note 3).
4. Re-suspend the cells at a concentration of 1×10^6 cells per mL in 10 mL of supplemented medium in a culture tube. The number and type of culture is dependent on the number of cells available (see Note 4). Carefully mix and incubate at 37°C.

3.2. Harvesting and Slide Preparation

3.2.1. Harvesting

1. Add 50 μL colchicine to each culture to be harvested.
2. After a specified time (see Note 5), centrifuge the sample for 5 min at 200 *g*.
3. If the sample is to be cultured in colchicine overnight, substitute pure colchicine with 0.2 mL dilute Colcemid/BrdU for 16 h.
4. Gently re-suspend the cell pellet and add ~10 mL 0.075 M KCl hypotonic solution to the culture tube (see Note 6).
5. After 5 min, re-spin the tube for 5 min at 200 *g* and gently re-suspend the pellet.
6. Keeping the pellet mobile using a vortex mixer (see Note 7), add fixative (3:1 methanol:acetic acid) drop-wise for the first millilitre, then generously to 5 mL. Top up with more fixative (to ~10 mL) without vortexing (see Note 8).
7. Centrifuge the sample again for 10 min at 200 *g* (see Note 9). Remove and discard the supernatant. As the cells are now fixed, the use of Virkon is no longer required.
8. Re-suspend the pellet and top up to ~10 mL with fresh fixative.
9. Repeat steps 6 and 7 a couple of times or until the sample no longer appears brown.
10. Leave the fixed cells at −20°C for a minimum of 1 h or preferably overnight before slide making (see Note 10). The cells may be stored long term at −20°C without deleterious effects, in which case they should be transferred to a small screw topped Eppendorf tube with the pellet re-suspended in ~1 mL of fixative.

3.2.2. Slide Preparation

The cell suspension obtained after harvesting is used to make slide preparations for both cytogenetic analysis and FISH.

1. Clearly label a slide with appropriate identification and handle only one patient sample at a time.
2. Wash the cells by repeating steps 6 and 7 above.
3. Repeat step 6 again, but this time dilute the re-suspended pellet with only a small amount of fresh fixative until the suspension appears slightly cloudy (see Note 11).

4. For FISH, drop 2 μL of cell suspension onto the slide using a p20 Gilson pipette and mark the area with a diamond pen (see Note 12).
5. For cytogenetic analysis, one large drop from a standard glass pipette is added to the slide (see Note 13).
6. Examine the slide under low power (×10) using a phase-contrast microscope. For best results, nuclei should appear grey (not phase bright) and flat, with no residual cytoplasm. Cells should be spread with minimal touching between nuclei but at a sufficiently high density to allow easy scoring on high power. For metaphase analysis, chromosomes should be grey (not phase bright) and not overlapping (see Note 14).
7. Slides should be aged for a minimum of 20 min on a hot plate at 60°C.

3.3. Cytogenetic Analysis of ALL Samples

3.3.1. Chromosome Banding

1. Prepare the following coplin jars in the order:
   (a) 1–2 mL trypsin solution in 50 mL saline
   (b) 50 mL saline (rinse)
   (c) 50 mL pH 6.8 buffer (rinse) × 2 (see Note 15)
   (d) 50 mL Giemsa stain
   (e) 50 mL pH 6.8 buffer (rinse) × 2
2. Dip the slide into the first coplin jar containing trypsin for 2–20 s (see Note 16). Rinse the slide in the saline and buffers.
3. Stain for 5 min in the Giemsa solution and rinse in the second buffer.
4. Mount the wet slide to examine the quality of the banding.
5. If satisfactory, remove the coverslip and carefully dry the slide on a hot plate. If further banding is required, repeat the procedure from step 2.
6. Mount the slide with a coverslip using mountant.

3.3.2. Microscope Analysis

A bright-field microscope is required for the analysis of G-banded metaphases, fitted with a low-power lens (×10) for slide scanning to locate the metaphases and a high-power oil immersion lens (×100) to undertake the detailed analysis.

1. Using the low-power lens, scan the slide to locate metaphases for analysis (see Note 17).
2. Add a drop of immersion oil and examine the metaphases using a high-power lens (×100) (see Note 18) (Fig. 1). First, count the chromosomes to determine the ploidy levels and whether there are chromosomal gains and/or losses. Methodically identify each chromosome pair within each metaphase to identify any numerical or structural chromosomal abnormalities.

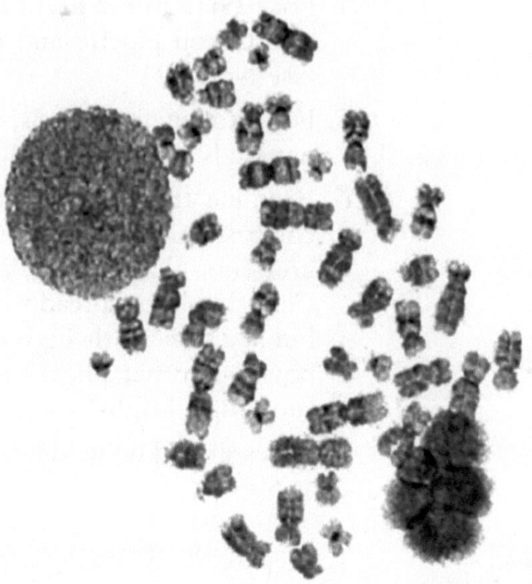

Fig. 1. A well spread, optimally banded metaphase from an ALL sample examined under high power (×100). Note the interphase nucleus on the left-hand side and debris on the right-hand side. There is no evidence of plasma membrane.

Only clonal abnormalities should be recorded. According to the International System for Human Cytogenetic Nomenclature (ISCN) (1), an abnormality is regarded as clonal if two or more cells contain the same chromosomal gain or structural change: three or more cells with the same chromosomal loss are required to define an abnormal clone.

3. If no abnormal cells are found, full analysis of 20 cells is required to assign a normal karyotype to an ALL sample. Less than 20 normal cells should be regarded as a fail. Although two (three) abnormal cells is sufficient to define an abnormal clone, it is advisable to analyse a minimum of 10 cells fully to indicate whether clonal evolution is present.
4. Karyotypes should be written according to the ISCN (1).

3.4. FISH Analysis

Due to the low mitotic index and poor chromosomal morphology in ALL, it is strongly recommended that FISH studies are carried out for the detection of those chromosomal abnormalities of prognostic significance, as detailed below (Subheading 3.5).

3.4.1. Probe Design

Three types of probe design described below are those most frequently used in the analysis of ALL. The individual probes are available from a number of companies.

3.4.1.1. Dual-Colour Breakapart Probes

These probes indicate disruption at a specific chromosomal breakpoint. Spectrum red (R) and SpectrumGreen (G) labelled probes are designed to hybridise to either side of a known

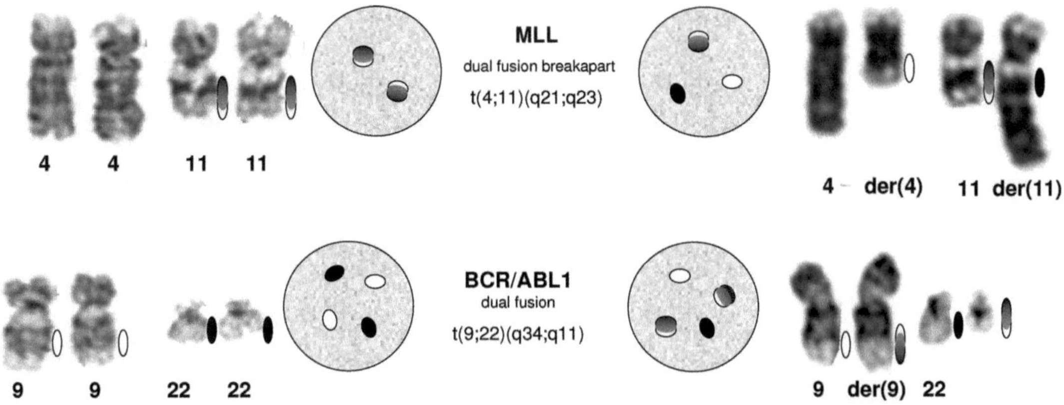

Fig. 2. A composite illustration of the breakapart and dual fusion probes and applications of FISH approaches, using *MLL* and *BCR-ABL1* rearrangements as examples. On the left-hand side, the normal configurations of the probes on metaphase chromosomes and in interphase cells are given, while on the right-hand side the expected corresponding abnormal signal patterns are shown. The black and white circles represent the red and green signals.

breakpoint cluster or to span a gene of interest. When the probe is hybridised to normal cells, the red and green signals are juxtaposed and appear as a yellow fusion (F) signal. A 0R0G2F signal pattern is produced in interphase, representing the intact locus on two normal homologous chromosomes. If a break occurs in the chromosome at this locus, the red and green signals become separated and appear as single signals. The resulting signal pattern is a "split" of one fusion signal into the component red and green parts (1R1G1F). This type of probe is particularly useful when the partner gene in a rearrangement is unknown, or when the target gene has multiple partners. The example of the breakapart probe given in Fig. 2 is the *MLL* locus located at chromosome band 11q23, which has more than 40 different partner genes.

3.4.1.2. Dual Fusion Probes

In dual fusion probes, both probes extend beyond the breakpoint of both chromosomes. Thus, when a translocation occurs, two fusions are formed, one on each derived chromosome as shown in the example given in Fig. 2 of the translocation, t(9;22)(q34;q11.2), which gives rise to the *BCR-ABL1* fusion gene. The standard abnormal signal pattern in this situation is 1R1G2F. The benefit of dual fusion probes over those which create only a single fusion is that they reduce the incidence of false-positive cells as dual fusion signal patterns rarely arise by chance.

3.4.1.3. Copy Number Probes

Centromeric probes are used to detect loss or gain of whole chromosomes and ploidy changes. They may be purchased, labelled with different fluorochromes, and should be applied in dual or triple colour combinations, e.g. in determination of high hyperdiploidy.

Unique sequence probes are used to identify deletion or gain of a particular gene. To detect the deletion of a gene of interest, a probe is selected to cover the gene (R). This is tested in a dual-colour hybridisation with another probe which identifies the chromosome on which this gene is localised, usually a centomeric probe (G). Two copies of the red and green probes (2R2G) indicate normal copy number, whereas 1R2G indicates a deletion of the gene of interest. Similarly extra copies of a specific gene may be identified.

3.4.2. FISH Method

The procedures described here refer to commercially available directly labelled probes, which are recommended for use in routine screening to provide consistency between laboratories. Probes should always be diluted according to the manufacturers' instructions. Most probes are used at a working dilution of one part probe to nine parts buffer mix. The total volume of probe mix used is dependent upon the size of the coverslip. For example, 3 μL is sufficient for a 13 mm^2 coverslip.

3.4.2.1. Hybridisation and Denaturation

The manufacturers' protocols should be employed; however, this standard protocol should apply to most probes.

1. Remove the probes from the freezer 5 min before use and leave at room temperature to thaw. Protect from direct light to prevent bleaching of the fluorochromes.
2. Pulse the probe containing tubes briefly in a micro-centrifuge to mix and collect all liquid together at the bottom of the tube.
3. Label a microtube with the name of each probe to be applied and aliquot the appropriate amount of probe into this tube taking care to avoid cross-contamination.
4. Pulse the probe mix as step 2 above.
5. Apply the appropriate volume of probe mix to each coverslip.
6. Invert the region of the slide to be hybridised onto the coverslip, aligning the coverslip with the spot of cell suspension. Surface tension will stick the coverslip to the slide when it is turned over.
7. Allow the probe mix to spread underneath the coverslip to the edges. This will usually require applying gentle pressure (see Note 19).
8. Seal the edges of the coverslip with rubber cement.
9. Place the slide in a hybridisation machine, set to the appropriate programme (see specific manufacturer's instructions) or use 72°C for 2 min to denature the probe and target followed by overnight hybridisation at 37°C (see Note 20).

3.4.2.2. Post-hybridisation Washes

1. Place a coplin jar containing stringent Wash 1 (see Notes 21–23) into a water bath and heat to 72°C.
2. Fill two more coplin jars: one with 2× SSC and one with Wash 2 and leave at room temperature.
3. When the solution in the first coplin jar reaches the required temperature, remove the slides one at a time from the hybridisation chamber. Using forceps, carefully remove the rubber cement.
4. Slide the coverslips off and place slides in 2× SSC. If the coverslips do not detach easily, place the slide into 2× SSC, where they will eventually soak off or become loose enough to detach in ~10 min.
5. Transfer the slides quickly to Wash 1 for 2 min. Ensure that the slides are in separate slots of the coplin jar (see Note 24).
6. Remove the slides from Wash 1 in the same order as they were inserted in order to keep the timings as similar as possible for each slide. Transfer slides to Wash 2 for 30 s to 2 min (see Note 25).

3.4.2.3. Mounting of Slides

1. Place the appropriate number of 24×50 mm^2 coverslips on absorbent paper and add 7 μL of slide mounting medium to each one.
2. For each slide in turn, drain off the excess liquid and dry the back of the slide on the paper.
3. Invert the slide onto the coverslip to come into contact with the slide mounting medium. Surface tension will stick the coverslip to the slide. Press firmly but carefully on the back of the slide whilst it is face down on the paper. This will allow excess fluid to be absorbed.
4. Turn the slide over and push out the air bubbles (see Note 26).
5. Store slides at 4°C in the dark until required.
6. For long-term storage, the edges of the coverslip should be sealed with nail varnish to prevent the slides from drying out and similarly stored in slide boxes at 4°C in the dark.

3.4.3. Interphase FISH Analysis

An epifluorescence microscope is required for analysis of the FISH slides. The minimum filter set needed for standard three-colour FISH is DAPI (counterstain) and appropriate filters for the red and green fluorochromes. A dual filter (which allows visualisation of both red and green fluorochromes at the same time) is frequently used for interphase analysis. It is recommended that results should be generated from the combined interpretation of two independent analysts. For a standard interphase FISH test on an ALL sample, 100 successful nuclei should be scored by each analyst, providing a total of 200 cells. If a sample is very sparse, a result from 100 cells may be acceptable with both analysts scoring at least 50 cells.

It is vital that the analyst has knowledge of the probe used and the expected signal patterns. All analysts should score the slides "blind," without knowledge of the other's results.

1. Select the correct area of cells to be analysed on a low-power lens (×10 or ×16) using the DAPI filter.
2. Using oil immersion, switch to a high-power lens (×100) and refocus the slide.
3. Use the dual filter to locate signals in the nuclei (see Note 27).
4. Determine the signal pattern and record it for each nucleus in turn. Avoid damaged nuclei and clumps of cells with indistinct boundaries.
5. Analyse 100 cells. A second analyst should analyse a further 100 cells from a different region of the area covered by the probe.

3.4.4. Calculation of Cut-Off Values

To guarantee the accuracy of the FISH result, it is important to determine cut-off values for each probe. There is a possibility of accidental juxtaposition of red and green signals in a normal cell or unexplained loss of a signal when no deletion is expected. These observations may be interpreted as a positive result in a normal case. To take such chance findings into account in the interpretation of results and to determine the false-positive level, a cut-off score should be calculated for every probe used as they can vary considerably. Five samples from normal samples should be evaluated. The sample type should be the same as the samples to be tested (bone marrow for leukaemia patients), and 200 nuclei should be scored by each analyst on all five cases (see Note 28). The false-positive rate for each and the mean false-positive rate for all five cases should be calculated. The cut-off level is the mean false-positive rate plus 3 standard deviations. This is expressed as a percentage and may differ between analysts. To maintain maximum accuracy, cut-off values should be revaluated periodically.

3.4.5. Interpretation of Interphase FISH Results

The FISH signal patterns and the number of cells with each pattern within the patient samples should be recorded independently by both analysts. This determines the number and types of cell populations present and their relative sizes. Any population occurring at a higher level than the cut-off value should be considered as positive (see Note 29). The results from the two analysts should be in close agreement. If they disagree, then a third independent analysis is required. The two sets of results which most closely agree after the third analysis should be used to provide the definitive result. Sometimes discrepant results may arise from poor probe hybridisation. In these circumstances, the test should be repeated or an alternative informative probe is used.

3.5. Important Chromosomal Abnormalities of Prognostic Significance in ALL

In both B- and T-lineage ALL, a large number of important chromosomal abnormalities have been reported. These are described in detail elsewhere (2). Although many have been shown to have prognostic significance, few are actually used in the risk stratification of patients for treatment. Here, we describe the small number of abnormalities in B-lineage ALL used to guide therapy, for which their accurate detection at the time of diagnosis is essential. Although cytogenetic analysis has always been the gold standard diagnostic approach, it is now regarded as insufficient as the sole test in these cases and additional complementary procedures (usually FISH or reverse transcriptase polymerase chain reaction, RT-PCR) are required for the complete genetic diagnosis. Here, we provide a brief description of these abnormalities and indicate the relative values of the techniques used for their detection as a guideline to the most appropriate diagnostic procedures (summarised in Table 1).

3.5.1. MLL Gene Rearrangements

Children with ALL and rearrangements of the *MLL* gene, particularly those under the age of 2 years, have a poor outcome. The most frequent abnormality is the translocation, t(4;11)(q21;q23). Although it is clearly apparent by cytogenetic analysis, the use of the MLL dual-colour breakapart probe (Fig. 2) confirms the involvement of *MLL* in these cases, as well as those involving other translocation partners, e.g. t(6;11)(q27;q23) and t(11;19)(q23;p13), which can be difficult to detect in poor quality preparations. This probe is useful to screen for the presence of cryptic abnormalities in patients with normal karyotypes and those which may be undetected in patients with a failed cytogenetic result (see Note 30). RT-PCR may be used as an alternative method, but is only informative in a multiplex approach including all fusion transcripts.

3.5.2. t(17;19)(q22;p13) with TCF3-HLF Fusion

The translocation, t(17;19)(q22;p13), which gives rise to the *TCF3-HLF* fusion, is a variant of t(1;19)(q23;p13), *TCF3-PBX1* fusion. Although very rare, patients with t(17;19) have an extremely poor outcome, with all reported cases having died. The abnormality is usually visible by cytogenetic analysis but may be confirmed using the dual-colour breakapart probe specific for *TCF3*. This probe can also, of course, be used for the detection of other translocations involving *TCF3*, primarily t(1;19).

3.5.3. t(9;22)(q34;q11.2) with BCR-ABL1 Fusion

Philadelphia chromosome (Ph) positive ALL has the translocation, t(9;22)(q34;q11.2), which gives rise to the *BCR-ABL1* fusion. Although rare in childhood, it is common in adult ALL. It is associated with a poor outcome in all age groups. Accurate identification is vital as patients with Ph positive ALL are now treated on specific protocols including imatinib treatment in addition to chemotherapy. Cytogenetic detection of t(9;22) is good; however,

Table 1
Chromosomal abnormalities used in risk stratification for ALL and the relative value of the techniques used for their detection

Abnormality	Prognosis	G-banding	FISH	RT-PCR	Notes
11q23 rearrangements t(4;11)(q21;q23) and multiple other translocation partners, involving *MLL* gene	Poor, particularly in children <2 years of age	Good but some translocations may be subtle and difficult to visualise	MLL dual-colour breakapart probe detects all translocations involving *MLL*, essential if cytogenetics has failed	Useful but multiplex approach required for detection of all partner genes	Care should be taken in interpretation of signal patterns indicating deletion of 3′ part of *MLL* probe
t(17;19)(q22;p13) with *TCF3-HLF* fusion	Extremely poor, although rare abnormality	Good	TCF3 dual-colour breakapart probe detects this translocation as well as t(1;19)(q23;p13) with *TCF3-PBX1* fusion, essential if cytogenetics has failed	Useful	
t(9;22)(q34;q11.2) with *BCR-ABL1* fusion	Poor	Good but some rearrangements may be cryptic	BCR-ABL1 dual-colour, dual-fusion probe detects cryptic abnormalities and deletions of der(9). Useful for disease monitoring	Effective, can distinguish between major and minor breakpoint	Q-PCR useful for monitoring of minimal residual disease
High hyperdiploidy (51–65 chromosomes).	Good	Good, but morphology may be too poor to identify individual trisomies	FISH using specific centromeric probes	Not applicable	DNA index by flow cytometry >1.16

Near-haploidy (25–29 chromosomes) and hypodiploidy (<44 chromosomes)	Poor	Good	FISH using specific centromeric probes can identify near-haploidy with a doubled population from classical high hyperdiploidy	Not applicable	DNA index can demonstrate near-haploid and the doubled populations Microsatellite analysis can distinguish between the doubled population and classic high hyperdiploidy
t(12;21)(p13;q22) with *ETV6-RUNX1* fusion	Good	Cytogenetically cryptic, but typical secondary abnormalities will be detected. Other techniques required for the detection of fusion	TEL-AML1 ES fusion probe is highly effective for the detection of fusion. It will also detect *ETV6* deletion, additional copies of chromosome 21 and der(21). Used to identify iAMP21	Effective for the detection of fusion, but will not provide information of secondary changes or diagnosis iAMP21	
iAMP21	Poor on standard therapy	Good for the detection of abnormal chromosome 21	TEL-AML1 ES fusion probe will indicate amplification of *RUNX1*. Chromosome 21 subtelomeric probe hybridised with *RUNX1* will distinguish iAMP21 from cases with multiple copies of entire chromosomes 21	Not applicable	Currently FISH with probes for *RUNX1* provide the only reliable detection method

a number of cryptic insertions have been described and three-way translocations may be difficult to interpret. FISH using the dual-colour, dual-fusion approach as shown in Fig. 2 provides a rapid and accurate detection method with a low false-positive rate. One advantage of this FISH approach is that deletions from the derivative chromosome 9 involving all or part of the reciprocal *ABL1-BCR* fusion may be simultaneously detected (3). RT-PCR for the BCR-ABL1 fusion transcript offers a good alternative detection method, which, with the appropriate primers, will distinguish between the major (M-BCR) and minor (m-BCR) breakpoint cluster regions within the *BCR* gene. The level of response to treatment may be measured by cytogenetics and FISH, which is a requirement of some of the imatinib trial protocols. However, the most accurate method to monitor the level of minimal residual disease following treatment is real-time quantitative PCR.

3.5.4. High Hyperdiploidy (51–65 Chromosomes)

In high hyperdiploidy, the majority of chromosomal gains are non-random; those most frequently occurring include trisomies of chromosomes 4, 6, 10, 14, 17, 18, 21, and gain of an X chromosome in both males and females. Multiple copies of chromosome 21 are often observed (4). Patients with high hyperdiploidy have a good prognosis. However, in US trials, the most favourable outcome is associated with the combined trisomies of chromosomes 4, 10, and 17 (triple trisomy) (5), while in UK trisomy 18 has shown the best outcome (4). Cytogenetic analysis will readily detect the presence of high hyperdiploidy in terms of chromosome number, but the poor chromosome morphology often precludes the identification of the gained chromosomes. FISH using selected centromeric probes accurately determines their origin. For example, three-colour hybridisation using centromeric probes specific for chromosomes 4, 10, and 17 shows the trisomies for these chromosomes in the same cells (see Note 31). There are no molecular techniques appropriate for the reliable detection of high hyperdiploidy. Determination of the DNA index at a level of >1.16 using flow cytometry will indicate the presence of a high hyperdiploid clone, but will provide no indication of the specific chromosomes gained.

3.5.5. Near-Haploidy (25–29 Chromosomes) and Hypodiploidy (<44 Chromosomes)

Near-haploidy provides an example of another numerical chromosomal change. Although rare, it is associated with an adverse outcome. In these cases, a haploid chromosome set with a single chromosome representing each pair is observed, typically with gains of chromosome X, 14, 18, and 21 (6). Cytogenetics provides a reliable method for detection. Near-haploidy is frequently associated with a population of cells with a doubling of the near-haploid chromosome number. Furthermore, in a number of cases, the doubled population is the only one seen at diagnosis. The presence of this doubled population does not improve the outcome, thus it is important that this presentation is distinguished from one with classical high hyperdiploidy. This is not possible from the determination of chromosome number alone. Thus, DNA index measurement may not be helpful unless the near-haploid population can also be detected. One problem is that often the doubled population is not an exact double of the near-haploid number. FISH may be helpful in accurate identification of the gained chromosomes. However, the most reliable method to distinguish high hyperdiploidy from a doubled near-haploid population is by microsatellite analysis. Markers for those chromosomes shown as diploid are more likely to be heterozygous and homozygous, respectively, in these two situations.

3.5.6. t(12;21)(p13;q22) with *ETV6-RUNX1* Fusion

This translocation is a common abnormality in young children with ALL, occurring at an overall incidence of ~22%. It was not discovered until the mid-1990s as the translocation is cryptic by cytogenetic analysis. FISH, using the dual-colour or extra signal approach as described for *BCR-ABL1* in Fig. 2, offers a highly effective detection method. RT-PCR for the detection of the fusion transcript is also a successful approach, but the advantage of FISH is that it will simultaneously identify the secondary chromosomal changes which are associated with the translocation, e.g. the gain of chromosome 21 (seen as an additional *RUNX1* (red) signal), an extra copy of the derived chromosome 21 (seen as an additional fusion (yellow) signal) and the deletion of the normal homologue of *ETV6* (seen as loss of the green signal specific for *ETV6*).

This probe design led to the identification of the abnormality described as intrachromosomal amplification of chromosome 21 (iAMP21) (7), which is associated with a poor outcome in children with ALL on standard therapy (8). Although the initiating mechanism of iAMP21 is unknown, patients are negative for the *ETV6-RUNX1* fusion, while all of them have at least three additional copies of the *RUNX1*gene located at the abnormal chromosome 21. To confirm the presence of this abnormality in interphase, the *RUNX1* probe is applied in a dual-colour hybridisation together with a probe specific for the subtelomeric region of chromosome 21. Patients with iAMP21 show normal or loss of copy number for this subtelomeric region. Therefore, a ratio of 5:1/2 *RUNX1*: subtelomere will distinguish iAMP21 from patients with multiple copies of intact chromosomes 21 as in high hyperdiploidy, in which the ratio will be 1:1. Currently, this FISH approach provides the only reliable detection method for iAMP21.

3.5.7. Notes

1. Samples are usually transported in a medium containing preservative-free heparin to support the bone marrow cells during transit and prevent clumping. Upon receipt, the cells are transferred to fresh complete medium to encourage growth.

2. 3% Virkon is used as an appropriate strength disinfectant against infectious bacteria.
3. To count cells, add 50 μL of bone marrow suspension into a small sterile container, e.g. an Eppendorf tube. Add 50 μL of red cell lysis reagent and shake gently to mix. Count the cells using a haemocytometer. Multiply the cell number by 2 (dilution factor) and then by 10^4 to obtain the number of cells/mL in the original sample.
4. Leukaemic blasts from patients with ALL must be cultured over short time periods as they cannot be maintained long term in vitro. The cultures should normally include one direct (harvested immediately) and one cultured overnight. However, if there are only sufficient cells for a single culture, the overnight is the best option. Spare cells should be spun down and stored in liquid nitrogen for future studies requiring DNA and RNA to be extracted.
5. Colchicine is a spindle poison, which arrests cells in metaphase. The incubation time varies according to laboratory practise, ranging from 15 min to 2 h. Sometimes overnight incubation is carried out; although this is likely to increase the mitotic index, many of the metaphases will have highly condensed chromosomes. Chromosome condensation can be reduced by adding dilute colchicine/BrdU to the overnight culture.
6. KCl may be pre-warmed to 37°C to prevent temperature shock to the cells and to promote chromosome spreading.
7. Vortexing during this first fixation is particularly important to prevent clumping of white cells. If the red blood cells have been removed prior to cell culture, the white cells become more fragile. Thus, keeping the pellet mobile as the fixative is added reduces the chance of the cells rupturing during the fixation process.
8. At this first fixation step, the suspension will appear brown due to the presence of lysed red blood cells.
9. Fixed cells require longer centrifugation to minimise cell loss.
10. Acetic acid softens the cell membrane; therefore, overnight fixation improves the spreading of the chromosomes as the cell membranes rupture more easily upon contact with the glass slide.
11. The concentration of the cells in suspension depends on the purpose for the slide being made. Under-dilution of a cell suspension may lead to poorly spread metaphases; over-dilution and subsequent over-spreading may lead to chromosome loss. If necessary, experiment to find the optimum dilution.

12. Marking the limits of the cell suspension "spot" helps to relocate the cells while the probe is being applied and later to read the result.
13. For FISH, it is possible to achieve good results from a very small aliquot of cell suspension: ~2 μL is usually adequate for most interphase and metaphase FISH. In contrast, larger quantities are required to make slides for conventional cytogenetic analysis and the cell suspension needs to be more dilute to allow the chromosomes to spread.
14. Slide making to produce good metaphase spreads from ALL blasts is difficult. It is susceptible to subtle variations in temperature, air current, and humidity. In general, warmer, moist, and still conditions produce the best slides. Conditions are less critical for slides made for interphase FISH as many probes are sufficiently robust to hybridise well to suboptimal slides. A number of steps may be added to improve metaphase quality of ALL samples.
    (a) The slide should be cleaned before use (wipe with a tissue soaked in a small amount of fixative).
    (b) Breathe on the slide to add a layer of condensation immediately prior to dropping the cell suspension.
    (c) Drop the cell suspension in a single motion.
    (d) Rotate the slide once after dropping the cells.
    (e) Add a second drop of fixative to the slide as the edges of the spot begin to dry: at the time that the Newton's rings become visible.
    (f) Add 3–4 drops of 70% acetic acid in methanol at the time that the Newton's rings become visible. Cover the spot and allow the acetic acid to remain on the slide and evaporate slowly. Leave for ~2–3 min, rotating the slide occasionally to ensure contact is maintained with the spot. Drain excess acetic acid from the slide onto a tissue and air dry.
    (g) Keep the slide motionless whilst drying to avoid unnecessary air currents.
    (h) Dry the slide slowly on a damp paper towel (to create a humid atmosphere).
    (i) Dry the slide by warming on the back of the hand.
15. It is important that the pH of the buffer is accurate.
16. The time in trypsin varies between samples and is also dependent on the ambient temperature. Therefore, it is advisable to dip one slide at a time to determine the optimum time to achieve acceptable banding patterns. When the chromosomes

appear too dark, a longer treatment time in trypsin is required. If they appear pale, restaining in Giemsa may be adequate.

17. In ALL, the metaphases are often of poor morphology; therefore, if the slide contains a mixture of cells of good and poor quality, it is likely that they represent normal and leukaemic cells, respectively. Initial analysis should be carried out on the lesser quality cells.
18. A green filter will render the pink Giemsa bands black and white for easier examination.
19. Air bubbles must be removed as they prevent efficient probe hybridisation. Use a pipette tip, blunt pencil or the barrel of a Gilson pipette to apply gentle pressure to the coverslip from the middle outwards.
20. Hybridisation of the slide for longer than 16 h (overnight) at 37°C may increase the levels of background staining.
21. Coplin jars break easily when exposed to sudden changes in temperature.
22. The detergent in Wash 1 removes background staining from the slides.
23. Do not overfill the jar as the addition of slides will displace the liquid.
24. Start the timer after the last slide has been added. This is a stringent wash; therefore, accurate timing is important. Add no more than five slides to each Coplin jar. The addition of each slide cools down the solution and it is important that the temperature is maintained at 72°C.
25. This wash reduces non-specific binding of probe and prevents the slide from drying out whilst preparing for the next stage.
26. Air bubbles must be removed as they can affect focussing of the microscope.
27. Signals may not be located within a single plane of focus, and therefore when the nuclei are scanned it is important to focus up and down through the nuclei.
28. It is important that cut-off values are determined for each analyst as interpretation of signal patterns can vary considerably between individuals. This exercise should demonstrate close agreement between analysts.
29. Although very low-level positive results, close to the cut-off value, are rare in diagnostic acute leukaemia samples, they may occur from time to time. The results should be scrutinised and if possible the sample should be tested with different probe or an alternative, appropriate molecular test.
30. Sometimes a FISH pattern of 0R1G1F, indicating deletion of the 3′ part of the *MLL* gene, may be observed. Although the

loss of the 3′ *MLL* signal may result from a concurrent translocation and deletion event, this observation should be interpreted with caution. A number of such cases have been shown to have no involvement of the *MLL* gene, with deletion of sequences 3′ of the gene covered by the probe. Such aberrant FISH signal patterns should be further investigated by alternative methods (9).

31. Hybridisation with centromeric probes specific for those chromosomes not usually gained in high hyperdiploidy, e.g. chromosomes 3 and 7, will distinguish high hyperdiploidy from the rare ploidy change of triploidy with gain of a complete haploid chromosome set.

References

1. Schafer, L., Slovak, M., and Campbell, L. (2009) *An International System for Human Cytogenetic Nomenclature*, S. Karger, Basel.
2. Harrison, C. J. and Johansson, B. (2009) Acute Lymphoblastic Leukaemia. In "Cancer Cytogenetics" (Third Edition). Eds S. Heim and F. Mitelman, John Wiley and Son Inc., NJ, USA.
3. Robinson, H. M., Martineau, M., Harris, R. L., Barber, K. E., Jalali, G. R., Moorman, A. V., Strefford, J. C., Broadfield, Z. J., Cheung, K. L., and Harrison, C. J. (2005) Derivative chromosome 9 deletions are a significant feature of childhood Philadelphia chromosome positive acute lymphoblastic leukaemia, *Leukemia* 19, 564–571.
4. Moorman, A. V., Richards, S. M., Martineau, M., Cheung, K. L., Robinson, H. M., Jalali, G. R., Broadfield, Z. J., Harris, R. L., Taylor, K. E., Gibson, B. E., Hann, I. M., Hill, F. G., Kinsey, S. E., Eden, T. O., Mitchell, C. D., and Harrison, C. J. (2003) Outcome heterogeneity in childhood high-hyperdiploid acute lymphoblastic leukemia, *Blood* 102, 2756–2762.
5. Sutcliffe, M. J., Shuster, J. J., Sather, H. N., Camitta, B. M., Pullen, J., Schultz, K. R., Borowitz, M. J., Gaynon, P. S., Carroll, A. J., and Heerema, N. A. (2005) High concordance from independent studies by the Children's Cancer Group (CCG) and Pediatric Oncology Group (POG) associating favorable prognosis with combined trisomies 4, 10, and 17 in children with NCI Standard-Risk B-precursor Acute Lymphoblastic Leukemia: a Children's Oncology Group (COG) initiative, *Leukemia* 19, 734–740.
6. Harrison, C. J., Moorman, A. V., Broadfield, Z. J., Cheung, K. L., Harris, R. L., Reza, J. G., Robinson, H. M., Barber, K. E., Richards, S. M., Mitchell, C. D., Eden, T. O., Hann, I. M., Hill, F. G., Kinsey, S. E., Gibson, B. E., Lilleyman, J., Vora, A., Goldstone, A. H., Franklin, I. M., Durrant, J., and Martineau, M. (2004) Three distinct subgroups of hypodiploidy in acute lymphoblastic leukaemia, *British Journal of Haematology* 125, 552–559.
7. Harewood, L., Robinson, H., Harris, R., Al Obaidi, M. J., Jalali, G. R., Martineau, M., Moorman, A. V., Sumption, N., Richards, S., Mitchell, C., and Harrison, C. J. (2003) Amplification of AML1 on a duplicated chromosome 21 in acute lymphoblastic leukemia: a study of 20 cases, *Leukemia* 17, 547–553.
8. Moorman, A. V., Richards, S. M., Robinson, H. M., Strefford, J. C., Gibson, B. E., Kinsey, S. E., Eden, T. O., Vora, A. J., Mitchell, C. D., and Harrison, C. J. (2007) Prognosis of children with acute lymphoblastic leukemia (ALL) and intrachromosomal amplification of chromosome 21 (iAMP21), *Blood* 109, 2327–2330.
9. Barber, K. E., Ford, A. M., Harris, R. L., Harrison, C. J., and Moorman, A. V. (2004) MLL translocations with concurrent 3′ deletions: interpretation of FISH results, *Genes Chromosomes Cancer* 41, 266–271.

Chapter 9

Cytogenetic Methods in Chronic Lymphocytic Leukemia

Claudia Haferlach and Ulrike Bacher

Abstract

In chronic lymphocytic leukemia of the B-lineage (B-CLL), cytogenetic alterations are highly relevant for prognosis and therapeutic decisions. With conventional techniques, chromosome banding analysis in CLL has been hampered by the low quality of metaphases and low rates of cytogenetic alterations due to a low in vitro proliferation rate of CLL cells. Thus, interphase fluorescence in situ hybridization (FISH) has become the standard technique for cytogenetic analysis in CLL. However, interphase FISH is not able to provide an overview on the whole karyotype. In order to improve chromosome banding analysis in CLL, specific stimulation techniques have been developed. These either use CD40 ligand or oligonucleotides (e.g., CpG) and IL-2 in combination. With the respective techniques, metaphase cultivation is successful in >90% of CLL cases and aberrant karyotypes can be detected in nearly 90% of CLL cases. This has allowed the detection of new clinically relevant subgroups (e.g., complex aberrant karyotype cases) and a more differentiated picture of distinct cytogenetic subtypes, e.g., cases with a 13q deletion. Efforts should continue to define the value of chromosomal banding in CLL focusing as well on the interaction with already established techniques such as interphase FISH or immunophenotyping.

Key words: Chronic lymphocytic leukemia, Cytogenetic alterations, Chromosomal banding, Culture techniques, Interphase FISH

1. Introduction

B-chronic lymphocytic leukemia (B-CLL) represents the most frequent mature lymphatic neoplasm and the most frequent leukemia in adults. It has an incidence of 3/100,000 per year in the western hemisphere. After the age of 70 years, the incidence increases up to almost 50/100,000 per year (1). The disease is characterized by sustained increase of peripheral blood lymphocytes $>5 \times 10^9$/L with a predominance of small mature lymphocytes in the blood smear. The immunophenotype of B-CLL is characterized by co-expression of CD5 and CD19 antigens. Expression of CD23 and the weak expression of surface light

Lynda J. Campbell (ed.), *Cancer Cytogenetics: Methods and Protocols*, Methods in Molecular Biology, vol. 730, DOI 10.1007/978-1-61779-074-4_9, © Springer Science+Business Media, LLC 2011

chain immunoglobulins allow us to distinguish most cases of B-CLL from other CD5-positive B-cell lymphoma (e.g., mantle cell lymphoma).

Clinically, CLL appears to be a highly heterogeneous disease. While some patients have a normal life expectancy, others show rapid progression with survival of a few months only. As a consequence of this clinical heterogeneity, therapeutic concepts range from a "watch and wait" strategy to allogeneic stem cell transplantation for selected younger or high-risk patients. Others may benefit from cytotoxic treatment or therapy with monoclonal antibodies against the CD20 antigen (rituximab). Thus, risk stratification is highly important in CLL. Diverse staging systems – lymphocyte doubling time, serum parameters such as beta-2 microglobulin, immunophenotype characteristics (e.g., ZAP-70 expression), or the immunoglobulin variable heavy chain (IgV_H) mutation status – play a central role in diagnostics and risk assessment (2).

In most patients with CLL, cytogenetic alterations can be identified and are highly relevant for prognosis (3). In contrast to the acute leukemias or myelodysplastic syndromes, only a few studies have investigated chromosome banding analysis in CLL due to the low in vitro proliferative activity even in the presence of B-cell mitogens. Either the generation of metaphases was completely hampered or clonal aberrations were detectable in 40–70% of cases only – due to the poor quality of the metaphases or as the normal hematopoietic cells showed in vitro proliferation while the CLL cells did not (4, 5).

Therefore, other methods were necessary to determine these recurrent abnormalities in CLL. Fluorescence in situ hybridization (FISH) and comparative genomic hybridization (CGH) on interphase nuclei do not require proliferating cells. With FISH, clonal aberrations were identified in >80% of CLL patients. Prognosis is worst in cases with 17p or 11q deletions, while other aberrations such as a sole deletion of 13q are associated with favorable outcomes (6).

However, both CGH and FISH have shortcomings. CGH detects genomic imbalances but misses balanced translocations. Interphase FISH is restricted to the genes/loci for which probes were selected. Thus, new approaches were needed to identify specific cultivation techniques capable of generating sufficient metaphases in CLL. Stimulation of CLL cells with CD40 ligand (a B-cell mitogen) or a combination of CpG-oligodeoxynucleotides and IL-2 was demonstrated to overcome the problem of established cultivation techniques in this disorder (7, 8). This chapter provides an overview of chromosome banding techniques using these new cultivation techniques in B-CLL.

1.1. Application of CD40

Conventional chromosome banding techniques used the B-cell mitogen 12-*O*-tetradecanoyl-phorbol-13-acetate (TPA). With this approach, the number and quality of metaphases were low in CLL cases. In lymph node proliferation centers in CLL, the environment protects the lymphatic cells from apoptotic and cytotoxic triggers. This environment is missing in in vitro culture. Prolonged in vitro CD40 stimulation of B-CLL cells was shown to result in upregulation of antiapoptotic Bcl-xL, A1/Bfl-1, and Mcl-1 proteins, and to mediate resistance to various classes of drugs, e.g., fludarabine or bortezomib. Thus, addition of CD40 was able to induce an antiapoptotic profile in CLL which was similar to the effects of *BCR-ABL1* in chronic myeloid leukemia (9). Using these antiapoptotic effects of CD40 application, generation of metaphases is markedly improved in CLL.

1.2. Application of Oligonucleotides and IL-2

B-CLL cells are arrested in G0/early G1 phase of the cell cycle and are characterized by a marked hyporesponsiveness toward a variety of polyclonal B-cell activators. Regulation of early cell cycle progression differs between B-CLL cells and normal B-cells. Costimulation with CpG-oligonucleotides and IL-2 can overcome this proliferative defect (10). Thus, another approach to induce in vitro stimulation of B-CLL cells is the above combination of the CpG-oligonucleotide (DSP30) and interleukin-2 (IL-2).

Briefly, for metaphase induction, peripheral blood mononuclear cells are cultured in RPMI 1640 medium with 20% fetal calf serum (FCS) in the presence of the immunostimulatory CpG-oligonucleotide DSP30 and interleukin 2 (IL-2). After 48 h, colcemid is added for another 24 h before chromosome preparation. The above procedure results, in most cases, in a resolution of 200–300 bands per haploid karyotype (11). This method has now become the standard technique in several laboratories and is very robust in a routine setting.

2. Materials

1. A fresh sample of peripheral blood or bone marrow is required (see Note 1). In the case of a peripheral blood leukocytosis, use 5–10 ml of heparinized peripheral blood (peripheral blood is preferred to bone marrow for karyotyping in CLL/malignant lymphoma, as contaminating dividing cells of normal hematopoiesis are less frequent). Alternatively, heparinized bone marrow may be used (see Note 2).
2. Medium: RPMI 1640 medium (Gibco, Gaithersburg, MD) with L-glutamine – 400 mL.

3. FCS (20%): 100 ml of FCS is added to 400 ml of RPMI; together with antibiotics/antimycotics. Store at 4°C temperature.
4. CpG-oligonucleotide DSP30: Order 1 μmol of DSP30: sequence: 5′- TsCsgsTsCsgsCsTsgsTsCsTsCsCsgs CsTsTsCs TsTsCsTsTsgsCsC, dilute 500 nmol in 5 mL distilled water to a concentration of 100 pmol/μL. Use 100 μL (=10 nmol) of DSP30 solution per 5 mL prepared medium (see Note 3).
5. Interleukin 2 (IL-2): Biological activity of IL-2: ≥1 × 10^7 U/mg (=10^5 U/10 μg). Add 10 μg IL-2 to 1,000 μL RPMI media 1640 (=100,000 U/1,000 μL = stock solution). Aliquot stock solution: 50 μL (=5,000 U) stock solution + 950 μL RPMI media 1640 into 20 Eppendorf tubes. Concentration of IL-2 for use: ~5,000 U/1,000 μL. Add 100 μL (=500 U) IL-2 per 5 mL prepared medium (see Note 3).
6. TPA (Phorbol-12-myristate-13 acetate).
7. Colcemid (0.15 g/mL) (Sigma, Munich, Germany).
8. Potassium chloride (KCl): 5.592 g KCl added to 1,000 mL distilled water. Store at 4°C temperature.
9. Carnoy's fixative: acetic acid (100%) – 50 mL; methanol – 150 mL. Storage temperature: –20°C, stability: 24 h (maximum 48 h).
10. Phosphate buffer: Di-Na-hydrogen phosphate – 11.4 g; K-dihydrogen phosphate – 4.9 g; add 1,000 mL distilled water. Store at 4°C temperature.
11. 2× SSC for chromosome banding: NaCl – 7.65 g; tri-sodium citrate dehydrate – 44.10 g; add 2,000 mL distilled water. Store at room temperature.
12. Giemsa Staining solution: 60 mL of phosphate buffer with 600 mL distilled water; remove 37.5 mL and add 37.5 mL of Giemsa stain.

3. Methods

3.1. Preparation of Samples for Cultivation of Metaphases

1. Measure the white blood cell (WBC) count of patient sample (Sysmex® nucleocounter); required cell count: 1 × 10^7 leukocytes/5 mL medium.
2. Add sample volume to culture tube with 5 mL of prepared medium.
3. Prepare two culture tubes: Add 100 μL of IL-2 solution and 100 μL of DSP30 solution to one culture tube. Add 50 μL TPA to the other culture tube (see Table 1).

Table 1
Supplements for cultivation

Supplements	Volume (μL)
Prepared medium	5,000
TPA	50
DSP30	100
IL-2	100

3.2. Cultivation of Metaphases for Cases with CLL

1. Incubate culture tubes for 48 h (37°C).
2. Add 150 μL colcemid (1.5 μg) to DSP30+Il2 culture and incubate for additional 24 h to generate a higher number of metaphases.
3. Incubate TPA culture for 72 h and then add colcemid for 2 h.

3.3. Harvesting of Cultures

1. Harvesting of cultures in our laboratory is done with the Tecan® pipetting robot.
2. In case the Tecan robot is not available or cannot be used due to technical problems (e.g., clots in the culture due to autoantibodies), the following manual procedure is used.
3. After centrifugation, remove the supernatant.
4. Suspend the sediment in 10 mL KCl (5.592 g/L).
5. After centrifugation and removal of supernatant, resuspend sediment in methanol/acetic acid (−20°C), wait for 30 min (room temperature) before proceeding.
6. Repeat step 5 twice without waiting period.
7. After centrifugation and removal of supernatant, resuspend sediment.
8. Drop suspension on glass slides and allow slides to dry.

3.4. Banding and Staining

1. Banding is carried out by incubation of dried slides at 60°C in 2× SSC solution.
2. Staining is performed using Giemsa stain diluted in distilled water and phosphate buffer for 3 min.

3.5. Evaluation of Metaphases

1. Capture metaphases using the metaphase finder Metafer (Metasystems®).
2. A minimum of 20 metaphases should be analyzed in CLL cases to establish a normal karyotype.
3. An abnormal karyotype should be correlated with the interphase FISH results for reporting (see Note 4).

4. Conclusions

Similar to other hematological malignancies, cytogenetic alterations are prognostically highly relevant in B-CLL. Chromosome banding analysis requires in vitro cell proliferation, while CLL cells show a low in vitro mitotic index (12). Therefore, with standard techniques, chromosome banding analysis shows high failure rates, and lower proportions of cases with aberrant karyotypes have been reported (5).

Interphase FISH analysis which does not require dividing cells has become the standard technique for diagnosis in CLL. Mostly, probes are used for the detection of trisomy 12 and deletions of 6q21, 11q22.3, 13q14, and 17p13. This allows the detection of aberrations in ~80% of CLL samples. Deletions of 13q are the most frequent alteration (in 55% of patients), being followed by 11q deletions, trisomy 12, and deletions of 6q and 17p (6, 13). However, in contrast to chromosome banding analysis, which gives an overview of all microscopically visible aberrations without previous knowledge of the affected regions, interphase FISH is able to detect limited patterns of aberrations only.

Thus, efforts continue to improve the karyotyping technique in CLL. One approach is the stimulation of the CLL cells with CD40 ligand or with specific cytokines. Using CD40-ligand stimulation, Buhmann et al. induced metaphases in >90% of cases in contrast to <80% with conventional methods. Cytogenetic aberrations were demonstrated in nearly 90% of cases in contrast to only 20–25% by conventional cytogenetic techniques (8). The approach of CD40-ligand stimulation, however, requires a labor-intensive, cellular co-culture system.

Alternatively, the CpG-oligonucleotide DSP30 and IL-2 can be combined for immunostimulation. With the above described culture techniques, stimulation of metaphases was successful in ~95% of cases in our laboratory. Chromosomal aberration rates were detected in >80% of cases. This corresponded to an almost twofold increase when compared with earlier chromosomal banding studies and was comparable to the aberration rate detected by FISH analyses (14, 15). Also, we found cytogenetic alterations in ~30% of cases with an apparently normal karyotype as assessed by FISH (Fig. 1). This was due to missing the respective alterations with a standard FISH panel. Only the use of CD40-ligand as a B-cell stimulus produced comparable results with respect to metaphase generation and chromosomal aberrations (7, 8).

Importantly three studies demonstrated that CpG/IL-2 does not induce clonal cytogenetic changes. The detected abnormalities

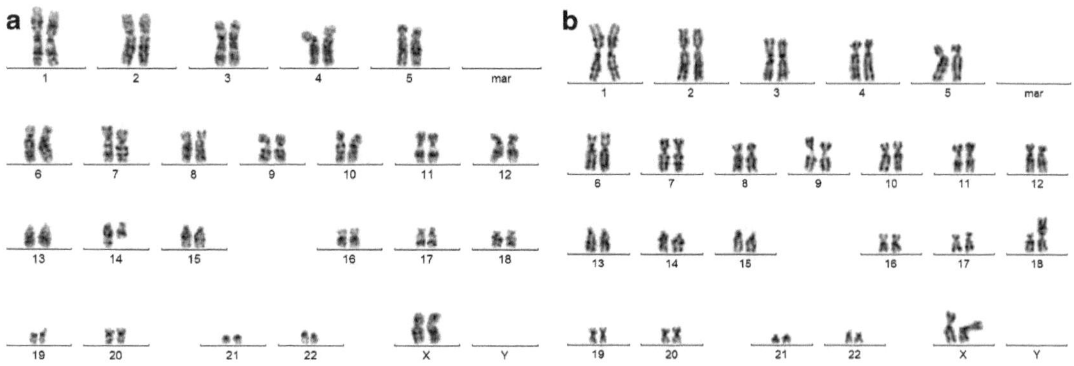

Fig. 1. Example of two cases in which the standard FISH panel did not detect any abnormality: (**a**) G-banded karyotype with interstial 14q deletion and (**b**) G-banded karyotype of a patient with an unbalanced rearrangement between chromosomes 2 and 18 leading to a trisomy 2p and a monosomy 18p.

hence represent true CLL-associated aberrations. Wu et al. cultured blood of healthy donors and did not observe clonal chromosome aberrations (16). Dicker et al. also showed that normal karyotypes were obtained after CpG/IL-2 stimulation of samples of some CLL patients and of healthy individuals and that sequential analysis yielded the same aberrations in two tested cases (11). Accordingly, Put et al. did not detect clonal aberrations in samples from five healthy donors, and most abnormalities in patient samples were found in both CpG/IL-2 and TPA cultures (17). Moreover, nonclonal aberrations were not more prevalent after CpG/IL-2 cultures compared with TPA cultures.

Chromosome banding analysis might contribute to prognostic predictions in CLL, and thus may lead to an improved understanding of the biological background of this heterogeneous disorder: one example was the description of reciprocal rearrangements in 34% of CLL patients by Mayr et al. Breakpoints clustered in the regions previously described as being deleted in CLL, e.g., 13q14, 11q21–1125, or 14q32. The occurrence of reciprocal translocations was an unfavorable prognostic parameter (7).

Previous studies demonstrated that increasing numbers of clonal cytogenetic abnormalities were associated with shorter survival (5, 7). We defined complex aberrant karyotypes by the combined occurrence of at least three clonal aberrations in accordance with the definition in other hematological malignancies and the cited studies in CLL (5, 18). In contrast to FISH data which were able to identify complex aberrant karyotypes in only ~3% of cases, complex aberrant karyotypes were detected in ~20%

of CLL cases with the above culture techniques. Similar to acute myeloid leukemia, there was an association between *TP53* deletion and the number of chromosome aberrations (14). Further, complex aberrant karyotypes were associated with an unmutated IgV_H status. Therefore, FISH analysis seems to underestimate the complexity and heterogeneity of cytogenetic aberrations in CLL when compared to modern chromosomal banding techniques (11, 15).

As an additional aspect, chromosome banding can reveal new subgroups within defined cytogenetic aberrations. In cases with a 13q deletion (according to interphase FISH), chromosomal banding can differentiate between interstitial deletions and reciprocal translocations (Figs. 2 and 3). Also, additional cytogenetic abnormalities were identified with chromosomal banding in >30% of cases which seemed to have a sole del(13q) with interphase FISH (15). Furthermore, clonal evolution can easily be followed by chromosome analysis (Fig. 4).

Finally, the use of automated procedures, such as pipetting or relocation of metaphases, allows us to improve the technical quality of chromosomal banding analysis, e.g., due to higher numbers of metaphases which can be screened for further evaluation. These novel options of automation in cytogenetics are able to contribute as well to the success of chromosomal banding in CLL cases.

In conclusion, stimulation with the oligonucleotide DSP30 and IL-2 is an easy and efficient stimulus for metaphase generation in CLL. Two recently published studies confirmed that this novel

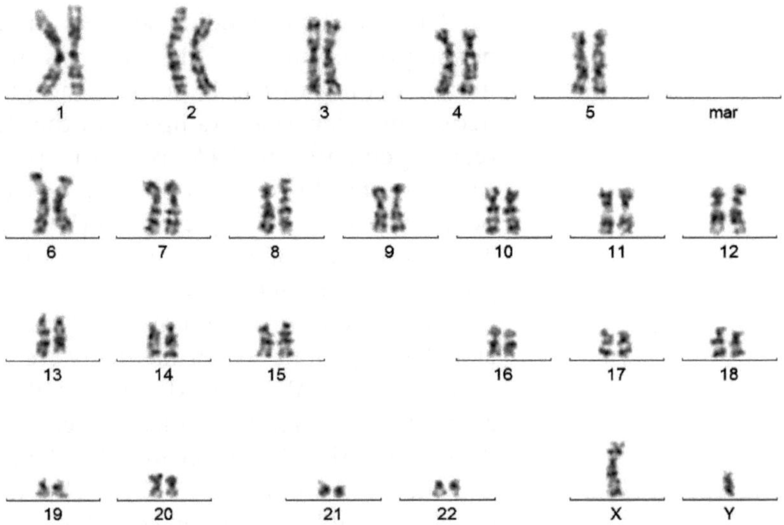

Fig. 2. Interstitial 13q deletion in CLL demonstrated by chromosome banding analysis.

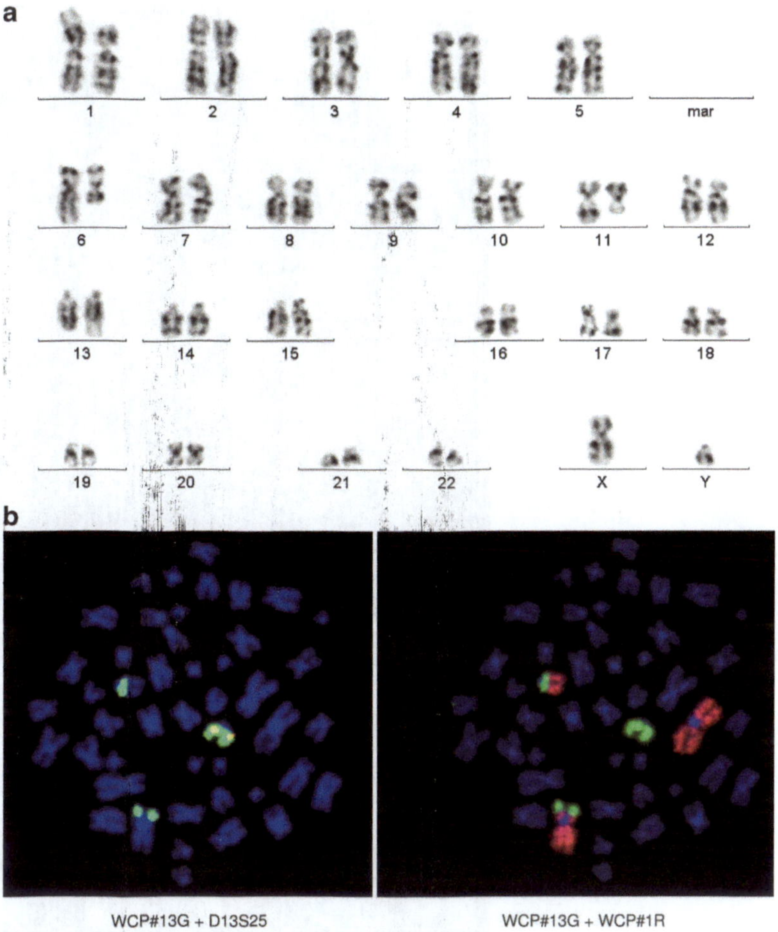

Fig. 3. Translocation between the short arm of chromosome 1 and the long arm of chromosome 13 associated with a loss of 13q14 material. (**a**) G-banded karyotype showing *t*(1;13) and a 6q-deletion, as well as an 11q-deletion. (**b**) Chromosome painting of the same patient demonstrating loss of 13q14 material by using a probe for the locus D13S25 which is missing on the derivative chromosome 1 as well as on the derivative chromosome 13 in conjunction with chromosome 13 WCP, and confirming the *t*(1;13) using chromosomes 1 and 13 WCP.

cultivation technique leads to an optimization of chromosome analysis in CLL (17, 19). This results in a more comprehensive genetic characterization of CLL, and insights in the whole genome improve. As demonstrated by the subgroup of complex aberrant karyotypes in CLL, new clinically relevant subgroups can be identified in CLL with chromosome banding. Future trials should evaluate the prognostic value of chromosome banding analysis in direct comparison to other clinically relevant parameters, e.g., interphase FISH or the IgV_H mutation status. Finally, it remains to be clarified which combination of techniques is necessary to optimize prognostic predictions in CLL.

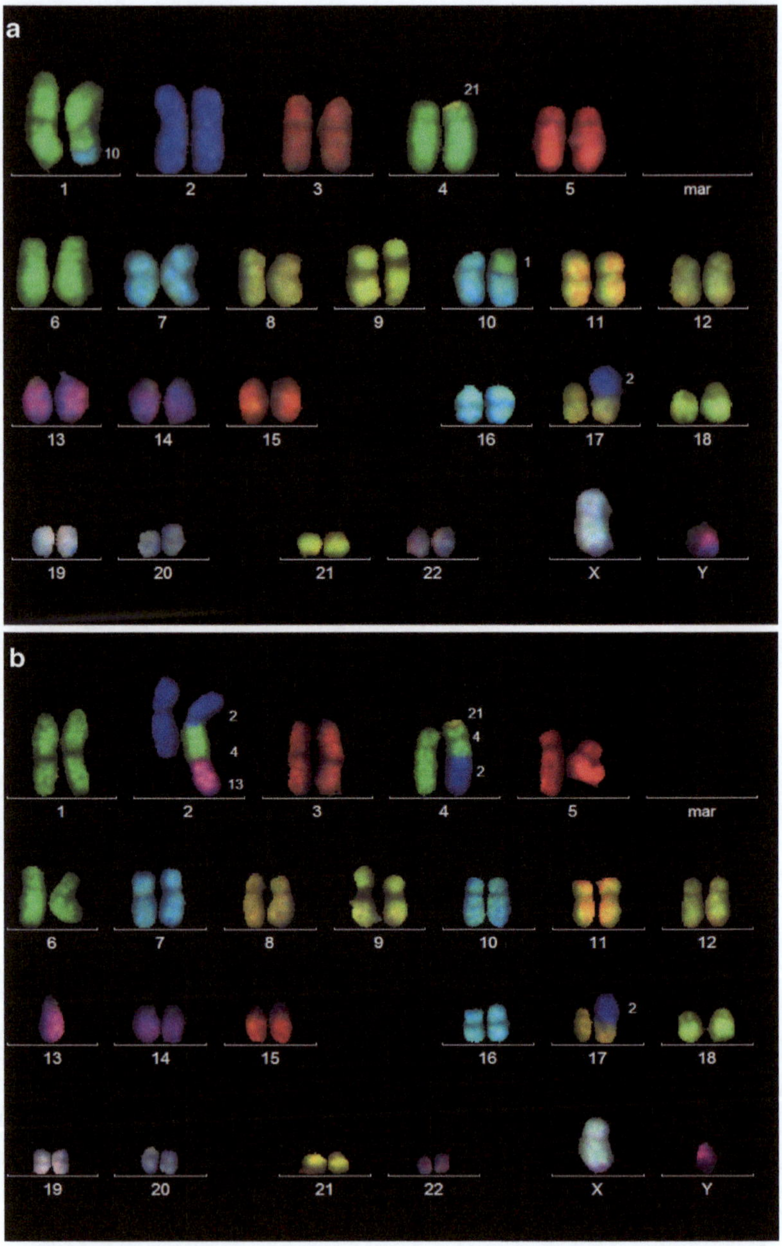

Fig. 4. (**a**, **b**) Two subclones in a CLL patient with clonal evolution – karyotype after 24-color FISH.

5. Notes

1. Laboratory entry of sample: optimal results are obtained if the sample is processed within 24 h after collection. The transport time is especially relevant for samples with high WBC counts.

2. In case of peripheral leukocytosis, peripheral blood samples provide better conditions for the stimulation of the clonal altered metaphases due to the higher amounts of contaminating nonmalignant cells in bone marrow samples.
3. New IL-2 and DSP30 batches should be tested for the stimulation of metaphases in parallel to already established batches.
4. If 13q deletions, as detected by interphase FISH, are not detectable with metaphase karyotyping, metaphase FISH with whole chromosome banding (WCP) and probes WCP#13 and D13S25 or D13S319 should be performed to distinguish between small cytogenetically cryptic deletions and failure of detection due to lack of proliferation of the aberrant clone.

References

1. Hallek M, Stahel RA, Greil R. ESMO Minimum Clinical Recommendations for diagnosis, treatment and follow-up of chronic lymphocytic leukemia. Ann Oncol 2005; 16 Suppl 1: i50–1.
2. Hallek M, Cheson BD, Catovsky D, Caligaris-Cappio F, Dighiero G, Döhner H et al. Guidelines for the diagnosis and treatment of chronic lymphocytic leukemia: a report from the International Workshop on Chronic Lymphocytic Leukemia (IWCLL) updating the National Cancer Institute-Working Group (NCI-WG) 1996 guidelines. Blood 2008; 111(12): 5446–56.
3. Shanafelt TD, Call TG. Current approach to diagnosis and management of chronic lymphocytic leukemia. Mayo Clin Proc 2004; 79(3): 388–98.
4. Oscier DG, Gardiner AC, Mould SJ, Glide S, Davis ZA, Ibbotson RE et al. Multivariate analysis of prognostic factors in CLL: clinical stage, IGVH gene mutational status, and loss or mutation of the p53 gene are independent prognostic factors. Blood 2002; 100(4): 1177–84.
5. Juliusson G, Oscier DG, Fitchett M, Ross FM, Stockdill G, Mackie MJ et al. Prognostic subgroups in B-cell chronic lymphocytic leukemia defined by specific chromosomal abnormalities. N Engl J Med 1990; 323(11): 720–4.
6. Döhner H, Stilgenbauer S, Benner A, Leupolt E, Kröber A, Bullinger L et al. Genomic aberrations and survival in chronic lymphocytic leukemia. N Engl J Med 2000; 343(26): 1910–6.
7. Mayr C, Speicher MR, Kofler DM, Buhmann R, Strehl J, Busch R et al. Chromosomal translocations are associated with poor prognosis in chronic lymphocytic leukemia. Blood 2006; 107(2): 742–51.
8. Buhmann R, Kurzeder C, Rehklau J, Westhaus D, Bursch S, Hiddemann W et al. CD40L stimulation enhances the ability of conventional metaphase cytogenetics to detect chromosome aberrations in B-cell chronic lymphocytic leukaemia cells. Br J Haematol 2002; 118(4): 968–75.
9. Hallaert DY, Jaspers A, van Noesel CJ, van Oers MH, Kater AP, Eldering E. c-Abl kinase inhibitors overcome CD40-mediated drug resistance in CLL: implications for therapeutic targeting of chemoresistant niches. Blood 2008; 112(13): 5141–9.
10. Decker T, Schneller F, Hipp S, Miething C, Jahn T, Duyster J, Peschel C. Cell cycle progression of chronic lymphocytic leukemia cells is controlled by cyclin D2, cyclin D3, cyclin-dependent kinase (cdk) 4 and the cdk inhibitor p27. Leukemia 2002; 16(3): 327–34.
11. Dicker F, Schnittger S, Haferlach T, Kern W, Schoch C. Immunostimulatory oligonucleotide-induced metaphase cytogenetics detect chromosomal aberrations in 80% of CLL patients: a study of 132 CLL cases with correlation to FISH, IgVH status, and CD38 expression. Blood 2006; 108(9): 3152–60.
12. Gahrton G, Robert KH, Friberg K, Zech L, Bird AG. Nonrandom chromosomal aberrations in chronic lymphocytic leukemia revealed by polyclonal B-cell-mitogen stimulation. Blood 1980; 56(4): 640–7.
13. Stilgenbauer S, Bullinger L, Lichter P, Döhner H. Genetics of chronic lymphocytic leukemia: genomic aberrations and V(H) gene mutation status in pathogenesis and clinical course. Leukemia 2002; 16(6): 993–1007.
14. Dicker F, Herholz H, Schnittger S, Nakao A, Patten N, Wu L et al. The detection of TP53 mutations in chronic lymphocytic leukemia independently predicts rapid disease progression

and is highly correlated with a complex aberrant karyotype. Leukemia 2009; 23(1): 117–24.

15. Haferlach C, Dicker F, Schnittger S, Kern W, Haferlach T. Comprehensive genetic characterization of CLL: a study on 506 cases analysed with chromosome banding analysis, interphase FISH, IgV(H) status and immunophenotyping. Leukemia 2007; 21(12): 2442–51.
16. Wu X, Nowakowski G, Smoley S, Arendt B, Peterson M, Van Dyke D et al. Cytogenetic analysis of normal human B cells following CpG stimulation: implications for interpretation of CpG induced CLL metaphase analysis. ASH annual meeting abstracts; Blood 2008; 112: #3124.
17. Put N, Konings P, Rack K, Jamar M, Van Roy N, Libouton JM et al. Improved detection of chromosomal abnormalities in chronic lymphocytic leukemia by conventional cytogenetics using CpG oligonucleotide and interleukin-2 stimulation: a Belgian multicentric study. Genes Chromosomes Cancer 2009; 48(10): 843–53.
18. Schoch C, Kern W, Kohlmann A, Hiddemann W, Schnittger S, Haferlach T. Acute myeloid leukemia with a complex aberrant karyotype is a distinct biological entity characterized by genomic imbalances and a specific gene expression profile. Genes Chromosomes Cancer 2005; 43(3): 227–38.
19. Struski S, Gervais C, Helias C, Herbrecht R, Audhuy B, Mauvieux L. Stimulation of B-cell lymphoproliferations with CpG-oligonucleotide DSP30 plus IL-2 is more effective than with TPA to detect clonal abnormalities. Leukemia 2009; 23(3): 617–9.

Chapter 10

Genetic Abnormalities in Non-Hodgkin's Lymphoma as Revealed by Conventional and Molecular Cytogenetics Methods of Analyses

Suresh C. Jhanwar and Ryan C. Denley

Abstract

Malignant non-Hodgkin's lymphoma (NHL) is a heterogeneous group of tumors, the histological classification of which based on morphologic evaluation alone is not always possible. Various technological advances in cytogenetics combined with molecular approaches have greatly enhanced our ability to identify genetic abnormalities in any given tumor type. The genetic abnormalities identified with the combination of these methods of analysis have resulted in various histological subtypes of NHL being linked with specific genetic abnormalities. Such a classification based on specific abnormalities has lead to the realization that the same abnormalities associated with initiation, transformation, and progression of the disease have also served as markers of diagnosis, prognosis, and predisposition to a given tumor type, and some abnormalities also served as markers for therapeutic targets. Results of such studies in NHL have not only identified the subsets of various histological types based on specific abnormalities, but, as is evident from recent literature, also set the stage for further evaluation using high-resolution array comparative genomic hybridization (CGH) and expression profiling.

Key words: Non-Hodgkin's lymphoma, Conventional cytogenetics, Molecular cytogenetics, FISH, CGH, SKY, RT-PCR

1. Introduction

Since the introduction of banding techniques in the early 1970s and the demonstration that the Philadelphia chromosome is highly diagnostic of chronic myelogenous leukemia and is derived from a reciprocal translocation between chromosomes 9 and 22, a vast extent of literature has accumulated. According to specific chromosome abnormalities in lymphohematopoietic systems and, to a lesser extent in solid tumors, the abnormalities not only correlate with specific histological subtypes and

Lynda J. Campbell (ed.), *Cancer Cytogenetics: Methods and Protocols*, Methods in Molecular Biology, vol. 730, DOI 10.1007/978-1-61779-074-4_10, © Springer Science+Business Media, LLC 2011

Table 1
Recurrent translocations in histological subtypes of non-Hodgkin's lymphoma (NHL), specific genes involved, and prognostic implications[a]

		Clinical relevance		
Histologic subtype	**Specific translocation**	**Diagnostic**	**Prognostic**	**Genes**
Follicular small cell NHL	t(14q;18q)	Yes	Yes	*IGH, BCL2*
Diffuse large cell NHL	t(3q;14q)	Yes	Yes	*BCL6, IGH*
Mantle cell lymphoma	t(11q;14q)	Yes	Yes	*CCND1, IGH*
Burkitt's lymphoma	t(8q;14q)	Yes	Yes	*MYC, IGH*
Anaplastic large cell NHL	t(2p;5q)	Yes	Yes	*ALK, NPM*

[a]Details of other specific translocations and rearrangements associated with various subtypes of NHL are described in detail in previous publications (18, 19)

immunophenotypes, but also with prognosis (Table 1). Such a close correlation led to the realization that the same specific genetic abnormalities which lead to the development of different cancers may also serve as their specific markers of prognosis, diagnosis, and predisposition (1–3).

Several technical innovations in molecular biology and their integration with conventional cytogenetics have allowed cytogeneticists to identify clinically relevant abnormalities with greater precision. The molecular methods of analysis which include fluorescence in situ hybridization (FISH), whole chromosome specific painting (WCP), comparative genomic hybridization (CGH), spectral karyotyping (SKY), and polymerase chain reaction (PCR) are methods that have revolutionized the way in which the genetic alterations associated with cancer are studied (2, 3).

Non-Hodgkin's lymphomas (NHLs) are mesodermal in origin and represent a heterogeneous group of tumors with various histological subtypes. The histological classification of NHL based on morphology alone is difficult. The 14q+marker chromosome detected in Burkitt's lymphoma was subsequently shown to be derived as a result of a translocation t(8;14)(q24;q32). It was not until 1982 that three independent groups showed the nature of the genes involved in the pathogenesis of this high-grade histological subtype of lymphoma. It was shown that the deregulation of the MYC gene was intimately associated with this subtype of lymphoma (4–8).

During the 1990s, the FISH technique became an indispensable procedure in many areas of research and clinical cytogenetics. FISH analysis in conjunction with karyotype not only has shown concordance between karyotype and interphase FISH results, but also demonstrates the utility of FISH analysis in identification of clinically relevant chromosome abnormalities in specimens that failed to yield metaphases for analysis or resulted in normal karyotypes. Results of such analyses also correlated with southern blotting and PCR studies. It is therefore feasible to identify a specific genetic abnormality in any given cancer by using one of these or several of these methods of analysis. In addition to FISH analysis of tumor chromosomes using specific probes (single copy genes, satellite probes, etc.), WCP analysis has further helped in identifying rearrangements not easily detectable in cases with less than optimum quality of metaphases or in those cases in which the origin of material in a marker chromosome could not be determined (2). Furthermore, analysis of previously unknown genetic alterations in cancer has become possible through the use of CGH, which is based on competitive binding of tumor and control DNA to normal metaphase chromosomes (2, 9). In recent years, CGH has been extensively used by several investigators to identify chromosomal abnormalities not easily detected by conventional cytogenetics in lymphohematopoietic as well as in solid tumors (9, 10). CGH is particularly informative in cases with low mitotic indexes such as Multiple Myeloma, CLL, or other cancers with low proliferative activity (9). Recent advances in methods for the visualization of complete chromosome constitution of tumor cells by SKY have revolutionized the way in which the complex genetic alterations often seen in tumors are identified (11–13). Such methods are being utilized to identify genetic alterations not easily detected by conventional cytogenetics, CGH, or WCP (3). Application of the above molecular methods of analysis in the identification of genetic abnormalities in a variety of cancers has been reviewed earlier (2).

Thus, the overall objectives are to provide a theoretical background through a series of specific examples representing some of the diagnostic categories outlined in Table 10.1, and to demonstrate the application of molecular/cytogenetic methods of analysis (FISH, CGH, SKY, and PCR) as an adjunct to conventional cytogenetics in the identification of clinically relevant genetic abnormalities in lymphoma that are highly relevant to determining diagnosis and prognosis, and are expected to help in the management of cancer patients.

2. Materials

2.1. Culture and Harvest

1. 15 mL conical tubes (Corning).
2. T_{25} culture flasks (Corning).
3. Culturing medium: RPMI 1640, 780 mL; fetal bovine serum (Gemini Bio Products), 200 mL; penicillin and streptomycin, 5,000 U/mL, 10 mL; L-glutamine, 200 mM, 10 mL.
4. RPMI wash solution: RPMI 1640, 980 mL; gentamicin and fungizone, 5,000 U/mL, 20 mL.

5. Colcemid (10 μg/mL) (Invitrogen).
6. Trypan Blue (Thermo Scientific).
7. KCl 0.075 M hypotonic solution (Invitrogen).
8. Carnoy's fixative (3:1 methanol:glacial acetic acid).

2.2. Chromosome Preparation and G-Banding Reagents

1. Trypsinizing solution: Trypsin-EDTA 10× (GIBCO), 10 mL; de-ionized distilled water, 35 mL; Hank's balanced salt solution (10×, without Ca^{2+} or Mg^{2+}) (GIBCO), 5 mL.
2. Giemsa Staining solution: Giemsa blood stain (J.T. Baker), 14 mL; Tris-Buffer solution (see Note 1), 12.5 mL; de-ionized distilled water, 37.5 mL.

2.3. Interphase FISH Preparation

1. 20× SSC (saline-sodium citrate) (Fisher), pH 5.3; dilute to 2× when necessary using de-ionized distilled water.
2. Protease digestion solution: Pepsin (Fisher), 25 mg; 10 mM HCl, 50 mL; prepare just before use and keep at 37°C.
3. Formaldehyde fixative solution: 1× PBS, 37 mL; 10% neutral buffered formalin (Fisher), 12.5 mL; 2 N $MgCl_2$ (Abbott Molecular), 0.5 mL.
4. 1× PBS (Fisher).
5. 100% Ethanol (Fisher).
6. Denaturing solution: formamide (Fisher), 343 mL; 20× SSC, 49 mL; de-ionized distilled water, 98 mL; pH 7.0–7.5.
7. 2× SSC/0.1% NP-40: 20× SSC, 100 mL; de-ionized distilled water, 850 mL; NP-40 (Abbott Molecular), 1 mL; pH 7.0–7.5, and bring volume to 1,000 mL with de-ionized distilled water.
8. 0.4× SSC/0.3% NP-40: 20× SSC, 20 mL; de-ionized distilled water, 950 mL; NP-40, 3 mL; pH 7.0–7.5, and bring volume to 1,000 mL with de-ionized distilled water.
9. Rubber cement.
10. DAPI II counterstain (Abbott Molecular).

2.4. Spectral Karyotyping

1. Spectrum Green-deoxyuridine triphosphate-dUTP (Abbott Molecular, Des Plaines, IL).
2. Spectrum Orange-dUTP (Abbott Molecular, Des Plaines, IL).
3. Texas Red-dUTP (Molecular Probes, Eugene, OR).
4. Biotin-16-dUTP (Boehringer Mannheim, Indianapolis, IN).
5. Digoxigenin-11-dUTP (Boehringer Mannheim, Indianapolis, IN).
6. Human cot-1 DNA (GIBCO/BRL, Gaithersburg, MD).
7. DAPI II.

2.5. Comparative Genomic Hybridization

1. Fluorescein-12-dUTP (NON-DuPont, Boston, MA).

2.6. Reverse Transcription-Polymerase Chain Reaction

1. Plasmids: home made.
2. TaqMan Universal PCR Master Mix (Applied Biosystems).
3. 40 U/μL RNase inhibitor (Invitrogen).
4. Screw-cap microcentrifuge tubes, autoclaved.
5. MicroAmp caps (ABI).
6. MicroAmp reaction tubes (ABI).
7. 96-Place 0.2 mL Amplification Tube Rack (USA Scientific).
8. TOPO TA Cloning Kit (Invitrogen).
9. QIAprep Spin Miniprep Kit (QIAGEN).
10. RNeasy Mini Kit (QIAGEN).
11. SuperScript® II Reverse Transcriptase (Invitrogen).
12. SuperScript® III Reverse Transcriptase (Invitrogen).
13. LB broth (Invitrogen).
14. Luria Agar (Invitrogen).
15. iCycler Optical Quality Sealing Tape (Bio-Rad).
16. iCycler PCR Plate, 96-well (Bio-Rad).
17. Primers (custom oligos synthesized by ABI) (working solution, 5 pmol/μL).
18. Probes (custom oligos synthesized by ABI with Reporter and Quencher) (working solution, 2 pmol/μL).
19. Pipette tips, with aerosol barrier.
20. Positive control cell line.
21. Negative controls (water, HL-60 cells).

3. Methods

3.1. Methods for Conventional Cytogenetic Study

3.1.1. Culturing of Cells Obtained from Solid Tissues

1. Using sterile techniques, place the lymph node or solid tissue in a petri dish and wash intact tissue three times with RPMI medium. Perform all the sterile procedures while working inside a type II HEPA filtered fume hood. After removing wash media, add several milliliters of RPMI to the tissue.
2. Dissociate the sample with two scalpels by gently scraping or patting the tissue until you begin to achieve a cell suspension. Collect the minced tumor and suspended cells in a 15 mL tube with 10 mL RPMI using a sterile pipette.
3. After a few seconds, the large debris will settle to the bottom of the tube. At this point, transfer the cellular supernatant into a fresh sterile 15 mL centrifuge tube. This suspension is now

in the appropriate state for having the number of non-red blood cells counted with a cell counter such as a Coulter Counter or by a hemocytometer. It is also advisable to assess the viability of the suspended cells using Trypan Blue (see Note 2).

4. Centrifuge suspension for 10 min at 600–1,000 *g*. Resuspend the resulting pellet in an appropriate volume of complete culturing medium in order to dispense specific number of cells. Typically, 4.0×10^7–6.0×10^7 cells are inoculated in a 10 mL culture flask. Set up the desired number of T_{25} flasks with culturing medium and cells and incubate for 24 h at 5% CO_2 and 37°C.

3.1.2. Cell Culture Harvest

1. At the completion of incubation, add 0.1–0.2 mL of Colcemid to each culture flask and swirl to mix (see Note 3). Incubate the cultures for 2 h at 4°C.
2. Transfer the contents of each flask into separate 15 mL tubes. Centrifuge the tubes for 10 min at 600–1,000 *g*. Remove the supernatant carefully without discarding or disrupting any of the pelleted material.
3. Once the supernatant has been removed, break up the pellet by tapping the bottom of the tube and resuspend the pellet in 10 mL of 0.075 M KCl hypotonic solution to assist in metaphase spreading. Incubate the solutions for 20–30 min in a 37°C water bath (see Note 4).
4. After the hypotonic incubation period, add ten drops of freshly prepared Carnoy's fixative (3:1 methanol:acetic acid) as a pre-fixation step into each test tube and invert gently to mix.
5. Centrifuge the tubes for 10 min at 600–1,000 *g*. Remove the supernatant carefully without discarding any of the pellet. Fully break up the cell pellet and resuspend it in 10 mL of freshly prepared, cold fixative while agitating (vortexing can be used) in order to keep white cells from clumping together. It is also advisable to allow the contents of the tube to incubate for at least 30 min to overnight after the initial fixation to ensure adequate fixation and lysis of the red blood cells. After 30 min, centrifuge again for 10 min at 600–1,000 *g*.
6. Repeat step 5 twice more or until the contents of the tube are clear and the cell pellet is white. It is not necessary to wait for 30 min after changing the fixative.

3.1.3. Preparation of Slides for Chromosome Analysis

1. Remove the supernatant very near to the surface of the cell pellet.
2. Resuspend the pellet and add fresh fixative until an ideal milky white concentration is obtained. If the cell suspension is too dense when the suspension is dropped onto the slide, there will

be significant crowding of nuclei and chromosome spreading will be inhibited. Too dilute suspensions result in a greater amount of time needed to find analyzable metaphases.

3. Drop the cell suspension from various distances onto a clean, dry slide taken from a dish of 70% ethanol. Tilt the slide in order to facilitate spreading of the solution across the entire surface of the slide. Allow the slide to dry (see Note 5).
4. Once dry, scan the slide under a phase microscope to assess for the presence of chromosomes and the quality of their spreading. If chromosome spreads are too tight to analyze, over-spread, or have a shiny appearance due to the persistence of cytoplasm, consider adjusting your technique.
5. Incubate all slides intended for G-band analysis in a 60°C incubator overnight. This is done to age the slides.

3.1.4. Giemsa-Banding

1. After 24 h, move the slides from a 60°C incubator to a 37°C incubator and allow them to cool.
2. Prepare and warm a trypsin banding solution to 37°C according to the preparation steps mentioned in Subheading 2.
3. Place a slide to be banded into the trypsin banding solution for 10–20 s. After 10–20 s, rinse the slide thoroughly with cold water to stop trypsin activity.
4. Place the slide in Tris-Buffer Giemsa Staining solution (prepared according to the steps mentioned in Subheading 2) for approximately 1 min. After 1 min, rinse the slide thoroughly with cold water to wash off the excess stain.
5. Blot or air dry the slide, place two drops of slide mounting medium on the slide surface, coverslip, and allow the mounting medium to dry.
6. Scan the slide for evaluation of banding and staining. Adjust trypsin and/or staining time according to the quality of chromosomal banding (see Note 6).

3.1.5. Analysis

1. Count and analyze a minimum of 20 metaphases when possible. This is most easily accomplished when metaphases are photographed and cut-out or computer image captured and digitally manipulated. Computer image capturing software often has provisions for digital cutting of chromosomes for the purpose of arranging them into karyograms.
2. A minimum of two metaphases should be karyotyped for each clone. A clone is defined as two or more cells possessing the same structural abnormalities or additional chromosomes indicating polysomy, and also as three or more cells with the same missing chromosomes indicating clonal monosomy (see Note 7).

3.2. Methods for Molecular Cytogenetic Study

3.2.1. Interphase FISH Slide Preparation

1. Obtain cells incubated for 24 h and harvested as per the protocols mentioned for conventional cytogenetic studies. This method produces optimal results, although it is possible to perform interphase FISH on cells that have not been cultured. Cells must, however, be harvested and fixed prior to FISH.
2. Prepare a slide from a fixed pellet as previously described. Considerations for chromosome spreading are not necessary unless metaphase FISH is being performed, although it is still advisable to minimize the amount of cytoplasm still contained in the cells. These slides are now ready for hybridization, but they can also be stored at –20°C.

3.2.2. Interphase FISH Slide Pretreatment

1. Place the slides intended for hybridization in a Coplin jar containing 2× SSC at 75°C for 2 min.
2. Remove the slide(s) from the 2× SSC and place them into the pepsin solution (prepared as described in Subheading 2) at 37°C for 10 min.
3. Remove the slide(s) from pepsin and place the slide(s) in 1× PBS for 5 min at room temperature.
4. Remove the slide(s) from the PBS and place the slide(s) in 1% formaldehyde for 5 min at room temperature.
5. Remove the slide(s) from the formaldehyde fixative and place the slide(s) in 1× PBS again for 5 min at room temperature.
6. After removing the slide(s) from 5 min in the final PBS solution, dehydrate the slide(s) in an alcohol gradient of 70, 85, 100% ethanol in that order for 1 min in each concentration at room temperature. Once dried, the slides have been pretreated and are ready for denaturation and hybridization (see Note 8).

3.2.3. Interphase FISH Slide Denaturation and Hybridization

1. Place the slide(s) in 70% formamide/2× SSC denaturing solution (prepared as described in Subheading 2) at 75°C for 3–5 min.
2. After 3–5 min, dehydrate the slide(s) in an ice-cold alcohol gradient of 70, 85, 100% ethanol for 1 min at each concentration in that order. This process will quickly arrest the DNA on the slide(s) in a single-stranded state.
3. Prepare the probe for hybridization according to a probe manufacture's protocol or according to the home brewer's preparation recommendations and denature the probe at 75°C for 5 min in a water bath. After 5 min, immediately place the probe on ice and allow to completely cool down.
4. Apply approximately 10 µL of probe mixture onto the slide in the area where hybridization is desired. It is often possible to apply two probes to a single slide on different sides.

5. Apply a 25 mm × 25 mm coverslip to the probe and seal the edges of the coverslip with rubber cement.
6. Place the slide in a dark prewarmed, humidified box or chamber, and incubate for 4–16 h at 37–39°C. During this hybridization period, the probe will anneal to the target DNA.

3.2.4. Interphase FISH Posthybridization Wash

1. Remove the rubber cement from the slide(s) with forceps or scalpel.
2. Place the slide(s) in 0.4× SSC/0.3% NP40 (prepared as described in Subheading 2) at 77°C for 2 min with agitation at first.
3. Remove the slide(s) after 2 min and immediately place the slide(s) in 2× SSC/0.1% NP40 (prepared as described in Subheading 2) at room temperature for 1 min, again with agitation at first.
4. Remove the slide(s) and allow them to air dry vertically and away from light.
5. Apply 10 μL of DAPI II counterstain onto the target area and overlay with a 25 mm × 25 mm coverslip.
6. Incubate slide(s) in a dark, −20°C environment for 30 min for optimal results and to activate DAPI antifading activity, or until needed for evaluation.

3.2.5. Interphase FISH analysis

1. Visualize a minimum of 100 to a maximum of 500 interphase cells under a fluorescent light microscope equipped with appropriate excitation filters.
2. Use a cell counter to enumerate cell populations (see Note 9).
3. Photograph a number of representative interphase cell images (both normal and abnormal) using digital imaging software.
4. Establish false-positive cut-off values for every probe analyzed which have been generated from known normal cases (see Note 10).

3.2.6. Spectral Karyotyping in Brief

1. Age metaphase slides prepared according to the procedures described earlier for 2–3 weeks prior to use.
2. Precipitate 22 ng of each of the differentially labeled chromosome painting probes in the presence of 50 μg of human Cot-1 DNA.
3. Hybridize for 2 days at 37°C.
4. Biotinylated probes are detected using avidin Cy5, and digoxigenin-labeled probes are detected with an antimouse digoxigenin antibody followed by goat antimouse secondary antibody conjugated to Cy5.
5. Counterstain chromosomes with DAPI II.

6. Imaging is performed using a SD200 Spectracube (Applied Spectral Imaging, Migdal Ha-Emck, Israel) mounted on a Leica DMIRBE microscope or a Nikon Eclipose E800 microscope using a custom designed optical filter (SKY-1) (Chroma Technology, Brattleboro, VT).
7. Using a Sagnac interferometer (Applied Spectral Imaging) in the optical head, an interferogram is generated at all image points, which is deduced from the optical path difference of the light which, in turn, depends on the wave length of the emitted fluorescence.
8. The spectrum is recovered by Fourier transformations, and the spectral information is displayed by assigning red, green, or blue colors to certain ranges. This RGB display renders chromosomes that are labeled with spectrally overlapping fluorochromes or fluorochrome combinations in similar colors.
9. Based on the measurement of the spectrum for each chromosome, a spectral classification algorithm is applied that allows the assignment of pseudocolors to all points in the image that have the same spectrum. This algorithm forms the basis for chromosome identification by SKY.
10. DAPI images are acquired from all metaphases analyzed using a DAPI-specific optical filter.
11. Breakpoints on the SKY-painted chromosomes are determined by examination of corresponding DAPI banding and by comparison with G-banded karyotypes of the same tumor. By this method, we are able to define the breakpoints on *add* and *der* chromosomes, but are unable to assign the precise breakpoints of chromosomal segments from partner chromosomes that generated the *add* or *der* chromosomes.
12. A breakpoint is considered recurring if identified in two or more cases by G-banding, SKY, or by both (see Note 11).

3.2.7. Comparative Genomic Hybridization in Brief

1. Extract tumor DNA.
2. Tumor (test) and normal (reference) DNAs are labeled by nick-translation with fluorescein-12-dUTP.
3. Equal amounts (200 mg) of tumor and normal DNAs are coprecipitated with 10 mg of human Cot-1DNA and resuspended in the hybridization mix before in situ hybridization to human metaphase chromosome spreads prepared from phytohemagglutinin-stimulated lymphocytes from normal individuals.
4. After hybridization, the slides are washed and the chromosomes are counterstained with DAPI II to enable identification of the chromosomes.

5. Fluorescent hybridization signals and DAPI-staining patterns are captured with a cooled charge-coupled device (CCD) camera (Photometrics, Tucson, AZ) attached to a Nikon Microphot-SA microscope and processed using an image analysis system.
6. The software performs a calculation of the green (tumor DNA) to red (normal DNA) fluorescent ratios along the length of each chromosome. The average of readings from eight metaphases are graphed for each chromosome and compared with the profile for the same chromosome in a reference DNA/reference hybridization to set the boundaries of gain and loss.
7. Ratios greater than 1.20 and less than 0.80 are considered to represent chromosomal gain and loss, respectively.
8. CGH detects DNA sequence copy numbers relative to the average copy numbers in the tumor but not to the ploidy level of the tumor. Chromosomal regions near the centromeres of 1, 9, 13–16, 21, and 22 are not scored for CGH analysis because of the highly repetitive sequences in these regions. Recurrence of a change is defined by its presence in two or more tumors (see Note 12).

3.3. Methods for Molecular Analysis

3.3.1. Construction of Plasmid Standard DNA Curves

1. Generate PCR fragments for target gene(s) and TATA binding protein (TBP). Set up qualitative RT-PCR using primers that are the same as those used for real-time PCR and analyze PCR products on 3% agarose gel.
2. Clone target RNA transcript and TBP PCR fragments into pCR 2.1-TOPO Vector and culture bacteria containing target gene fragment or TBP PCR fragment using TOPO TA Cloning Kit (see the manufacturer's kit manual for suggested protocol in detail).
3. Purify plasmid DNA using QIAprep Spin Miniprep Kit (see the manufacturer's kit manual for suggested protocol in detail). Usually 5 μg of plasmid DNA can be generated by using one QIAprep Spin Miniprep column.
4. Dilute plasmid DNA in sterile water at the concentrations of 1,000,000 copies/10 μL, 100,000 copies/10 μL, 10,000 copies/10 μL, 1,000 copies/10 μL, 100 copies/10 μL, and 10 copies/10 μL according to molecular weight of the plasmid. Use 10 μL from each concentration as PCR template for the standard curves.

3.3.2. RT-PCR Amplification

1. Total RNA may be extracted using an RNeasy Mini Kit which is capable of optimal RNA extractions from multiple tissue types including bone marrow and lymph node samples. Extract the total RNA as described in the kit (see the manufacturer's kit manual for suggested protocol in detail).

2. Determine the number of samples to be amplified including controls. Positive controls are RNA from cell lines. It is advisable to include negative controls derived from HL-60 cells for the normal transcripts, a reagent control where no RNA is added into the reaction, and an RT (–) control where Superscript III Reverse Transcriptase is omitted. HL-60 and water controls are to be run with each set of amplification reactions. The RT (–) control is to be run with each sample.
3. Prepare the sample for reverse transcription (RT) as described below:

RNA (5 μg) + water	10 μL
Primer TBPR + A2R (2 pmol each)	2 μL
10 mM dNTP	1 μL

At this point, denature the mix at 65°C for 5 min and put on ice immediately. Then proceed by adding the following reagents.

5× RT Buffer	4 μL
RNase inhibitor (40 U/μL)	1 μL
DTT (0.1 M)	1 μL
SuperScript III (200 U/μL) (see Note 13)	1 μL
Total volume	20 μL

4. The RT reaction is run at 55°C for 60 min, then heated to 75°C for 5 min; at which time, add 80 μL of water for each reaction, mix and store it at –70°C.
5. Prepare the mix for the PCR as described below:

Taqman Universal PCR Master Mix	25 μL
Primer 1 (15 pmol)	3 μL
Primer 2 (15 pmol)	3 μL
Probe (10 pmol)	5 μL
Water	4 μL
Total volume	40 μL

6. Multiply the volume of each component by the number of total tubes plus 10% to make a master mix for the entire set of reactions. Run duplicates for RT (+) samples and single reactions for RT (–) samples.
7. Add 40 μL of PCR master mix to each reaction tube or plate. Duplicates are required for patient samples at this point. Pipette carefully at a slight angle to avoid splashing the solution and add 10 μL of template from the above RT reaction or water for negative controls and plasmid DNA for the

standard curve. Place cap on each tube or sealing tab for the plate and do a brief spin to remove air bubbles at the bottom of the tube.

8. Place the tubes or plate into Bio-Rad iCycler and run the PCR at the following conditions:

95°C × 8.5 min	1 cycle
95°C × 0.5 min, 62°C × 1 min, 72°C × 0.5 min	50 cycles

9. The PCR amplification program will be completed in approximately 2–3 h. After the step of amplification, discard the tubes or plates without opening them. Analyze the file generated by the machine during the real-time PCR and calculate copy number for each sample. Divide the copy number of target gene transcripts by the copy number of TBP for the sample. The ratio represents the expression level for this sample (see Note 14).

3.4. Summary Comments

It is clear that technological advances combined with molecular approaches have greatly enhanced our ability to identify genetic abnormalities in any given tumor type. The genetic abnormalities identified with the combination of these methods of analysis have resulted in the realization that the same abnormalities associated with initiation, transformation, and progression of the disease have also served as markers of diagnosis, prognosis (14), and predisposition to a given tumor type. Results of such studies in NHL have not only identified the subsets of various histological types based on specific abnormalities, but, as is evident from the recent literature, also set the stage for further evaluation using high-resolution array CGH and expression profiling.

In fact, the global gene expression profiles have further demonstrated the presence of several subtypes in diffuse large cell lymphomas based on c-Rel amplification and BCL-2 expression (15, 16). Finally, in recent years, investigators have also considered these specific genetic abnormalities as the markers for targeted therapy in lymphoma (17).

4. Notes

1. Components for Tris-Buffer stock solution: Trizma Base Tris hydroxymethyl aminomethane (Sigma), 24.2 g; Maleic Anhydride (Sigma), 23.3 g; de-ionized distilled water, 1,000 mL; pH solution to 5.6 with sodium hydroxide. It is advisable to wear a mask when working with maleic anhydride.

2. Evaluate the viability of the cells in regards to the number of cells which do not absorb the blue stain. These cells have a working active transport system and are considered viable.
3. The Colcemid works to depolymerize microtubules separating the chromosomes and maximizes the number of metaphases that can be obtained in a cell culture.
4. This is a variable in the harvesting procedure that is changed at the user's discretion. If experience at a certain hypotonic time results in broken metaphases after slide preparation, decrease the hypotonic time. If metaphases are too tight and not well spread, it is advisable to increase the hypotonic time.
5. Environmental conditions greatly influence the drying of the cell suspensions in fixative and the rate of drying greatly impacts on the chromosomes ability to spread. This is due to surface tension and atmospheric pressure pressing down on the individual nuclei as the fixative layer dries. Drying of the fixative too quickly does not permit enough pressure on the nuclear membrane to burst and, therefore, expel the cytoplasm and adequately spread the chromosomes. There is great consensus that the optimum conditions for slide dropping are 25°C and 40% humidity. Where possible, a controlled environment such as a chamber is recommended for slide dropping. If this is not available, experimentation with multiple techniques such as hot plates, steam baths, and variations in the height of the drop can assist in optimizing the slide preparations.
6. Assessment of the quality of banding and staining is rather subjective; however, the goal of any cytogeneticist is to maximize the number of visible bands on a chromosome while preserving appropriate contrast. Under-trypsinization results in darkly stained chromosomes with too much protein, while over-trypsinized chromosomes appear ghostly, and G-bands are difficult to distinguish. Inappropriately trypsinized chromosomes may appear to have ambiguous, faint bands if they were not exposed to stain long enough, and may have merging, indistinct bands if left in stain for too long. A balance of all variables must be met in order to optimize a preparation.
7. Karyotype analysis is the gold standard when examining genetic abnormalities of any NHL tumor. It is a precise method, but suffers from high failure rates (5–50%). It is especially useful in identifying translocations (see Fig. 1), deletions, aneuploidy, and segment amplification.
8. The following steps are optional, however, they optimize the quality of the FISH preparation by removing protein, decreasing background, and making DNA more accessible. In addition, all the following steps through the postwash hybridization are described as being performed on a single

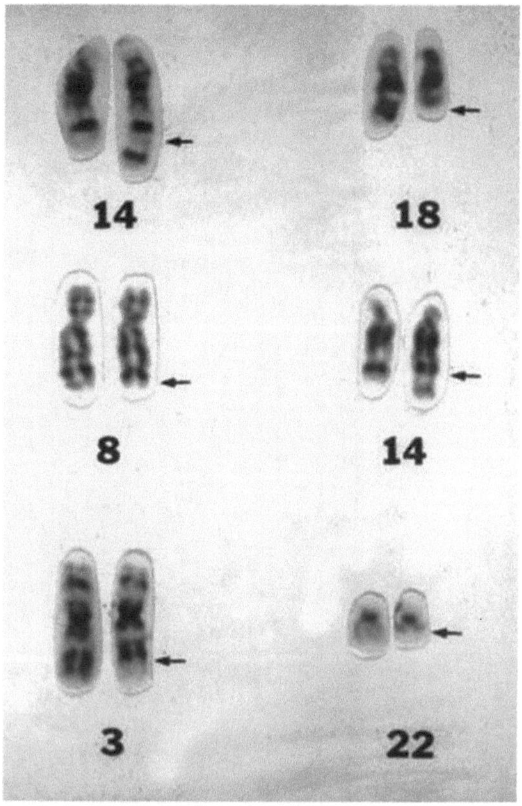

Fig. 1. Representative G-banded partial karyotypes to show recurrent translocations: t(14;18)(q32;q21) in follicular lymphoma (*top*), t(8;14)(q24;q32) in Burkitt's lymphoma (*middle*), and t(3;22)(q27;q11) in diffuse large cell lymphoma (bottom).

slide; however, multiple slides can be processed at once provided the number of slides does not greatly vary the ambient temperature for any single step.

9. When considering a cell for evaluation, the cell must not be overlapping another cell. In addition, enumeration must be performed on signals that are separated by at least the distance of a single intact signal. This is also important when establishing the state of a break-apart probe. The probe signals must be separated by the distance of a single signal in order to be considered separate.

10. FISH analysis is a less precise method of genetic study due to the existence of false positives and false negatives but it is nonetheless a very valuable quantitative technique performed with relative ease. Some benefits of FISH are that it does not require dividing cells, requires only intact nuclei, can be performed expeditiously, large numbers of cells can be examined in a short period of time, and it can be performed on embedded and archival tissues or on routine smears if necessary (see Fig. 2).

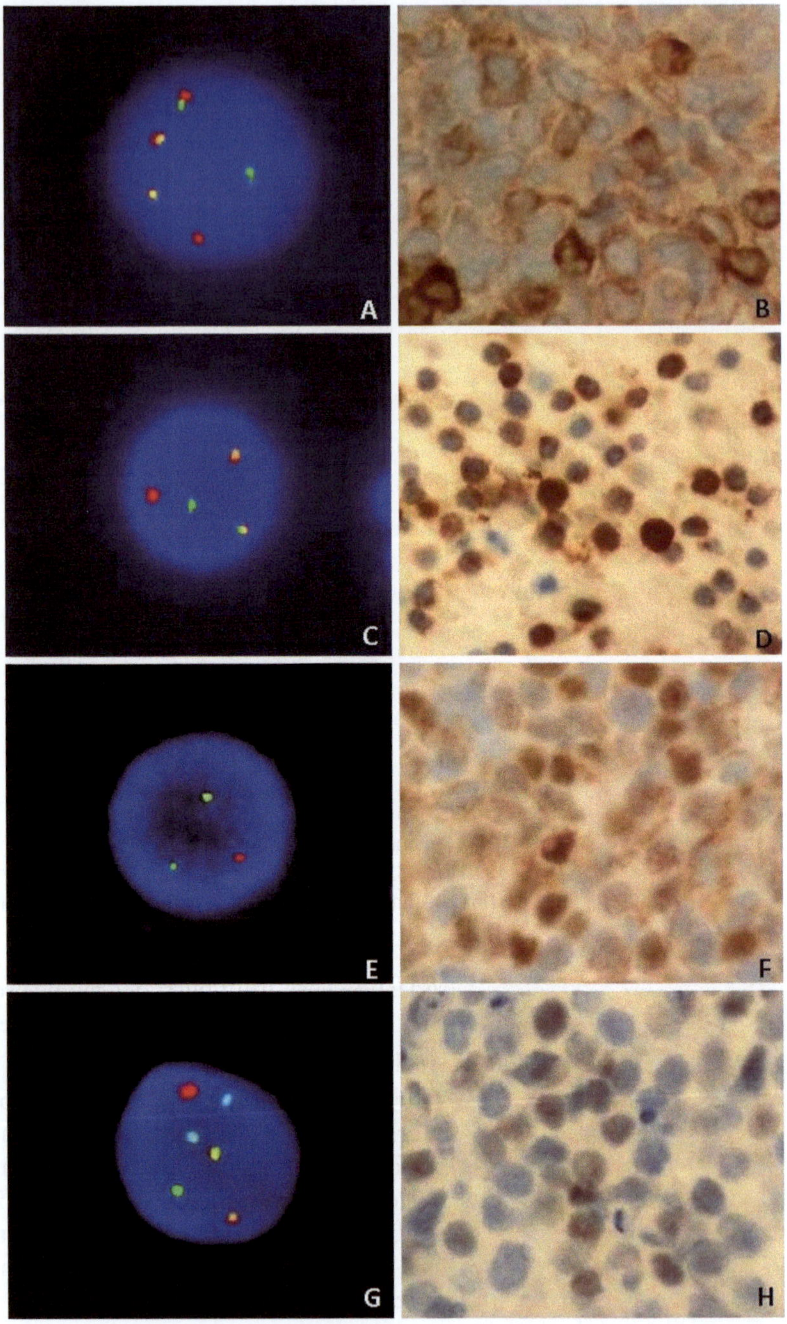

Fig. 2. Interphase FISH and immunostaining to show a close correlation between specific translocations and deregulated protein expression. (**a**) interphase FISH showing a t(14;18)(q32;q21) and (**b**) BCL-2 expression in follicular lymphoma; please note that the extra fusion signal seen in the interphase FISH image represents an additional copy of the derived chromosome 18; (**c**) interphase FISH showing a t(11;14)(q13;q32) and (**d**) CCND1 expression in mantle cell lymphoma; (**e**) interphase FISH showing BCL-6 rearrangement and (**f**) BCL-6 expression in diffuse large cell lymphoma; (**g**) interphase FISH showing a t(8;14)(q24;q32) and (**h**) MYC expression in Burkitt's lymphoma.

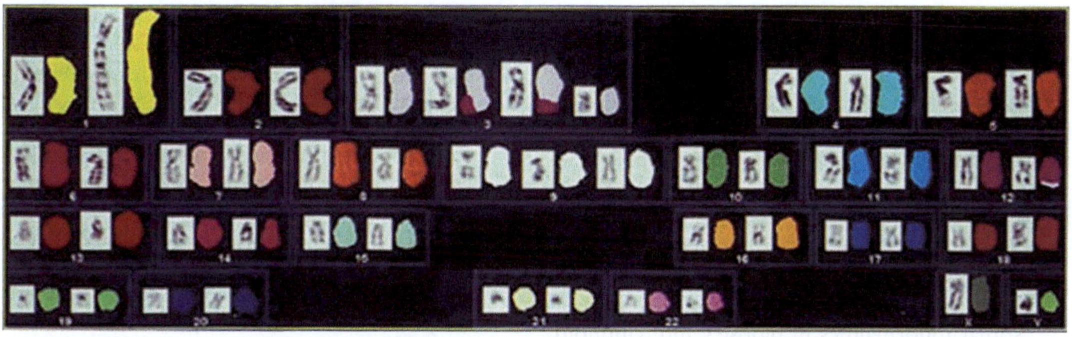

Fig. 3. G-banded and SKY karyotypes from metaphases obtained from a lymph node culture of a patient with the histopathological diagnosis of diffuse large cell lymphoma. Please note that unbalanced rearrangements affecting chromosomes 3 and 12, among others, could not be characterized based on G-banding alone, but were identified by SKY. The rearrangement at the 3q region was subsequently shown by interphase FISH to be due to BCL-6 rearrangement which is highly associated with large cell lymphoma.

11. The SKY method involves several steps. First, all the chromosomes are labeled with a different combination of fluorescent dyes. The spectral imaging system then measures the full visible light spectrum emitted from the sample in order to acquire a spectral image of the chromosomes. Finally, SKY View, the analysis software, analyzes the results of the spectral image and displays the metaphase and a karyotype image where each pair of chromosomes appears with a unique color. This allows one to easily identify translocations and components of marker chromosomes since the abnormal chromosomes are multicolored. It is a relatively new technique, but its utility exists in its ability to detect subtle translocations (see Fig. 3).
12. CGH is a method in which parameters are not yet fully defined, but is potentially of great value. Two of its limitations are that it requires over 20% tumor cells in a specimen in order to register losses and gains, and that it cannot identify balanced structural rearrangements or ploidy levels.
13. In lieu of SuperScript III when preparing the RT samples, add 1 μL of water for the RT (–) control.
14. RT-PCR is a precise, nonquantitative test, although its biological significance is in question. It requires multiple primers for the many described genetic abnormalities.

References

1. Chaganti, R.S.K. and Jhanwar, S.C. (1985) Chromosomal models of cancer etiology. *In*: Genetics in Clinical Oncology. (Eds. Chaganti, R.S.K. and German, J.L.), Oxford University Press, New York, 60–79.
2. Jhanwar, S.C. (1999) Chromosome painting and comparative genomic hybridization: application in identification of structural and numerical abnormalities in lymphohematopoietic and solid tumors. *In*: An Introduction to

Fluorescent In Situ Hybridization. (Eds. Andreeff, M. and Pinkel, D.), John Wiley & Sons Press, New York.

3. Salman, M., Jhanwar, S.C., and Ostrer, H. (2004) Will the new cytogenetics replace the old cytogenetics? *Clin Genet.* **66**, 265–275.
4. Manolov, G. and Manolova, Y. (1972) Marker band in one chromosome 14 from burkitt lymphomas. *Nature.* **237**, 33–34.
5. Zech, L., Haglund, U., Nilsson, K., and Klein, G. (1976) Characterization chromosomal abnormalities in biopsies and lymphoid cell lines from patients with burkitt and non-burkitt lymphomas. *Int J Cancer.* **17**, 47–56.
6. Neel, B.G., Jhanwar, S.C., Chaganti, R.S.K., and Hayward,W.S. (1982) Two human c-*onc* genes are located on the long arm of chromosome 8. *Proc Natl Acad Sci.* **79**, 7842–7846.
7. Taub, R., Kirsch, I., Morton, C., Lenoir, G., Swan, D., Tronick, S., et al. (1982) Translocation of the c-*myc* gene in the immunoglobulin heavy chain locus in human Burkitt lymphoma and murine plasmacytoma cells. *Proc Natl Acad Sci.* **79**, 7837–7841.
8. Dalla-Favera, R., Bregni, M., Erikson, J., Patterson, D., Gallo, R.C., and Croce, C.M. (1982) Human c-*myc onc* gene is located on the region of chromosome 8 that is translocated in Burkitt lymphoma. *Proc Natl Acad Sci.* **79**, 7824–7827.
9. Cigudosa, J.C., Rao, P.H., Calazanz, M.J., Odero, M.D., Michaeli, J., Jhanwar, S.C., et al. (1998) Characterization of nonrandom chromosomal gains and losses in multiple myeloma by comparative genomic hybridization (CGH). *Blood.* **91**, 3007–3010.
10. Rao, P., Houldsworth, J., Dyomina, K., Parsa, N.Z., Cigudosa, J.C., Louie, D.C., et al. (1998) Chromosomal and gene amplification in diffuse large cell lymphoma. *Blood.* **92**, 234–240.
11. Rao, P.H., Cigudosa, J.C., Ning, Y., Calasanz, M.J., Lida, S., Tagawa, S., et al. (1998) Multicolor spectral karyotyping (SKY) identified new recurring breakpoints and translocations in multiple myeloma. *Blood.* **92**, 1743–1748.
12. Nanjangud, G., Rao, P.H., Hegde, A., Terayu-Feldstain, J., Donnelly, G., Qin, J., et al. (2002) Spectral karyotyping identifies new. Rearrangements, translocations, and clinical associations in diffuse large B-cell Lymphoma. *Blood.* **99**, 2554–2561.
13. Dyomin, V.G., Palanisamy, N., Lloyd, K.O., Dyomina, K., Jhanwar, S.C., Houldsworth, J., et al. (2000) MUC1 is activated in a B-cell lymphoma by the t(1;14)(q21;q32) translocation, and is rearranged and amplified in B-cell lymphoma subsets. *Blood.* **95**, 2666–2671.
14. Offit, K., LoCoco, F., Louie, D.C., Parsa, N.Z., Leung, D., Portlock, C., et al. (1994) Rearrangement of the Bcl-6 gene as a prognostic marker in diffuse large cell lymphoma. *N Engl J Med.* **331**, 74–80.
15. Rosenwald, A., Wright, G., Chan, W.C., Connors, J.M., Campo, E., Fisher, R.I., et al. (2002) The use of molecular profiling to predict survival after chemotherapy for diffuse large-B-cell lymphoma. *N Engl J Med.* **346**, 1937–1947.
16. Lossos, I.S. (2005) Molecular pathogenesis of diffuse large B-cell lymphoma. *J Clin Oncol.* **23**, 6351–6357.
17. O'Connor, O.A. (2005) Targeting histones and proteasomes: new strategies for the treatment of lymphoma. *J Clin Oncol.* **23**, 6429–6436.
18. Chaganti, R.S.K., Nanjangud, G., Schmidt, H., and Teruya-Feldstein, J. (2000) Recurring chromosomal abnormalities in non-hodgkin's lymphoma: biologic and clinical significance. *Semin Hematol.* **37**, 396–411.
19. Nanjangud, G., Rao, P.H., Terayu-Feldstein, J., Donnelly, G., Qin, J., et al. (2007) Molecular cytogenetic analysis of follicular lymphoma (FL) provides detail characterization of chromosomal instability associated with the t(14;18)(q32;q21) positive and negative subsets and histologic progression. *Cytogenet Genome Res.* **118**, 337–344.

Chapter 11

Metaphase Cytogenetic Techniques in Multiple Myeloma

Jeffrey R. Sawyer

Abstract

Metaphase chromosome studies in multiple myeloma (MM) are performed as part of the diagnostic workup, as surveillance to monitor the therapeutic response, and at relapse to help direct therapy. Unfortunately, the abnormal clones in many patients have a low proliferative activity and therefore, in many cases, the analyzable metaphase cells are derived from normal hematopoiesis. This limitation has been overcome in part by the use of fluorescence in situ hybridization (FISH) of interphase nuclei, which is an important adjunct to metaphase analysis. However, the metaphase karyotype remains the primary cytogenetic tool used at our institute for diagnostic and prognostic purposes. To maximize the possibility of detecting abnormal cells in conventional metaphase studies, we routinely employ at least three different cell harvest techniques on each marrow specimen. These include a direct harvest, a 24 h culture, and either a synchronized or unsynchronized 48 h culture. Recently, fine needle aspirates of solitary plasmacytomas guided by magnetic-resonance imaging have also been shown to provide diagnostic metaphase karyotypes. Regardless of the origin or quality of the specimen, some uninformative cytogenetic results may simply be due to insufficient number and type of cultures being initiated and subsequently examined.

Key words: Multiple myeloma, Cytogenetics, Cell culture techniques

1. Introduction

Multiple myeloma (MM) is a B-cell malignancy of the plasma cells characterized by complex cytogenetic aberrations. In myeloma, plasma cells proceed through a series of phases during malignant transformation, including a non-proliferative phase, an active phase with a small percentage of proliferating cells, and a fulminant phase with an increase in plasmablastic cells. In patients with active myeloma, only about a third of patients will show an abnormal karyotype, which is usually complex with multiple structural and numerical aberrations (1–3). In some respects, metaphase cytogenetics can be regarded as a biological test. In early myeloma, the cells are believed to be stroma-dependent;

Lynda J. Campbell (ed.), *Cancer Cytogenetics: Methods and Protocols*, Methods in Molecular Biology, vol. 730,
DOI 10.1007/978-1-61779-074-4_11, © Springer Science+Business Media, LLC 2011

therefore, taking the myeloma cells out of their supportive microenvironment and placing them in tissue culture will result in apoptosis of the cell and no informative mitoses. As myeloma cells become stroma-independent in the advanced stages of the disease, taking the myeloma cell out of the microenvironment does not result in cellular apoptosis. The myeloma cells not only survive, but also they proliferate in tissue culture and give rise to abnormal mitoses. Therefore, finding abnormal mitoses in a sample can, in a certain respect, be regarded as a surrogate marker for stroma-independent cells and more advanced disease (4).

The specificity of chromosome abnormalities associated with various stages of MM is a recognized tool in the differential diagnosis and treatment of this disease. In general, myeloma karyotypes can be divided into two groups based on the pattern of chromosome aberrations (5). Approximately 55–60% of patients have a hyperdiploid karyotype which usually includes the most common trisomies of 3, 5, 7, 9, 11, 15, 19, and 21. The second group is classified as non-hyperdiploid or hypodiploid. These patients are more likely to present with IgH translocations including a t(14;16)(q32;q23) or t(4;14)(p16;q32) translocation. The hypodiploid patients commonly show chromosome losses involving deletions or monosomies for chromosomes 13, 14, 16, and Y.

Chromosome 1 aberrations are common in the more proliferative clones and associated with aggressive disease. During the cytogenetic progression of myeloma, chromosome 1 aberrations can occur in both hyperdiploid and hypodiploid karyotypes and coincide with very complex karyotypes and genomic instability. The most common aberrations include interstitial deletions of 1p (6), interstitial duplications of the 1q12–23 region, and whole-arm jumping translocations of 1q. Jumping translocations of 1q (JT1q) are large, unbalanced rearrangements associated with low-level gene amplification for 1q and, in certain cases, the loss of material on the receptor chromosome (7–9). JT1q can occur either as telomeric or centromeric translocations and most commonly occur with receptor chromosomes 16 or 19, resulting in a net gain of the genes on 1q and net loss of genes on 16q or 19q.

The application of metaphase cytogenetic analysis in myeloma has traditionally been hampered by low mitotic index and poor quality of banded karyotypes. To increase the likelihood of obtaining abnormal metaphase spreads, it is important to initiate multiple harvests of each specimen. The procedures most commonly used include direct harvest, 24, 48, and 72 h cultures, depending on specimen quantity and arrival times. Following specimen processing, all samples undergo Giemsa-banding and, when needed, a combination of adjunct fluorescence in situ hybridization (FISH) and spectral karyotyping (SKY) methods.

When adequate sample is available, we routinely set up a direct harvest, in addition to 24 and 48 h cultures. The direct harvest procedure is initiated promptly upon specimen arrival. The bone marrow is treated with a solution containing trypsin, hypotonic, and Colcemid (THC) (10). Hypotonic cell swelling commences immediately, while trypsin helps break up cell clusters and alters cell membranes and appears to facilitate spreading of chromosomes due to general proteolytic effects on membranes. Colcemid arrests mitotic cells among the proliferating portion of the population. As opposed to tissue culture, the immediacy of the direct harvest THC treatment, especially with regard to the action of Colcemid in the hypotonic solution, gives a more representative picture of the mitotic index and proportion of various cell types in the marrow at the time of aspiration. Another advantage of the direct protocol is the speed with which karyotype analysis can be performed. Slides may be ready for analysis within a few hours after receiving the specimen if slides are baked 1–2 h immediately following the harvest. In some patients, chromosomally abnormal mitoses may be found only in the direct harvest, or alternatively only in the 24, 48, or 72 h cultures. Therefore, in addition to direct harvest, subsequent tissue culture of bone marrow is also necessary. When possible, a 48 h synchronized cell culture, in addition to the asynchronous cultures, can result in a harvest with a higher mitotic index and morphologically superior metaphase spreads than those observed in routine unsynchronized preparations (11). After an initial 48 h asynchronous growth period, methotrexate (a cell cycle blocking agent) may be added to the cultures to induce synchrony. Methotrexate is an analog of folic acid (vitamin B6) which binds to dihydrofolate reductase at the same site as the normal substrate, dihydrofolate. This binding blocks the conversion of dihydrofolate into tetrahydrofolate, which is necessary for the incorporation of thymidine into the DNA molecule during synthesis. Addition of methotrexate to cultures therefore provides a block which stops the cells at the G1/S phase of the cell cycle. After 17 h exposure to methotrexate, the cells are released from the block by washing once with unsupplemented medium and then resuspended in thymidine-rich medium. A peak mitotic index is observed when the cultures are harvested 4.5–5.5 h after release.

An additional source for cytogenetic analysis is fine needle aspiration (FNA) of osteolytic lesions. Osteolytic lesions presumably occur as a consequence of both the activation of osteoclasts and the inactivation of osteoblasts, mediated by the interaction of myeloma cells with the bone marrow microenvironment. Magnetic-resonance imaging is an established technique for the diagnosis of solitary plasmacytoma of bone in MM. Karyotypes derived from these osteolytic lesions have also been shown to have diagnostic and prognostic implications in myeloma (12). The FNA specimens are prepared for harvest by pretreatment with a collagenase solution and then harvested by standard methods.

2. Materials

1. 15 mL Plastic centrifuge tubes (Corning).
2. 16 mL Sterile polystyrene round-bottom tubes (Falcon).
3. 5 mL Sterile serological pipettes.
4. 10 mL Sterile serological pipettes.
5. 5.75 in. Pasteur pipettes (glass) (Fisher).
6. Transfer pipettes (disposable polyethylene) (Fisher).
7. Microscope slides, 25 × 75 mm, super frost.
8. 25 cm^2 Tissue culture flasks.
9. RPMI 1640 media.
10. Complete RPMI Culture Medium (also used as transport medium):
    (a) 500 mL RPMI 1640 (Sigma).
    (b) 125 mL Fetal bovine serum (Irvine).
    (c) 10 mL L-Glutamine/penicillin–streptomycin solution (Invitrogen).
    (d) 250 μL Sodium heparin (1,000 μL/mL).
11. Hypotonic solution – 0.075 M potassium chloride (KCL) (Sigma): 1.1 g/200 mL distilled reagent grade water.
12. Fixative: five parts methanol, absolute, acetone free (Fisher) to two parts glacial acetic acid (Fisher). Standard fixative: Carnoy's fixative, three parts absolute methanol to one part glacial acetic acid.
13. Trypsin–EDTA (1×) (0.25%) (Life Technologies).
14. Colcemid (Irvin Scientific) 10 μg/mL: dilute 10 μg/mL Colcemid 1:1 with Hank's balanced salt solution for 5 μg/mL.
15. Hank's balanced salt solution – HBSS (Sigma).
16. Ethidium bromide solution (Sigma 10 mL) stock solution is 10 mg/mL. To make working solution: 9 mL of HBSS per 1 mL ethidium bromide stock solution (Store at 2–8°C) (possible carcinogen).
17. Collagenase CLS II (Worthington Biomedical) 100 mg: reconstitute each vial with 12.5 mL of complete RPMI 1640 media and transfer to sterile 30 mL vial via syringe with 0.2 μm filter. Store at 2–8°C.
18. Methotrexate solution (Bedford Labs): to make stock solution (10^{-4} M), combine 2 mL vial (50 mg) of sterile isotonic solution and 98 mL HBSS. To make working solution (10^{-5}) dilute 1 mL of stock methotrexate (10^{-4}) with 9 mL HBSS. Add 100 μL to each culture.

19. Thymidine – MW 242.23 (Sigma 1G) stock solution (10^{-2} M): add 0.25 g thymidine to 100 mL HBSS. To make working solution (10^{-3} M), add 2.0 mL of 10^{-2} stock solution to 18.0 mL of HBSS. Add 100 μL of 10^{-3} solution to each culture.
20. Sterile petri dishes 35 × 10 mm (Falcon).

3. Methods

3.1. Specimen Collection

Specimens are collected in the Bone Marrow Clinic. Typically, 4 mL of bone marrow is drawn and placed into 6 mL complete RPMI 1640 transport media. Specimens are transported to the laboratory and kept at room temperature until cultures are initiated. Specimens are centrifuged for 10 min at 200 *g*. Supernatant is aspirated leaving approximately 3 mL of specimen to be divided between different cultures.

3.1.1. Bone Marrow Specimen Wash

All bone marrow specimens not received in Cytogenetics transport media must be washed.

1. Place bone marrow in a sterile tube containing 6 mL of 1640 unsupplemented media.
2. Centrifuge for 10 min at 200 *g*.
3. Supernatant is aspirated leaving approximately 3 mL of specimen to be divided between different cultures.

3.2. Procedure for Direct Harvest (see Note 1)

THC Preparation (Trypsin, Hypotonic, Colcemid)
Prepare tubes:

1. Place 1.0 mL sterile trypsin–EDTA into a 16 mL sterile round-bottom tube.
2. To this, add 9 mL KCl prepared for direct culture with the addition of sodium heparin and prewarmed to 37°C (may also be used at room temperature). The KCl flask should be inverted before aliquotting to assure even distribution of sodium heparin into solution.
3. Using a sterile Pasteur pipette, add 1.0 mL of bone marrow to each tube as prepared above.
4. Add 50 μL of 10 μg/mL Colcemid. Invert to mix.
5. Incubate THC tubes in 37°C water bath for 60 min.
6. Remove from water bath and add 10–15 drops of freshly prepared methanol/acetic acid 5:2 fixative. Mix well.
7. Centrifuge for 10 min at 200 *g*.
8. Aspirate supernatant leaving approximately 1 mL above the cell pellet.

9. Carefully re-suspend the cell pellet with a transfer pipette. Keep the cell suspension in the lower third of the pipette. With a clean pipette, slowly add 2 mL of freshly prepared methanol/acetic acid 5:2 fixative drop-by-drop with constant agitation. Add an additional 4 mL of fix and mix thoroughly (see Note 2).
10. Let stand at room temperature for 20 min.
11. Centrifuge 10 min at 200 *g* and aspirate supernatant.
12. Re-suspend the pellet with a transfer pipette, keeping the suspension in the lower third of the pipette. Add 6 mL of fresh fix and mix thoroughly with pipette.
13. Repeat steps 11 and 12 for a total of three washes.

3.3. Set-Up Procedure for 24 and 48 h Unsynchronized Flask Harvest

1. Using aseptic technique, aliquot 10 mL of Complete RPMI 1640 into one to two tissue culture flasks. If no direct harvest was performed, a 72 h culture should also be initiated.
2. Label one flask 24 h harvest, one flask 48 h harvest, and (if specimen did not receive a direct culture) one 72 h harvest.
3. Add 1 mL of bone marrow with sterile Pasteur pipette to each flask. Incubate at 37°C for 24, 48, or 72 h.

3.3.1. Harvest Procedure for 24, 48, and 72 h Cultures

1. After appropriate incubation, thoroughly agitate cultures by rocking flasks back and forth. To each flask, add 50 μL of 10 μg/mL Colcemid. Return to incubator for 1 h.
2. Gently agitate flask to re-suspend cells. Pour contents of culture flask into a labeled centrifuge tube and centrifuge for 10 min at 200 *g*.
3. Aspirate supernatant taking care to leave 1 mL above the cell pellet.
4. Gently re-suspend the cell pellet with 2 mL warm (37°C) KCl hypotonic solution. Mix gently with pipette then slowly add an additional 6 mL of hypotonic solution. Mix again gently, avoiding bubbles.
5. Place in water bath at 37°C for 15 min.
6. Add 10–15 drops of freshly prepared methanol/acetic acid 5:2 fixative. Mix well by inverting the tube.
7. Centrifuge at 200 *g* for 10 min and aspirate the supernatant.
8. Re-suspend the cell button with a Pasteur pipette keeping the suspension in the lower one-third of the pipette. Then, with a clean pipette, slowly add 2 mL of freshly prepared methanol/acetic acid 5:2 fixative drop-by-drop with constant agitation. Add an additional 4 mL of fix and mix thoroughly.
9. Stand at room temperature for 20 min.

10. Centrifuge at 200 *g* for 10 min.
11. Aspirate supernatant and add 6 mL of fresh fixative. Re-suspend cell pellet and mix thoroughly.
12. Centrifuge at 200 *g* for 8 min.
13. Repeat steps 11 and 12 for a total of three washes.

3.3.2. Slide Making

For each of the cultures, after the harvest procedure is completed, the slides are prepared by applying diluted fixed cells in a drop-wise manner to pre-cleaned glass microscope slides. The aim is to achieve at least two slides per culture from the initial dropping for a total of six slides. In most cases, there will be three cultures per specimen warranting at least two slides per culture. If there are fewer than three cultures for the specimen, the number of slides dropped per culture will need to be adjusted to achieve a total of six slides. Additional slides may be prepared later if additional material is needed to complete the study.

To prepare slides:

1. Centrifuge fixed specimen at 200 *g* for 8 min and discard supernatant.
2. Re-suspend the pellet in fresh 5:2 fixative. The amount of fixative added depends upon the size of the pellet. The final suspension should appear slightly cloudy.
3. Drop 3–5 evenly spaced drops from a Pasteur pipette onto a pre-cleaned slide. Various techniques may be employed to prevent over or under spreading of mitoses (see Note 3). Label each slide with the patient's last name, the specimen number, and the preparation date. Also label each slide with an accession number corresponding to its harvest time (e.g., D, 24, 48). After slides are prepared, add enough fixative to bring the total volume of the cell suspension to at least 3 mL. Centrifuge at 200 *g* for 8 min and seal caps with Para-film. Fixed cell pellets are stored at −15 to −20°C.
4. Slides are baked over-night in a 65°C oven.

3.4. Procedure for Synchronized 48 h Bone Marrow Cultures

1. Aseptically aliquot 10 mL of prepared complete RPMI 1640 medium into a tissue culture flask.
2. Label the flask with appropriate culture time, 48 or 72 h.
3. Add 1 mL of washed bone marrow specimen.
4. Cap flask and place in a 37°C incubator.

3.4.1. Establishing the Methotrexate Block

1. After 24 h (around 3:00 p.m.) (or 48 h for 72 h culture), add 100 μL of methotrexate working solution (final concentration of 10^{-5} M). Mix gently.
2. Cap and place back into the 37°C incubator.

3.4.2. Releasing the Cells from the Block

Release the cells 17 h after adding methotrexate (around 8:00 a.m.) as follows:

1. Mix the culture by gently agitating the flask.
2. Pour the contents of the flask into a sterile screw-top 16 mL round-bottom polystyrene tube.
3. Centrifuge at 200 *g* for 10 min.
4. Aspirate supernatant.
5. Add 10 mL pre-warmed RPMI 1640 unsupplemented media and re-suspend the pellet gently by capping the tube and inverting several times.
6. Centrifuge at 200 *g* for 8 min.
7. Aspirate supernatant to within 1 mL of cell pellet in round-bottom tube.
8. Add 9 mL complete RPMI 1640 medium and 100 μL thymidine working solution to the culture. Gently resuspend by capping and inverting the tube several times.
9. Pour into a clean sterile culture flask. Label 48 h (or 72 h).
10. Return to the 37°C incubator for 4 h.

3.4.3. Harvesting the Culture

1. Four hours after release, add 100 μL of ethidium bromide to the flask. Return flask to the 37°C incubator for 20 min.
2. Add 20 μL of Colcemid (10 μg/mL) to the flask. Return to the 37°C incubator for 60 min.
3. Proceed with harvest as described for non-synchronized bone marrow culture (Subheading 3.3.1, step 2).

3.5. Fine Needle Aspirate Procedure for Solitary Plasmacytoma

FNA specimens are collected under sterile conditions, placed in cytogenetics 6 mL RPMI 1640 transport medium, and delivered promptly at room temperature.

3.5.1. Procedure for Culture Set-Up

1. Centrifuge specimen at 200 *g* for 10 min.
2. Under a laminar flow hood, aspirate supernatant leaving approximately 0.5 mL of media above the cell pellet.
3. Transfer cell pellet to a 30 mm petri dish and very finely mince with scalpels.
4. Add 1 mL of prepared collagenase solution and incubate 2–3 h. (Even if there is no visible tissue prior to centrifugation, collagenase should be added to the cell pellet in the petri dish and incubated.)
5. After 2 h at 37°C, wash the specimen by adding 3–5 mL unsupplemented medium to the petri dish and aspirate up and down with a 5 mL serological pipette to help dissociate any remaining clumps.

6. Centrifuge at 200 *g* for 8 min and remove supernatant.
7. FNA's are cultured in 24 and 48 h flasks and placed in the incubator with the caps loosened. If there is not a visible piece of tissue in the tube, only a single 48 h flask is set up.

3.5.2. Harvest of FNA

See Subheading 3.3.1.

4. Notes

1. Direct cultures are only set up on bone marrow specimens received on the day of specimen collection. Bone marrow specimens received late in the day that do not undergo a direct harvest should have a 72 h flask culture set up in addition to the 24 and 48 h flasks.
2. It is very important to execute the fixation step of the cell harvest properly, since a poor quality fixation may result in unwanted excess cytoplasmic background and poor spreading. Care should be taken to thoroughly re-suspend the cells following hypotonic in order to optimize the fixation. The fixative should be added in a drop-by-drop fashion with constant agitation of the tube. Different geographical regions experience different levels of local humidity. In our area with high humidity, the standard cytogenetic fixative, Carnoy's fixative (three parts methanol to one part acetic acid), can result in overspreading of chromosomes. Under high humidity conditions, alternative ratios of methanol and acetic acid can be tested to achieve the desired spreading of chromosomes. The routine fixative used in our laboratory is five parts methanol to two parts acetic acid.
3. For slide making, we have assembled our own Plexiglas slide dropping chambers (3×2×2 ft) which are modeled after the environmentally controlled slide dropping chambers sold commercially. We use ultrasonic humidifiers inside these chambers when necessary to adjust the humidity. Typically, 72–76°F and 55–60% relative humidity produces the best quality slides. Variations on drying of slides at different angles may be tried. Begin by drying the slide at a 45° angle. The slide should dry evenly with no moisture allowed to pool on the slide. After the cells have been washed in fixative at least three times, the final cell suspension can be altered to manipulate the spreading of the chromosomes. If the chromosomes appear to be under spread, the concentration of acetic acid in the final cell suspension can be increased, and the drying of the slide can be slowed by lowering the drying angle of the slide. Increasing the humidity level at the station work is also effective. Alternately,

if the sample is spreading too much, the technologist may slightly increase the concentration of methanol in the specimen and/or increase the drying time of the slide by raising the angle of the slide to an almost upright position.

References

1. DeWald G.W., Kyle R.A., Hicks G.A., and Griepp P.R. (1985) The clinical significance of cytogenetic studies in 100 patients with multiple myeloma, plasma cell leukemia or amyloidosis. *Blood* **66**, 380–90.
2. Sawyer J.R., Waldron J.A., Jagannath S., and Barlogie B. (1995) Cytogenetic findings in 200 patients with multiple myeloma. *Cancer Genet Cytogenet* **82**, 41–9.
3. Sawyer J.R. and Singhal S. (2002) Cytogenetics of plasma cell disorders. In *Myeloma*, Eds. J. Mehta and S. Singhal. Martin Dunitz, London, 65–80.
4. Zhan F., Sawyer J., and Tricot G. (2006) The role of cytogenetics in multiple myeloma. *Leukemia* **20**, 1484–6.
5. Chng W.J., Glebov O., Bergsagel P.L., and Kuehl W.M. (2007) Genetic events in the pathogenesis of multiple myeloma. *Best Pract Res Clin Haematol* **20**, 571–96.
6. Shaughnessy Jr., J., Zhan F., Burington B., Huang Y., Colla S., Hanamura I., Stewart J., Kordsmeier B., Randolph C., Williams D., Xiao Y., Xu H., Epstein J., Anaissie E., Krishna S., Cottler-Fox M., Hollmig K., Mohiuddin A., Pineda-Roman M., Tricot G., van Rhee F., Sawyer J., Alsayed Y., Walker R., Zangari M., Crowley J., and Barlogie B. (2007) A validated gene expression model of high-risk multiple myeloma is defined by deregulated expression of genes mapping to chromosome 1. *Blood* **109**, 2276–84.
7. Sawyer J.R., Tricot G., Mattox S., Jagannath S., and Barlogie B. (1998) Jumping translocations of chromosome 1q in multiple myeloma: Evidence for a mechanism involving decondensation of pericentromeric heterochromatin. *Blood* **91**, 1732–41.
8. Sawyer J.R., Tricot G., Lukacs J.L., Lichti-Binz R., Tain E., Barlogie B., and Shaughnessy J. (2005) Genomic instability in multiple myeloma: Evidence for jumping segmental duplications of chromosome 1q. *Genes Chromosomes Cancer* **42**, 95–106.
9. Hanamura I., Stewart J.P., Huang U., Zhan F., Santra M., Sawyer J.R., Hollmig K., Zangarri M., Pineda-Roman M., van Rhee F., Cavall F., Burington B., Crowley J., Tricot G., Barlogie B., and Shaughnessy Jr., J.D. (2006) Frequent gain of chromosome band 1q21 in plasma cell dyscrasias detected by fluorescence in situ hybridization: Incidence increases from MGUS to relapsed myeloma and is related to prognosis and disease progression following tandem stem cell transplantations. *Blood* **108**, 1724–32.
10. Hozier J. and Lindquist L. (1980) Banded karyotypes from bone marrow: A clinical useful approach. *Hum Genet* **52**, 205–9.
11. Yunis J.J. (1981) New techniques in the study of human neoplasia. *Hum Pathol* **12**, 540–9.
12. Walker R., Barlogie B., Tricot G., Haessler J., Anaissie E., Shaughnessy J., Epstein J., van Hemert R., Erdem E., Hoering A., Crowley J., Ferris E., Hollmig K., van Rhee F., Zangari M., Pineda-Roman M., Mohiuddin A., Yaccoby S., Sawyer J.R., and Angtuaco E.J. (2007) Magnetic resonance imaging in multiple myeloma: Diagnostic and clinical implications. *J Clin Oncol* **25**, 1121–28.

Chapter 12

Detection of Chromosome Abnormalities Using Cytoplasmic Immunoglobulin Staining and FISH in Myeloma

Bruce R. Mercer and Kathleen C. Rayeroux

Abstract

The low proliferation rate of myeloma cells in vitro can result in a normal cytogenetic karyotype with the abnormal cell population not being detected. Because plasma cell myeloma is a patchy disease, conventional FISH is also hampered by normal cell contamination. Identification of plasma cells by cytoplasmic immunoglobulin staining in combination with FISH (cIg FISH) can ensure that only the cells of interest are analyzed, and thus the results obtained are a more accurate reflection of the plasma cell population karyotype. Current literature suggests that probes for t(4;14), t(14;16), and del(17)(p13) should be used in routine diagnostic testing; however, this technique can be used for any probes of interest. In this chapter, we present the techniques and methods used in our laboratory for the detection of abnormalities in plasma cells by cIg staining in conjunction with FISH.

Key words: Plasma cell myeloma, Cytoplasmic immunoglobulin staining, Fluorescence in situ hybridization, FGFR3/IGH, IGH/MAF, TP53

1. Introduction

This protocol is provided for the identification of plasma cells in bone marrow samples using simultaneous staining of cytoplasmic immunoglobulin (cIg) with FISH. This test is designed for use in patients with plasma cell myeloma (PCM) or other plasma cell disorders.

PCM is a multifocal bone marrow-based plasma cell neoplasm, usually associated with an overproduction of a paraprotein or Bence Jones protein in serum or urine (1). PCM includes asymptomatic (smoldering) myeloma, nonsecretory myeloma, and plasma cell leukemia. Monoclonal gammopathy of undetermined significance is considered a precursor to PCM.

Chromosomal abnormalities are detected by conventional cytogenetics in approximately 30% of PCM cases (2–4). Many genetic

Lynda J. Campbell (ed.), *Cancer Cytogenetics: Methods and Protocols*, Methods in Molecular Biology, vol. 730, DOI 10.1007/978-1-61779-074-4_12, © Springer Science+Business Media, LLC 2011

abnormalities have been found to be associated with myeloma. Hyperdiploidy, with trisomies of chromosomes 3, 5, 7, 9, 11, 15, 19, and 21 is a common finding. Hypodiploidy, less common than hyperdiploidy, is generally considered an unfavorable prognostic indicator (4–6). Monosomy 13 or deletion of 13q by conventional karyotyping is also considered a poor prognostic indicator (7). The low proliferation rate of cultured plasma cells, however, may result in a normal cytogenetic karyotype without the abnormal cell population being detected.

Conventional FISH studies are used to detect a range of molecular abnormalities, but do not distinguish between the abnormal plasma cell population and other hemopoietic cells in the bone marrow. This can lead to a normal FISH result, particularly when there is a low level of plasma cells in the sample, which may not reflect the myeloma karyotype. Monoclonal plasma cells are usually greater than 10% of cells in the bone marrow of patients with PCM, but some patients may have fewer than 10% plasma cells in their marrow aspirate. The European Myeloma Network recommendations for FISH in myeloma (2007) state that it is not acceptable to report FISH results in myeloma without either plasma cell enrichment or employing some means of plasma cell identification, so that only plasma cells are scored.

As plasma cells express either kappa or lambda antibodies (see Note 1), cIg staining can be used with an anti-kappa or anti-lambda antibody to facilitate identification. Using this method, an overall abnormality rate of greater than 80% has been reported in myeloma bone marrow samples (8, 9).

FISH testing of PCM cases generally incorporates the use of FISH probes with a proven link to prognosis. Currently, recommended probes are the *FGFR3/IGH* dual fusion probe to detect the cryptic t(4;14), the *IGH/MAF* dual fusion probe to detect the t(14;16), and a *TP53/*D17Z1 dual color probe to identify deletions of the short arm of chromosome 17, del(17)(p13) (7). All of these abnormalities are associated with a poor prognosis in PCM and are not necessarily visible by conventional cytogenetics. The role of FISH for deletions of chromosome 13q has recently been questioned by studies suggesting that loss of 13q is prognostically most significant when detected by conventional cytogenetics (7, 10, 11).

FISH for a number of other chromosome abnormalities may play a role in the investigation of PCM in the future, but these abnormalities are currently either under review or of unclear prognostic significance. Of note, amplification of 1q21 has been seen in 30–40% of patients (12–14). It is considered to be an event associated with disease progression that often accompanies the 4;14 or 14;16 translocations; however, its significance remains to be determined (2, 12). Other translocations of interest involve the cyclin D group, *CCND1* at 11q13, *CCND2* at 12p13, and

CCND3 at 6p21, and other genes of the MAF group, *MAFA* at 8q24.3 and *MAFB* at 20q12. Rapidly evolving therapeutic options for PCM mean that this is an area in a state of dynamic flux, and recommendations are likely to continue to be revised and updated.

2. Materials

All reagents are potentially hazardous. It is advisable to use appropriate safety procedures, including personal protective equipment, when handling these materials and to avoid contact with skin and mucous membranes. Before commencing testing, it is necessary to check the patient's immunoglobulin status (see Note 2) and it is also important to note the plasma cell percentage (see Note 3).

2.1. Specimens: Cytogenetic Fixed Cell Suspension

The following method is optimized for use on PCM specimens that have been cultured and harvested using standard cytogenetic methodology (not described in this chapter) (see Note 4). These protocols result in the specimen being fixed in Carnoy's fixative.

2.2. Pretreatment of Slides

1. Centrifuge.
2. Phase contrast microscope.
3. 95°C Waterbath.
4. Shandon Cytoslides (Thermo Scientific, Cheshire, WA), with pre-marked circle.
5. 70% Ethanol: add 30 mL distilled water (dH_2O) to 70 mL of 100% ethanol. Store at room temperature.
6. 80% Ethanol: add 20 mL dH_2O to 80 mL of 100% ethanol. Store at room temperature.
7. 95% Ethanol: add 5 mL dH_2O to 95 mL of 100% ethanol. Store at room temperature.
8. 10 mM Citrate buffer: add 1 g of citric acid and 1.45 g tri-sodium citrate to 500 mL dH_2O, pH 6.0. Make to 1 L with dH_2O. Store at 4°C.
9. Phosphate buffered saline (PBS).
10. Coplin jars.

2.3. Hybridization

1. Probes:
   (a) LSI FGFR3/IGH dual color dual fusion probe (Vysis, Downers Grove, IL).
   (b) LSI TP 53/CEP17 dual color probe (Cytocell, Cambridge, UK).
   (c) LSI IGH/MAF dual color dual fusion probe (Vysis, Downers Grove, IL).

2. Hybridization buffer (supplied with commercial probes) (Vysis, Downers Grove, IL).
3. Sterile dH_2O.
4. 0–10 μL Pipette.
5. Pipette tips.
6. 12-mm Circular coverslips (Menzel-Glaser, Braunschweig, Germany).
7. Fixogum (Marabu, Bietigheim-Bissingen, Germany).
8. Hotplate (minimum 80°C).
9. Humidified chamber (see Note 5).
10. 37°C Incubator.

2.4. Washing

1. 20× SSC: dissolve 175.3 g NaCl and 88.2 g sodium citrate in ~600 mL dH_2O and make up to 1 L. Check pH is between 7 and 7.5. Adjust if required with NaOH or HCl. Store at room temperature.
2. NP40 (Calbiochem, San Diego, CA).
3. 0.4× SSC/0.3% NP40: mix 20 mL of 20× SSC with 980 mL dH_2O. Add 3 mL of NP40. Mix well. Store at room temperature.
4. 2× SSC/0.1% NP40: mix 100 mL of 20× SSC with 900 mL dH_2O. Add 1 mL of NP40. Mix well. Store at room temperature.
5. 73°C Waterbath.
6. PBS.

2.5. cIg Staining

1. AMCA Anti-Human Kappa Chain (Vector, Burlingame, CA) (α-K). Store at 4°C.
2. AMCA Anti-Human Lambda Chain (Vector, Burlingame, CA) (α-Λ). Store at 4°C.
3. AMCA Anti-Goat Ig (H+L) (Vector, Burlingame, CA) (α-G). Store at 4°C.
4. Parafilm (Pechiney, Menasha, WI).
5. Vectashield Mounting Medium for Fluorescence (Vector, Burlingame, CA) (see Note 6). Store at 4°C.
6. 22×50-mm coverslips (Menzel-Glaser, Braunschweig, Germany).
7. Humidified chamber.
8. 37°C Incubator.
9. PBS.
10. 1.5-mL Eppendorf tubes.

11. 0–10 μL pipette.
12. 20–200 μL pipette.

2.6. Analysis

1. Fluorescent imaging microscope. The following method is based upon using a Fluorescent Axioplan 2 imaging microscope (Zeiss, Jena, Germany) with appropriate objectives and filter sets (see Note 7).
2. Immersol™ 518F: immersion oil for fluorescence (Zeiss, Jena, Germany).
3. Analysis software. The following method describes using the ISIS image analysis software (Metasystems, Altlussheim, Germany).

3. Methods

See Fig. 1 for an illustrated summary of the method.

3.1. Pretreatment of Slides

1. Assemble six coplin jars; fill one with citrate buffer, two with PBS, and one each with 70, 80, and 95% ethanol, respectively.
2. Before commencing the pretreatment process, warm the coplin jar containing citrate buffer to 95°C in a waterbath.
3. Centrifuge the Carnoy's fixed bone marrow suspension at 200 *g* for 5 min and carefully remove the supernatant. Resuspend to 5 mL with 95% ethanol and centrifuge again at 200 *g* for 5 min (see Note 8).
4. Resuspend the pellet in 95% ethanol to produce a slightly cloudy suspension (see Note 9).
5. Using a 10-μL pipette, drop 10 μL of the fixed cells within the pre-marked circle on the slide and allow to air-dry.
6. Assess cell density under the phase contrast microscope. Ideally there should be a minimum of 20 cells per field of view under a 10× objective (see Note 10).
7. Transfer the slides into the pre-warmed citrate buffer at 95°C for 10 min.
8. Remove the coplin jar from the waterbath and leave to cool at room temperature for 30 min. Place the coplin jar directly on the bench top to hasten cooling.
9. Transfer the slides to a PBS solution at room temperature for 2 min. Repeat the wash for a further 2 min in a second coplin jar.
10. Dehydrate the slides in an ethanol series of 70, 80, and 95% at room temperature for 1 min in each coplin jar and then air-dry the slides.

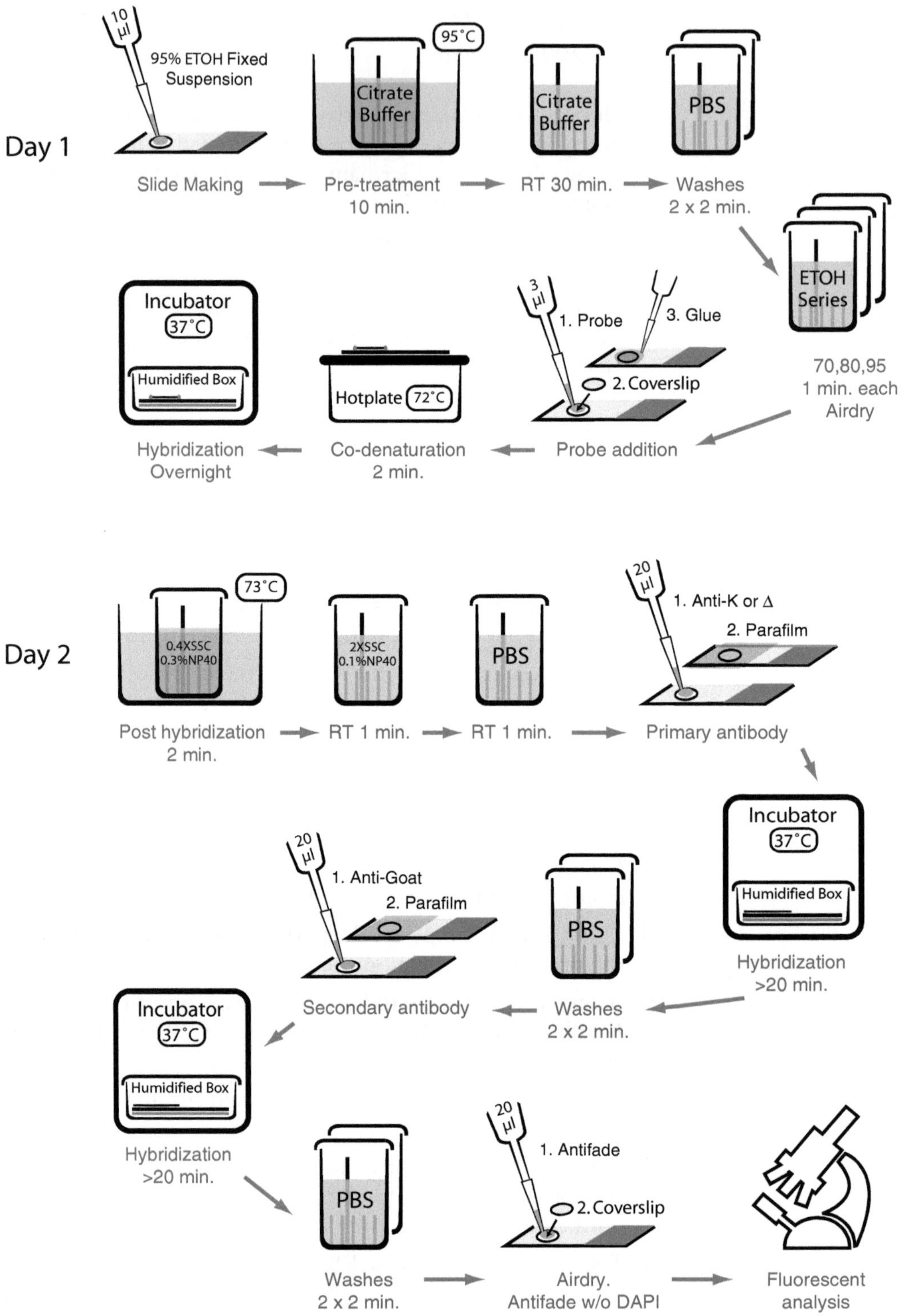

Fig. 1. Illustration of the step-by-step cIg FISH procedure.

3.2. Hybridization

1. Set the hotplate to 72°C and pre-warm a humidified chamber in a 37°C incubator.
2. Prepare probe mixtures according to the manufacturer's protocol (see Note 11).
3. Apply 3 μL of probe mixture within the pre-marked circle on the slide, and cover with a 12-mm circular glass coverslip. Ensure that the probe mixture spreads evenly under the coverslip without any air bubbles. If bubbles are present, gently press on the coverslip to push them out. Seal the coverslip with Fixogum.
4. Place slides on the hotplate for 2 min to denature both target and probe DNA and transfer to the pre-warmed humidified chamber (see Note 12).
5. Hybridize overnight or for a minimum of 14 h in a 37°C incubator.

3.3. Post-hybridization Washes

1. Pre-warm a coplin jar of 0.4× SSC/0.3% NP40 to 73°C in a waterbath (see Note 13).
2. Carefully remove Fixogum and the coverslips from the slides and wash for 2 min in the 0.4× SSC/0.3% NP40 solution.
3. Wash the slides in 2× SSC/0.1% NP40 solution at room temperature for 1 min, then wash in PBS for 1 min.

3.4. cIg Staining

1. For each slide dilute 2 μL of either AMCA Anti-Human Kappa or Lambda Chain antibody, depending on the patient's immunoglobulin status (see Note 14), with 18 μL of PBS in an Eppendorf tube.
2. Place the slides flat in the humidified chamber, apply 20 μL of the diluted antibody onto each slide, overlay with a sheet of parafilm large enough to cover all the slides in the chamber, and gently push out any air bubbles.
3. Incubate slides at 37°C for a minimum of 15 min.
4. Wash slides twice in PBS at room temperature for 2 min.
5. For each slide dilute 1 μL of AMCA Anti-Goat Ig (H+L) antibody with 19 μL of PBS in an Eppendorf tube.
6. Place the slides flat in the humidified chamber, apply 20 μL of the diluted antibody onto each slide, overlay with parafilm, and gently push out any air bubbles.
7. Incubate slides at 37°C for a minimum of 15 min.
8. Wash the slides twice in PBS at room temperature for 2 min (the PBS from step 3 can be reused).
9. Air-dry the slides.
10. Pipette 20 μL of Vectashield mounting medium (without DAPI) onto each slide and overlay with a coverslip.

3.5. Analysis

1. Turn on the fluorescent microscope and allow the bulb to warm up for 5–10 min.
2. Use the 10× objective to find the focal plane of the area of interest.
3. Apply a drop of immersion oil to the coverslip and examine under the 100× objective. Use the DAPI filter (see Note 15) to check that the labeled plasma cells can be distinguished from unlabeled cells (see Notes 16 and 17).
4. Identify the plasma cells by the blue AMCA cytoplasmic staining and distinctive plasma cell morphology (see Note 18).
5. Using the triple and/or individual filters, check that FISH signals are present in the nucleus of the plasma cells (see Note 19).
6. Only the clearly AMCA-positively stained nonoverlapping cells should be scored for FISH signals. If unsure whether a cell is a plasma cell or not, it is advisable *not* to score the cell in question.
7. Ensure that all signals in each labeled cell are scored by focussing up and down, as the three-dimensional nature of the cells may result in signals being in different focal planes.
8. Check a few non-cIg-stained cells to confirm that they contain a normal signal pattern.
9. Score a total of 100 plasma cells between two scientists (50 each) (see Note 20) for each probe tested.
10. Capture a minimum of five labeled cells for each abnormal clonal population observed and a minimum of two labeled cells for those with a normal signal pattern.
11. If a low number (less than 50) of labeled plasma cells are available for analysis, the first scorer should capture images of all the cells scored. The second scientist can then assess the captured images and if necessary relocate the cells down the microscope to confirm the signal pattern. It is recommended that a minimum of ten images are captured (see Note 21).
12. Record the results.

3.6. Interpretation of Results

The majority of plasma cells should be identifiable by cIg labeling. In our experience, the labeling strength and plasma cell morphology vary significantly between specimens, and practical experience is invaluable in correctly identifying plasma cells. FISH signals should be visible in the cIg-labeled cells for each of the probe mixtures used. A population is generally considered to be abnormal if more than 20% of the labeled cells display the same abnormal signal pattern. In reality, abnormal FISH signal patterns are usually present in the majority of plasma cells due to the preferential expansion of the myeloma cells over any residual normal plasma cells.

The non-cIg staining cells generally display a normal FISH signal pattern and should be checked as an internal control (see Note 22). If there are no signals visible in the non-cIg staining cells, the probe may not be performing appropriately and caution should be taken in interpreting the plasma cell signals.

4. Notes

1. Immunostaining for kappa or lambda light chains is recommended as it gives a stronger result than CD138.
2. If the Ig status of the patient is unknown, a combination of both kappa and lambda light chain antibodies may be used (see Note 14).
3. The percentage of plasma cells will give some indication of the level of difficulty of scoring. It is advisable not to test many patients with very low plasma cell percentages in the same experiment, as scoring of these cases is extremely time consuming.
4. To obtain sufficient cells for cIg FISH we recommend that, at the time of setting up cultures for conventional cytogenetic analysis, an additional 24-h culture be set up (if sufficient sample is available) with ~1.5–2 times the cell concentration of the standard 24-h culture. As samples processed for cIg FISH are no longer useful for conventional cytogenetics analysis, the original 24-h culture can be retained for this purpose.
5. A humidified chamber can be any sealed plastic container with a cloth or sponge dampened with 2× SSC placed inside on the base.
6. It is important that the mounting media does *not* contain DAPI, as the blue of the DAPI will interfere with the identification of blue AMCA cIg-stained cells.
7. The appropriate filters are vital for optimal visualization, scoring, and image capture. For viewing of the AMCA fluorochrome to identify the plasma cells, our laboratory uses an excitation filter 365/10, Dichroic 400 DCLP, Emission 460/50. For scoring of FISH signals that are labeled with either SpectrumOrange or SpectrumGreen, single filters with excitation/emission wavelengths of 559/588 nm or 497/524 nm respectively, are recommended. Alternatively, a triple filter set will allow simultaneous visualization of all three fluorochromes.

8. For optimal plasma cell staining, the bone marrow suspension should be fixed in 95% ethanol. Carnoy's fixed samples produce pale and diffuse plasma cell staining, which makes identification of the individual plasma cells difficult.
9. From our experience, 95% ethanol fixed samples can be stored long term at –80°C. Long-term storage in Carnoy's fixative at this temperature is also satisfactory.
10. Cells will frequently appear phase bright and clumped. The clumping is a side effect of the ethanol fixation and does not seem to be reduced even when resuspending with continuous vortexing or agitation.
11. Some probes are provided ready to use, containing both the hybridization buffer and probe, while others require the user to prepare a probe mixture from separate probe and hybridization buffer vials. If preparing a probe mixture, ensure the probe is well mixed before adding to the hybridization buffer.
12. Place slides on a 37°C hotplate following co-denaturation and before placing into the humidified chamber, if large numbers of slides are being probed.
13. The manufacturer's conditions for the post-hybridization washes should be followed.
14. If the Ig status of the patient is unknown, an antibody mixture of one part of α-K, one part of α-Λ, and eight parts of PBS should be used.
15. Labeled cells may have variable staining intensity. In some cases, the plasma cells may be clearly distinguishable on triple filter; alternatively, the DAPI filter may be required.
16. If the AMCA staining is unsatisfactory or absent, there may be a number of possible causes and remedies:
    (a) The patient sample may have a low plasma cell percentage with few or no plasma cells present on the slide. Check the plasma cell percentage and concentrate the sample for a repeat test. If the plasma cell percentage is extremely low, a result may not be possible.
    (b) If the sample was collected more than 24 h prior to culture, the plasma cells may have degraded and a repeat, freshly collected bone marrow sample may be required. Also, ensure that the culture with the shortest incubation time is used for FISH testing, ideally 24 h or less. Although the plasma cells will survive in most cultures for 48–72 h, at least one failed result in our hands appeared to be due to no 24 h culture being available.
    (c) If the labeling is poor for no identifiable reason, it is advisable to reapply the cIg staining steps. Remove the coverslip and rinse the slide in PBS for 2 min. Repeat the

cIg staining process and increase the incubation time to 30 min for both of the antibody steps.

(d) The wrong antibody or antibody dilution may have been used. Check the Ig light chain status of the patient and ensure that the correct dilutions of α-kappa, α-lambda, and α-goat antibodies in PBS were used.

(e) The fixation may be incorrect. Ensure that the specimen is fixed in 95% ethanol and not in 3:1 Carnoy's fixative. Refix the specimen with freshly made 95% ethanol.

17. If all of the cells appear to be labeled, there may be a number of possible causes and remedies:

(a) Antifade with DAPI may have been used. Repeat the test using Vectashield mounting medium without DAPI.

(b) The post antibody attachment washes may not have been carried out. Repeat the wash steps in PBS.

(c) The antibody concentration may be too strong, leaving residual antibody on the slide or nonspecific labeling. Check the dilutions of the antibody mixtures. If additional washes in PBS do not remove the excess antibody, it may be necessary to repeat the test with reduced concentrations of antibodies.

18. Not all cases will have classic plasma cell morphology. In some instances, there may be reduced cytoplasm around the nucleus and/or cIg staining over the entire cell (some examples of variable plasma cell morphology and cIg staining are shown in Fig. 2).

19. There may be a number of problems associated with the FISH signals:

(a) FISH signals may be present only in unlabeled cells. The plasma cells may be dead. Check the morphology of the labeled cells for evidence of cell degradation. Ensure that the shortest culture is used. For old or very poor samples, obtaining a repeat sample may be required.

(b) If there are no signals in any cells, this may point to a probe, hybridization, or post-hybridization wash problem. Check that the probe has been correctly made and applied, the denaturation and hybridization performed at the correct time, and temperatures and the post-hybridization washes performed at the correct concentration and temperature.

(c) High inter/intracellular probe background could point to a post-hybridization wash problem. The temperature may be too low, the salt concentration too high, or there was insufficient incubation in the post-hybridization solution. Rewash slides according to the manufacturer's recommendations.

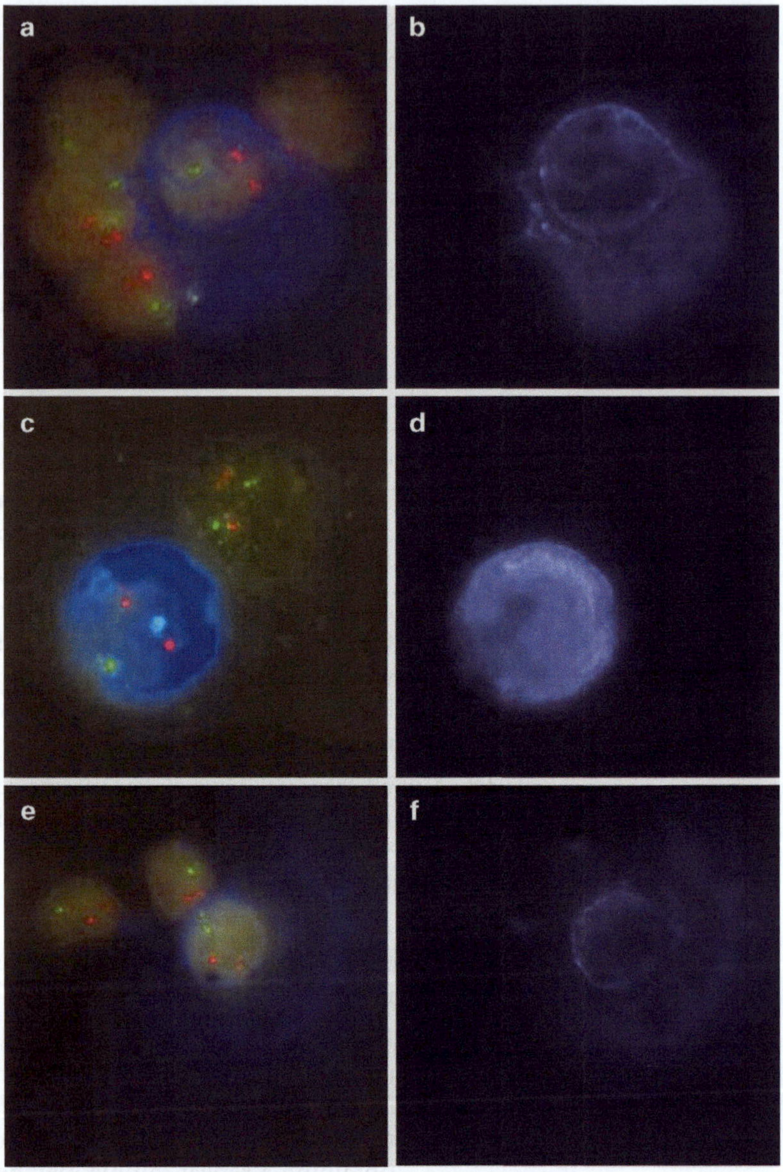

Fig. 2. cIg FISH images demonstrating the variable morphology and cIg staining of plasma cells in PCM. (**a**) cIg FISH showing an aberrant signal pattern for the LSI *FGFR3/IGH* dual fusion probe (Vysis) in a cIg-positive cell; (**b**) the corresponding AMCA image demonstrates the classic plasma cell morphology with a large amount of eccentric cytoplasm; (**c**) cIg FISH showing a normal signal pattern for the LSI *FGFR3/IGH* probe in a cIg-positive cell; (**d**) the corresponding AMCA image demonstrates nonclassical plasma cell morphology with strong cIg staining over the entire cell; (**e**) cIg FISH showing aberrant signal patterns in both cIg-positive and negative cells for the LSI *FGFR3/IGH* probe (where there is weak staining of the plasma cells as demonstrated, scanning and identification of cIg-positive cells on a triple color filter may not be possible); (**f**) in the corresponding AMCA image, the weakly stained plasma cell is easier to identify.

20. If there is a discrepancy between the results of the two scientists scoring FISH, a further 50 cells should be scored by a third scientist.
21. Results are only interpretable from <50 cells if >75% have an identical signal pattern.
22. A small number of our cases have demonstrated an abnormal cell population in the non-plasma cells, but not in cIg-positive cells. This unusual result may represent a malignant clone of cells devoid of cIg light chain expression, either primitive plasma cells, possibly malignant stromal cells, or a secondary hematological malignancy, such as a myelodysplastic syndrome.

Acknowledgements

We would like to thank Adrian Zordan, Crisoula Batzios, Kathy Somana and Lynda Campbell for their help with the development of this technique in our laboratory.

References

1. Swerdlow SH, Campo E, Harris NL, Jaffe ES, Pileri SA, Stein H, Thiele J, Vardiman JW. (Eds.). WHO Classification of Tumours of Haematopoietic and Lymphoid Tissues. Lyon: IARC; 2008.
2. Lai JL, Zandecki M, Mary JY, et al. Improved cytogenetics in multiple myeloma: a study of 151 patients including 117 patients at diagnosis. *Blood* 1995;85:2490–7.
3. Dewald GW, Kyle RA, Hicks GA, Greipp PR. The clinical significance of cytogenetic studies in 100 patients with multiple myeloma, plasma cell leukemia, or amyloidosis. *Blood* 1985;66:380–90.
4. Sawyer JR, Waldron JA, Jagannath S, Barlogie B. Cytogenetic findings in 200 patients with multiple myeloma. *Cancer Genet Cytogenet* 1995;82:41–9.
5. Smadja NV, Bastard C, Brigaudeau C, Leroux D, Fruchart C. Hypodiploidy is a major prognostic factor in multiple myeloma. *Blood* 2001; 98:2229–38.
6. Debes-Marun CS, Dewald GW, Bryant S, et al. Chromosome abnormalities clustering and its implications for pathogenesis and prognosis in myeloma. *Leukemia* 2003;17:427–36.
7. Stewart AK, Bergsagel PL, Greipp PR, et al. A practical guide to defining high-risk myeloma for clinical trials, patient counseling and choice of therapy. *Leukemia* 2007;21:529–34.
8. Chen Z, Issa B, Huang S, et al. A practical approach to the detection of prognostically significant genomic aberrations in multiple myeloma. *J Mol Diagn* 2005;7:560–5.
9. Fonseca R, Blood E, Rue M, et al. Clinical and biologic implications of recurrent genomic aberrations in myeloma. *Blood* 2003;101:4569–75.
10. Stewart AK, Fonseca R. Prognostic and therapeutic significance of myeloma genetics and gene expression profiling. *J Clin Oncol* 2005; 23:6339–44.
11. Avet-Loiseau H, Attal M, Moreau P, et al. Genetic abnormalities and survival in multiple myeloma: the experience of the Intergroupe Francophone du Myelome. *Blood* 2007;109: 3489–95.
12. Chng WJ, Glebov O, Bergsagel PL, Kuehl WM. Genetic events in the pathogenesis of multiple myeloma. *Best Pract Res Clin Haematol* 2007;20:571–96.
13. Fonseca R, Van Wier SA, Chng WJ, et al. Prognostic value of chromosome 1q21 gain by fluorescent in situ hybridization and increase CKS1B expression in myeloma. *Leukemia* 2006;20:2034–40.
14. Hanamura I, Stewart JP, Huang Y, et al. Frequent gain of chromosome band 1q21 in plasma-cell dyscrasias detected by fluorescence in situ hybridization: incidence increases from MGUS to relapsed myeloma and is related to prognosis and disease progression following tandem stem-cell transplantation. *Blood* 2006; 108:1724–32.

Chapter 13

Cytogenetics of Solid Tumours

Robyn Lukeis and Mary Suter

Abstract

The study of chromosome abnormalities in solid tumours provides valuable information for the diagnosis and prognostic stratification of a variety of tumour subtypes. Technical challenges are encountered in tissue culture, harvesting and finally the interpretation of complex chromosome abnormalities. Molecular cytogenetic studies complement karyotype analysis, but the ability to assess all chromosome abnormalities simultaneously underscores the value of conventional cytogenetic analysis. This chapter provides detailed methods for solid tumour cytogenetics from tissue culture to banding.

Key words: Solid tumour, Chromosome, Karyotype, Tissue culture

1. Introduction

Techniques that allow the detection of clinically relevant genetic abnormalities have a growing role in the diagnosis and prognostic prediction of a variety of solid tumours. Methodologies used to detect them may vary depending on laboratory resources and tissue availability. The increasing knowledge of the significance of the abnormalities refocuses attention on conventional cytogenetics as a tool for genome-wide detection of clinically important abnormalities.

Solid tumour cytogenetics is often viewed as challenging, particularly in the setting of the routine laboratory. There is sufficient published data to allow the description of non-random karyotypic abnormalities (1) for all subtypes of solid tumours, whether common or rare, benign or malignant. The utility of solid tumour cytogenetics in the diagnostic setting is, however, typically limited to a subset of tumour types. Molecular genetic and molecular cytogenetic methods increasingly complement conventional cytogenetics, but the ability to view a karyotype is

Lynda J. Campbell (ed.), *Cancer Cytogenetics: Methods and Protocols*, Methods in Molecular Biology, vol. 730,
DOI 10.1007/978-1-61779-074-4_13, © Springer Science+Business Media, LLC 2011

still advantageous for cases where there are differential diagnoses, or where other methods are difficult to interpret. Reciprocal translocations are readily identified, including common and variant partner chromosomes. Cytogenetics has an important role in the diagnosis of childhood tumours (2, 3).

Several factors contribute to the challenge in achieving and interpreting a tumour karyotype. Optimal in vitro culture conditions vary between tumour types. Solid tumours of different histological origin will have different optimal conditions for growth. Time and resource factors usually limit the ability to tailor conditions for all the different subtypes of tumour that may be referred. Relevant original references should be consulted for specific applications.

Solid tumours do not contain pure tumour cell populations. Tumour biopsies typically contain varying amounts of normal stromal and haemopoietic cells. During tissue disaggregation, these can establish as a contaminating normal cell line.

Solid tumours can contain significant inter- and intra-tumour heterogeneity, with complex karyotypes and chromosome instability. Multiple abnormal clones and significant non-clonal intratumour heterogeneity can exist as a feature of some types of solid tumours. There is evidence to suggest that some tumours are in fact polyclonal rather than monoclonal in origin (4). This limits the utility of conventional cytogenetics for some common tumour types, for example, breast and lung cancers. Longer term tissue culture can lead to the selection of subclones suited to in vitro culture, and new abnormalities can be acquired, with the resultant karyotype not necessarily representative of the original tumour. The study of short-term cultured tumour cells is necessary to identify the primary karyotypic abnormalities (5).

Tumour karyotypes can test the limits of sensitivity of G-banding in describing clonal abnormalities, with numerous and complex structural rearrangements. Karyotypic complexity and instability are characteristic of some tumours, and may be useful markers for future therapies once the genetic underlying mechanisms are understood (6–8). Rearrangements are often non-reciprocal, leading to partial gain or loss of material. Karyotypic variation between patients with the same diagnosis is also common in some tumour types, making description of common genetic changes difficult. This has led to the reference to common regions of gain or loss, where the chromosomal mechanism may be different between patients, but the net effect the same. ISCN 2009 allows the use of "inc" to denote incomplete karyotypes, enabling description of the pertinent abnormalities while demonstrating the limitations of the investigation.

1.1. Application of Solid Tumour Cytogenetics

Conventional cytogenetics is a routine test for childhood small round-cell tumours, including neuroblastoma and Ewing/primitive neuroectodermal tumours (PNET). FISH studies for important markers are common in these tumours (see Table 1) (3, 9).

Table 1
Cytogenetic abnormalities and FISH for solid tumours

Solid tumour type	Cytogenetic abnormality	Incidence (%)	Gene rearrangement	Probe type: Break-apart (BA)	Probe type: Locus-specific	Probe type: Translocation probe
Medulloblastoma (17)	idic(17)(p11.2)	50	Relative loss of TP53: 17q		TP53 (17p13)/ MPO (17q21–23)	
Peripheral nerve sheath tumour: Schwannoma	del(22q)					
Neuroblastoma (3)	Amp MYCN Del(1p36.3) Relative gain of 17q Deletion 11q		Amp MYCN Del(1p36.3) Relative gain of 17q	 MLL	MYCN/CEP2 1p36/CEP1 TP53 (17p13)/ MPO (17q21–23)	
Myxoid liposarcoma (9)	t(12;16)(q13;p11)		FUS/DDIT3 fusion gene	DDIT3 FUS		
Well-differentiated and de-differentiated liposarcoma (9)	r(12)(q13–15)		Amp MDM2 Amp 5′DDIT3	DDIT3	MDM2/CEP2	
Synovial sarcoma (9)	t(X;18)(p11.2;q11.2) Monophasic Biphasic	80	SS18/SSX2 SS18/SSX1 SS18/SSX1 SS18/SSX4 (rare)	SS18 SS18 SS18 SS18		
Alveolar rhabdomyosarcoma (2)	t(2;13)(q35;q14) t(1;13)(p36;q14)	60 20	PAX3/FOXO1 PAX7/FOXO1	FOXO1 FOXO1		PAX3/FOXO1 DC SF PAX7/FOXO1 DC SF

(continued)

Table 1 (continued)

Solid tumour type	Cytogenetic abnormality	Incidence (%)	Gene rearrangement	Probe type		
				Break-apart (BA)	Locus-specific	Translocation probe
Clear cell sarcoma (malignant melanoma of soft parts) (9)	t(12;22)(q13;q12)	Characteristic	ATF1/EWSR1	EWSR1		
Small round-cell tumcurs (PNET, Ewing's, Askin's) (2)	t(11;22)(q24;q12)	90	FLI1/EWSR1	EWSR1		
Posterior uveal/choroidal melanoma (18)	+8 or i(8)(q10) –3 dup(6p) del(6q)	40 50–60 25 25		 BCL6	 CEP3	MYC/IGH/CEP8

BA break-apart rearrangement probe, *DC DF* dual colour dual fusion translocation probe, *DC SF* dual colour single fusion translocation probe

Cytogenetic analysis is useful in distinguishing subtypes of soft tissue sarcomas, where reciprocal translocations are a feature. The laboratory may also expect to receive biopsies from tumours where the pathological findings are not definitive and where knowledge of the karyotypic abnormalities expected in the differential diagnoses is important. Non-random genetic changes and prognostic markers are described in the common tumours, including breast, lung, brain, colon, and prostate cancers, but it is unlikely that a karyotype will be requested due to difficulties in culturing and the complexity of the karyotypes. FISH or MLPA testing are used to detect important prognostic genetic features of breast (HER2 and EGFR) and brain tumours (1p/19q ratio in oligodendrogliomas), while FISH can be used to detect tumour cells in bronchial washes (lung cancer) or urine (bladder cancer).

In the research setting, it is useful to obtain a karyotype at the initiation of a cell line, as continuous passaging of cells can lead to selection of subclones and accumulation of abnormalities that may not be representative of the initial tumour population.

2. Materials and Equipment

2.1. Initiation of Cultures

1. Tumour wash (high-risk tumours only): 50-mL RPMI 1640, 5-mL Hank's 10× calcium- and magnesium-free balanced salt solution (BSS, Gibco), 500-μL gentamicin (10 mg/mL, Gibco), 50-μL fungizone (Amphotericin B, Gibco) made up to 100 mL with sterile water. Store at 4°C and discard after 1 month (see Note 1).
2. Culture flasks: 12.5- and 25-cm² sterile culture flasks, 0.22-μm vented cap, canted neck (BD Falcon Greiner).
3. Complete culture medium (CCM): 150-mL RPMI 1640, 40-mL foetal bovine serum, 1-mL gentamicin (10 mg/mL), 2-mL L-glutamine (200 mM), 2-mL 7.5% w/v sodium bicarbonate solution, 2-mL HEPES (1 M) (Gibco).
4. Giant cell tumour conditioned medium (GCTCM) (Irvine Scientific).
5. Insulin–Transferrin–Selenium (Gibco).
6. Collagenase Type II (Sigma-Aldrich): for overnight disaggregation, use at a concentration of 200–400 U/mL.
7. Sterile transfer pipettes, sterile Petri dishes, sterile disposable scalpels.
8. Sterile 10-mL centrifuge tubes.
9. Blunt-ended needles.

10. Sharps bins/contaminated bins for disposal of biohazardous material.
11. 37°C 5% CO_2 incubators.

2.2. Harvest of Solid Tumour Cultures

1. Spindle inhibitor: Colcemid (10 µg/mL in Hank's balanced salt solution, Gibco).
2. Trypsin–EDTA (10×, Gibco).
3. Hypotonic solution: 0.075 M Potassium chloride (KCl) solution: 5.59 g KCl (Merck)/litre sterile water. Ensure hypotonic is pre-warmed to 37°C before use.
4. 4% Aqueous glacial acetic acid: 1-mL Glacial acetic acid (Labserv Analar) + 24-mL deionised water.
5. Carnoy's fixative: 3:1 (v/v) Methanol (Mallinckrodt ChromAR HPLC):glacial acetic acid.

2.3. Slide Preparation

1. High-quality microscope slides.
2. Absolute alcohol.
3. Sterile water.
4. Freshly prepared Carnoy's fixative (3:1 methanol:glacial acetic acid) fixative.

2.4. GTW Banding

1. Nine Coplin jars.
2. Trypsin stock (Difco 1:250).
3. Hank's Ca- and Mg-free BSS (10×, Gibco).
4. Foetal calf serum.
5. 70% Ethanol.
6. 95% Ethanol.
7. pH 6.8 Buffer, prepared as directed (LabChem).
8. Wright's Stain (Sigma-Aldrich).

3. Methods

The major challenge in solid tumour cytogenetics is to convert solid tumour tissue into a single cell suspension. There are additional inherent difficulties associated with the culture of solid tumours, due to factors such as the frequent presence of necrosis, and their exhibiting widely variable cell cycle times, both of which affect adversely the chances of successful tissue culture and cytogenetic analysis.

For safety reasons, latex gloves and a coverall with elasticised wrists should be used at all times when initiating and harvesting solid tumour cultures. These specimens are potentially biohazardous, and hazardous reagents are used in the harvest procedure.

3.1. Transport of Samples

All biopsy specimens should be placed immediately into transport medium to ensure they remain hydrated and transported to the laboratory without delay for immediate processing. If delay is unavoidable, specimens should be maintained hydrated at 4°C for no more than 12 h.

3.2. Touch Imprints

1. Touch imprints are a valuable resource for subsequent FISH analysis (see Note 2). If no touch imprints have been sent with the tumour, label 4× positively charged (silanised) slides with the patient's name, medical record number (MRN), and date of collection.
2. Using a sterile blunt-ended needle, gently touch each slide three times with the tumour specimen.
3. Allow the slides to air-dry in the biohazard safety cabinet.
4. Fix all touch imprints for 15 min in Carnoy's fixative.
5. Allow to air-dry in the Biohazard Safety Cabinet, before wrapping fixed slides in foil, labelling the outside of the foil appropriately and storing at –20/–30°C.

3.3. Initiation of Cultures

1. Assess the tumour biopsy specimen to set up the size and quantity of culture flasks: if the specimen is small (~2 mm × 2 mm), only one 12.5-cm^2 culture flask should be set up. If the specimen is >2 mm × 2 mm, 2× 12.5-cm^2 culture flasks should be set up. If the specimen is ~1 cm × 1 cm, 2× 25-cm^2 cultures should be set up. If the specimen is very large (>2 cm), multiple cultures should be set up (see Note 3).
2. In a biohazard Type II cabinet and using sterile techniques, set up culture flask/s as follows:

   12.5-cm^2 Volume flask: 5-mL CCM.

   25-cm^2 Volume flask: 10-mL CCM.
3. Incubate all flasks at 37°C for at least 25 min.
4. Label the top of a sterile 35 mm × 10 mm Petri dish with the patient's name. Using a sterile transfer pipette, transfer the tumour biopsy sample to the Petri dish. If the specimen is very large, transfer it to a larger 90 mm × 15 mm Petri dish. Ensure the specimen is covered by 1–2 mL of accompanying transport medium.
5. If transport medium is cloudy, it may contain valuable viable cells: transfer it to a sterile 10-mL centrifuge tube and centrifuge at 160 *g* for 10 min.

6. Assess the tumour, according to its:
   (a) Appearance: Homogeneous or heterogeneous, consisting of several areas with different properties; if heterogeneous, separate flasks should be set up for each distinct area.
   (b) Colour: Red indicates good blood supply; greyish/dark brown indicates tissue may be necrotic; white-tan suggests healthy tissue.
   (c) Consistency: Tough or crunchy indicates connective tissue, requiring overnight disaggregation with Collagenase II; friable indicates a suspension culture should be set up as well as a monolayer culture and Collagenase II may not be required.
   (d) Note also if the tumour is suspected of being a recurrence. The provisional diagnosis is only a guide and may change once further anatomical pathology tests have been performed. For this reason, both suspension and monolayer cultures should be set up whenever possible to optimise chances of obtaining a representative karyotype (see Note 4).
7. Add Collagenase II to the required culture flask/s: 300 μL is required for each 12.5-cm^2 flask and 600 μL for each 25-cm^2 flask (see Note 5).
8. Return the flask to the 37°C incubator until required.
9. To culture flasks without Collagenase II, add GCTCM and ITS as follows:

   12.5-cm^2 Volume flask: 100 μL GCTCM + 100 μL ITS.

   25-cm^2 Volume flask: 200 μL GCTCM + 200 μL ITS.
10. Open the scalpel blades carefully to ensure the blades do not touch the surface of the cabinet. Press gently onto the specimen with the flat surfaces of the scalpel blades, to release any cell suspension. When this is complete, mince the tumour with both scalpel blades until all remaining tissue is <0.5 mm in size. Discard used scalpels into a sharps container in the biohazard cabinet.
11. Divide minced tissue between culture flasks (both with and without Collagenase II) and place in 37°C 5% CO_2 incubator.
12. If tumour is particularly cellular or if a high proliferation index has been reported, check the tumour culture at the end of the set-up day and split if required.
13. On the day following set-up, examine all culture flasks for culture depletion. Split or feed cultures if the medium looks severely depleted (yellow). Cultures should be split or fed with CCM + GCT + ITS (see Subheading 3.3).
14. Examine all culture flasks under an inverted microscope for adherence and for the presence of potentially harmful

non-viable cells or necrotic tissue. At each examination, assess the attachment, flattening, and proliferation of tumour cells. Mitotic activity may be observed as cells rounding up into spheres, or under high magnification, by the presence of mitotic figures. Excess non-viable cells or necrotic tissue can induce apoptosis, due to cell-lysing enzymes in the necrotic cells. Remove these with a sterile Pasteur pipette and replace the volume with CCM + GCT + ITS. If fibroblasts appear to be outgrowing the tumour cells, harvest immediately (see Note 6).

3.4. Monitoring and Splitting of Cultures

1. For each collagenased culture, warm 10-mL RPMI in the 37°C incubator for 25 min.
2. Set up a fresh culture flask (12.5 or 25 cm^2), depending on the flask volume set up originally.
3. Add 5-mL CCM + 100 μL GCTCM + 100 μL ITS to each 12.5-cm^2 flask and add 10-mL CCM + 200-μL GCTCM + 200-μL ITS to each 25-cm^2 flask.
4. Warm all flasks in a 37°C incubator for at least 25 min.
5. Monitor tissue dissociation in the collagenased flask under the inverted microscope. Cells on the edges of the tumour pieces should appear to be "falling off". With a sterile transfer pipette, gently pipette the tumour fragments to facilitate dissociation.
6. With the sterile transfer pipette, transfer the collagenase-containing medium and tissue pieces into a sterile 10-mL centrifuge tube.
7. Centrifuge the collagenased specimen at 160 *g* for 20 min.
8. Discard the supernatant and add the warmed RPMI.
9. Centrifuge the tube at 160 *g* for 20 min.
10. Remove the supernatant and add the tissue pieces to the warmed flask/s.
11. Re-incubate in the 37°C incubator till the following day.
12. If sufficient cells are present, harvest the suspension culture flask.
13. Check any non-collagenased monolayer tumour cultures for cell attachment with the inverted microscope.
14. On each subsequent day, all monolayer cultures must be examined daily under an inverted microscope until the cultures show attachment.
15. Cells usually go into log phase growth after approximately 72 h in culture (see Notes 7 and 8).
16. Harvest when the surface of the flask is 50–70% confluent (see Note 9).

3.5. Harvest of Solid Tumour Cultures

Prepare Carnoy's fixative immediately before use.

3.5.1. Suspension Culture Harvest (see Note 10)

1. Warm 0.075 M KCl (10 mL for each culture to be harvested) in the 37°C waterbath. Add 25-μL colcemid (10 μg/mL) to 12.5-cm² flask or 50-μL colcemid (10 μg/mL) to 25-cm² flask (final colcemid concentration = 0.05 μg/mL) (see Notes 11 and 12).
2. Return flask to the 37°C incubator for 120 min.
3. With a plastic transfer pipette, transfer the contents of the flask to a centrifuge tube and centrifuge at 160 *g* for 10 min.
4. Remove all but 0.5-mL supernatant, resuspend the pellet, and add pre-warmed 0.075 M KCl to a final volume of 9 mL (see Note 13).
5. Incubate in the 37°C waterbath for 10 min.
6. Invert each tube carefully and re-incubate in the 37°C waterbath for a further 15 min.
7. With a plastic Pasteur pipette, add 1 mL 4% aqueous glacial acetic acid.
8. Invert the tubes to mix and leave in the biohazard cabinet for 15 min.
9. Centrifuge at 160 *g* for 10 min.
10. Remove all but 0.5-mL supernatant, resuspend the pellet, and add initially dropwise freshly prepared Carnoy's fixative. When 4 mL fixative has been added, add fixative to a total volume of 9 mL.
11. Store at 4°C until ready to prepare slides.

3.5.2. Monolayer Culture Harvest

1. Warm 0.075 M KCl (10 mL for each culture to be harvested) in the 37°C waterbath.
2. Add 25-μL colcemid (10 μg/mL) to 12.5-cm² flask or 50-μL colcemid (10 μg/mL) to 25-cm² flask.
3. Return to the 37°C incubator for 120 min.
4. After 120 min, remove the culture medium and floating cells to a 10-mL centrifuge tube.
5. With a plastic transfer pipette, add 1 mL trypsin–EDTA to a 12.5-cm² flask or 2 mL to a 25-cm² flask and return the flask to the incubator for 10 min.
6. Remove the flask from the incubator and tap the flask firmly on the floor of the biohazard cabinet until the monolayer cells start to lift off the lower surface of the flask.
7. Examine under an inverted microscope to ensure all adhered cells have lifted off the flask.
8. Return the contents of the centrifuge tube to the flask and mix gently (the foetal bovine serum in the culture medium will stop the action of the trypsin).

9. If cells do not lift from the base of the flask, remove the trypsin–EDTA to the centrifuge tube and add a fresh 1-mL (or 2-mL) aliquot of trypsin–EDTA to the flask. Tap the flask firmly and re-incubate for a further 10 min.
10. With a plastic transfer pipette, transfer the contents of the flask to a centrifuge tube/s and centrifuge at 160 *g* for 10 min.
11. Remove all but 0.5-mL supernatant, resuspend the pellet, and add pre-warmed 0.075 M KCl to a final volume of 9 mL.
12. Incubate in the 37°C waterbath for 10 min.
13. Invert each tube carefully and re-incubate in the 37°C waterbath for a further 15 min.
14. With a plastic Pasteur pipette, add 1 mL 4% aqueous glacial acetic acid. Invert the tubes to mix and leave on the floor of the biohazard cabinet for 15 min.
15. Centrifuge at 160 *g* for 10 min.
16. Remove all but 0.5-mL supernatant, resuspend the pellet, and add initially dropwise freshly prepared Carnoy's fixative. When 4-mL fixative has been added, add fixative to a total volume of 9 mL.
17. Store at 4°C until ready to prepare slides.

3.6. Slide Preparation and Banding

This method describes the preparation of high-quality slides with adequately spread mitoses, showing minimum cytoplasm. Both factors are crucial to obtaining crisp GTW- or other banding patterns. Use gloves and a protective gown when cleaning and preparing slides to minimise exposure to ethanol solutions.

3.6.1. Slide Cleaning

1. Soak slides for a minimum of 2 h in AR grade ethanol.
2. Transfer slides to a container of tap water and leave for 10 min.
3. Empty water and refill with deionised water. Leave slides in deionised water until ready to clean.
4. Wipe each slide clean with a lint-free tissue and store in absolute ethanol until ready to use.

3.6.2. Slide Making

1. When *three* fixation changes of the fixed cell suspension have been completed, discard supernatant, resuspend cell pellet, and slowly add sufficient fixative to give a slightly cloudy suspension.
2. Prepare three Coplin jars as follows: Coplin jar 1 – containing 100% ethanol; Coplin jar 2 – containing deionised water; Coplin jar 3 – containing 5 mL freshly prepared Carnoy's fixative.

3. Dip each labelled slide once in Coplin jar 1, then dip in Coplin jar 2, until the slide is coated in a uniform layer of water (approximately ten dips).
4. Using a disposable transfer pipette, drop three drops of cell suspension along the slide ending at the labelled end (see Note 14).
5. Immediately afterwards, place three drops of fresh fixative from Coplin jar 3 on top of the cell suspension and drain the slide *along one long side*. If humidity is <40%, place the slide on a slide tray to dry at room temperature. If humidity is >40%, place the slide on a slide warming tray, warmed to 30°C, and allow to dry (see Note 15).
6. Check the slide under the microscope for an acceptable cell concentration. If too thick, dilute the cell suspension, or if too dilute, re-centrifuge the suspension and reconstitute in a smaller volume.
7. When the slides are dry, place in a 60°C oven overnight.

3.6.3. GTW Banding

1. Set up eight 50-mL Coplin jars as follows:
   (a) 1-mL Trypsin stock + 49-mL Hanks BSS (adjust to pH 7.0).
   (b) 2-mL Foetal calf serum + 48-mL Hanks BSS (adjust to pH 7.0).
   (c) 50-mL 1×Hank's BSS (adjust to pH 7.0).
   (d) 70% Alcohol.
   (e) 95% Alcohol.
   (f) pH 6.8 Working buffer.
   (g) Wright's stain (20%).
   (h) Sterile water.
2. Allow solutions to reach room temperature before use.
3. Dip aged slide in Coplin jar 1 for the chosen trypsin time (~5 s).
4. Transfer slide to Coplin jar 2 for 10 s.
5. Transfer slide to Coplin jar 3 for 60 s.
6. Drain briefly then dip slide three times in 70% alcohol.
7. Dip slide three times in 95% alcohol.
8. Drain briefly, then dip 10–12 times in pH 6.8 working buffer, until the alcohol is rinsed off.
9. Stain in Wright's stain for 2–3 min.
10. Rinse slide *briefly* in purified water and air-dry.
11. Coverslip slide, which is now ready for analysis (see Note 16) (5, 10–18).

4. Notes

1. Depending on the biopsy site, tumours are susceptible to infection and it is prudent to exercise caution by pre-treating high-risk samples (lung and gastrointestinal tract tumours) with an antibiotic-fungicide wash.
2. Pathologists responsible for solid tumour biopsy collections should be encouraged to supply 3–4 touch imprints from the biopsy material on positively charged (silanised) slides, labelled appropriately with the patient's name, medical record number, and date of collection. These should be transported to the Cytogenetics laboratory in a sealed container and should be handled in a biohazard type II safety cabinet with care appropriate to biohazardous material.
3. The success rate in obtaining a karyotype is dependent on the quality of the initial tumour biopsy material received for processing. Tumour biopsies can represent a heterogeneous population of cells, which often contain a significant number of non-tumour cells (e.g. haemopoietic and stromal) and necrotic tissue. Cytogenetics also requires an adequate amount of material to establish multiple cultures. This is often problematic, particularly when the initial biopsy may be small.
4. Whenever sample size allows, anchorage-independent and monolayer cultures need to be initiated, both of which require daily monitoring for medium depletion and determination of peak mitotic activity.
5. Minced tumour tissue may be disaggregated over the weekend using half the concentration of collagenase used overnight (11).
6. If the first culture harvest yields few mitoses with extremely low resolution (<<300 band level), subsequent cultures should be harvested with a lower colcemid concentration (10-μL colcemid/12.5-cm^2 flask or 20-μL colcemid/25-cm^2 flask) for a longer colcemid exposure time (e.g. 4 h).
7. Less than 20% tumours are ready to harvest before 72 h of culture. Ensure that rounded cells, which have not yet attached or just have a round morphology, are not confused with dividing cells. Cultures are only ready to harvest when cells begin to divide.
8. If a culture is greatly overgrown but is still dividing, perform a shake-off culture. Add colcemid as usual and then lightly trypsinise the culture (for 45 s). Tap the flask gently on the base of the biohazard cabinet to remove *only the mitotic* cells. Proceed with the harvest as described above.

9. Overgrowth by normal (diploid) stromal fibroblasts may occur in long-term cultures, and it is prudent to harvest all tumour cultures by 10 days following culture initiation.
10. Suspension harvests are crucial for the following tumour types:
    (a) All small round blue cell tumours, especially neuroblastomas.
    (b) Tumours which are friable and have shed cells into the transport medium.
    (c) Cells with a particularly high mitotic index.
11. When multiple cultures have been set up, colcemid concentrations and exposure times should be varied amongst culture harvests to optimise chances of successful tissue culture.
12. Variable colcemid exposure times (up to overnight) and/or concentrations may be required. Protocols should be adapted to suit laboratory individual conditions.
13. Variant hypotonic solutions are favoured is some protocols. A more dilute than standard KCl hypotonic solution, 0.067 M KCl, has been reported to reduce cytoplasm, thereby improving subsequent G-banding of the chromosomes, as well as enabling reduced incubation time in hypotonic (11). A 1:1 0.075 M KCl:1% w/v sodium citrate has also been used successfully in large-cell tumours, such as melanomas.
14. If over-spreading of metaphases occurs with the wet-slide technique, the humidified technique can be used:
    (a) Humidify the slide by either breathing on the surface of the slide, or by holding the slide over a steam bath.
    (b) Using a disposable transfer pipette, from a height of 1–3 cm, drop a few drops of suspension evenly along the slide ending at the labelled end. Blow gently across the slide and dry it by placing immediately on a wet towel. When dry, place the slides overnight in the 60°C oven.
15. If metaphases are under-spread with the wet-slide technique, the cold 60% acetic acid technique may be useful:
    (a) Store clean slides in 60% (w/w) aqueous solution of glacial acetic acid at 4°C.
    (b) On the bench, drain the slide briefly and drop two drops of fixed cell suspension onto the slide from a height of about 3 cm. Drain excess acetic acid from the slide, allow to dry, and then store overnight in a 60°C oven.
16. Clonal abnormalities can be more easily identified in complex heterogeneous cases by using a table. Abnormalities for each cell are listed for each chromosome. Patterns of chromosome gain and loss for groups of patients are best presented as ideograms, with regions of gains and loss identified on either side.

References

1. Mitelman F., Johansson B. and Mertens F. (eds) (2010) Mitelman Database of Chromosome Aberrations and Gene Fusions in Cancer. http://cgap.nci.nih.gov/Chromosomes/Mitelman.
2. Slater O. and Shipley J. (2007) Clinical relevance of molecular genetics to paediatric sarcomas. *J. Clin. Pathol.* **60**:1187–1194.
3. Ambros P.F., Ambros I.M., Brodeur G.M., Haber M., Khan J., Nakagawara A., Schleiermacher G., Speleman F., Spitz R., London W.B., Cohn S.L., Pearson A.D.J. and Maris J.M. (2009) International consensus for neuroblastoma molecular diagnostics: Report from the International Neuroblastoma Risk Group (INRG) Biology Committee. *Br. J. Cancer* **100**:1471–1482.
4. Parsons B.L. (2008) Many different tumor types have polyclonal tumor origin: Evidence and implications. *Mutat. Res.* **659**:232–247.
5. Pandis N., Bardi G. and Heim S. (1994) Interrelationship between methodological choices and conceptual models in solid tumor cytogenetics. *Cancer Genet. Cytogenet.* **76**:77–85.
6. Ma S. and Imreh S. (2007) Non-random genomic instability in cancer: A fact, not an illusion. *Semin. Cancer Biol.* **17**:1–90.
7. Heng H.H.Q., Liu G., Bremer S., Ye K.J., Stevens J. and Ye C.J. (2006) Clonal and non-clonal chromosome aberrations and genome variation and aberration. *Genome* **49**:195–204.
8. Carvajal R.D., Tse A. and Schwartz G.K. (2006) Aurora kinases: New targets for cancer therapy. *Clin. Cancer Res.* **12**:6869.
9. van de Rijn M. and Fletcher J.A. (2006) Genetics of soft tissue tumors. *Annu. Rev. Pathol.* **1**:435–466.
10. Lundgren R., Mandahl N., Heim S., Limon J., Hendrikson H. and Mitelman F. (1992) Cytogenetic analysis of 57 primary prostatic adenocarcinomas. *Genes, Chromosomes Cancer* **4**:164.
11. Fletcher J. (2000) Metaphase Harvest and Cytogenetic Analysis of Solid Tumor Cultures in Dracopoli N.C. (ed), *Current Protocols in Human Genetics*, John Wiley & Sons, New York.
12. Mandahl N. (1992) Methods in Solid Tumour Cytogenetics in Rooney D.E. and Czepulkowski B.H. (eds), *Human Cytogenetics Volume II: Malignant and Acquired Abnormalities, A Practical Approach*, IRL Press, Oxford, pp. 155–187.
13. Barch M.J., Knutsen T. and Spurbeck J.L. (eds) (1991) *The AGT Cytogenetics Laboratory Manual*, 3rd edition, Lippincott-Raven, Philadelphia, p. 279.
14. Phillips C. pers. comm. Oncology Cytogenetics Laboratory, Emory University Hospital, Atlanta, Georgia, USA.
   Slovak M. pers. comm. City of Hope National Medical Center, Cytogenetics Laboratory, Duarte, California, USA.
15. Hoglund M., Gisselsson D., Hansen G.B. and Mitelman F. (2004) Wilms tumors develop through two distinct karyotypic pathways. *Cancer Genet. Cytogenet.* **150**:9–15.
16. Bown N., Cotterill S.J., Roberts P., Griffiths M., Larkins S., Hibbert S., Middleton H., Kelsey A., Tritton D. and Mitchell C. (2002) Cytogenetic abnormalities and clinical outcome in Wilms tumor: A study by the U.K. Cancer Cytogenetics Group and the U.K. Children's Cancer Study Group. *Med. Pediatr. Oncol.* **38**:11–21.
17. Lamont J.M., McManamy C.S., Pearson A.D., Clifford S.C. and Ellison D.W. (2004) Combined histopathological and molecular cytogenetic stratification of medulloblastoma patients. *Clin. Cancer Res.* **16**:5482–5493.
18. Damato B. and Coupland S.E. (2009) Translating uveal melanoma cytogenetics into clinical care. *Arch. Ophthalmol.* **127**:423–429.

Chapter 14

Fluorescence In Situ Hybridization on Formalin-Fixed, Paraffin-Embedded Tissue Sections

Adrian Zordan

Abstract

Although in situ hybridization has been in use for over 30 years, its application to the study of solid tissue has only recently been adopted. Despite the numerous reports of the viability of formalin-fixed, paraffin-embedded (FFPE) tissue for fluorescence in situ hybridization (FISH) testing, this technique has not been universally implemented in the routine diagnostic setting. This is most likely due to the perception that the process is more technically demanding than FISH using conventional cytogenetic samples. FFPE FISH does, however, enable retrospective analysis of archived tissue samples and is helpful in the diagnosis of morphologically difficult cases such as Burkitt-like lymphoma, diffuse large B-cell lymphoma, and mantle-cell lymphoma.

Key words: Formalin-fixed, paraffin-embedded, Fluorescence in situ hybridization, Pretreatment, Tissue digestion, Proteinase K

1. Introduction

Formalin-fixed, paraffin-embedded (FFPE) tissue represents a vast and largely untapped resource for both research and diagnostic cytogenetic studies. Interphase fluorescence in situ hybridization (FISH) has traditionally been performed on fresh or cultured cells but, in the absence of these, FFPE tissue constitutes an attractive alternative (see Note 1) (1–3). Millions of oncology specimens are routinely processed and stored as FFPE tumor blocks annually. FISH on FFPE tissue is not a novel concept; however, due to its technically demanding nature, it has not been universally adopted (4).

Essentially, there have been two technical strategies used for FISH analysis of FFPE specimens. The first of these requires extraction of intact nuclei from thick (40–60 μm) sections of the tumor block prior to hybridization assays being performed on the resulting cell suspension (see Note 2). The second approach utilizes thin

Lynda J. Campbell (ed.), *Cancer Cytogenetics: Methods and Protocols*, Methods in Molecular Biology, vol. 730, DOI 10.1007/978-1-61779-074-4_14, © Springer Science+Business Media, LLC 2011

(2–6 μm) sections of the whole paraffin-embedded block for the FISH assay. Both methods have inherent advantages and limitations, but the major point of differentiation is that the processing of thick sections to obtain intact nuclei will destroy the tissue architecture. Conversely, the thin-section technique preserves tissue morphology and allows for direct comparison of the FISH results with the morphological details observed on adjacent hematoxylin and eosin (H&E)-stained sections. Using thin sections for FISH, it is therefore possible to confidently distinguish tumor nuclei from stromal and other normal nuclei in the section.

There are numerous factors affecting the outcome of FISH on FFPE tissue, including heterogeneous fixation conditions, the age of section, tissue type, degree of fibrosis, and thickness of section, to name a few. Essentially, however, the success of FISH on FFPE tissue is directly dependent on the accessibility of the target DNA within the cell nuclei (5). As such, the most important process in the FFPE FISH protocol is the pretreatment. This involves processing the sample through sodium bisulfite, a protease solution, phosphate-buffered saline (PBS), and an ethanol series. This sequence of washes acts to remove nuclear, cytoplasmic, and extracellular matrix proteins as well as eliminating the protein cross-links created by formalin fixation. All of these steps act to ensure efficient probe infiltration for FISH (6, 7).

The many compounding variables associated with FFPE tissue FISH mean that protocol customization is required for optimal results. Consequently, protease digest times should be tailored for each individual sample using parallel experiments. This approach will minimize the analysis time and ensure that the best possible results are achieved. Control FFPE tissue samples of the same age and cell type should always be processed along side any diagnostic specimens. If this is not possible, the nontumor cells in the diagnostic sample (as identified by the H&E-stained slide) may be used as internal negative controls for hybridization.

Once the FFPE tissue has been processed with the aforementioned pretreatment regime, it is suitable for a conventional FISH assay. Our experience with this has shown that a standard co-denaturation protocol will provide reliable and reproducible results. There is, however, one recommended amendment to the co-denaturation protocol, and this involves raising the denaturation temperature to 80°C or even as high as 95°C. This increase in temperature is to counter the tissue's added resistance to denaturation, which is a result of prior formalin fixation.

The advent of commercially available all-in-one FFPE FISH kits has helped to simplify the implementation of this technique in a cytogenetic setting. Whether processing with an all-in-one kit or using an in-house protocol, however, the general principles of the processing and the idiosyncrasies encountered are similar. When dealing with FFPE tissue, it is not possible to use a one-size-fits-all approach and each assay, regardless of the source of the solutions, will need to be tailored for each FFPE section. Indeed, it could be said that there is an art to the successful application of the FFPE pretreatment.

2. Materials

Note that many of the reagents listed below are hazardous or dangerous goods. Where possible, handling of these solutions should be conducted in a fume hood or other appropriate fume extraction system. Where this is not possible, ensure that the laboratory area is ventilated well. Use appropriate safety procedures including personal protective equipment and avoid contact with skin and mucous membranes. Dispose of waste or empty containers in accordance with relevant legislation.

2.1. FFPE Slide Pretreatment

1. 50-mL Coplin jars.
2. Waterbath with lid (45°C).
3. Magnetic stirrer.
4. Sterile distilled H_2O (dH_2O).
5. Xylene.
6. Ethanol (Analytical grade): prepare v/v dilutions of 100% ethanol with dH_2O. Store at ambient temperature and discard stock solutions after 6 months. 70% Ethanol: add 30 mL dH_2O to 70 mL of 100% ethanol; 80% ethanol: add 20 mL dH_2O to 80 mL of 100% ethanol; 90% ethanol: add 10 mL dH_2O to 90 mL of 100% ethanol; 100% ethanol.
7. 20× SSC: dissolve 175.3 g NaCl and 88.2 g Na citrate in 600 mL dH_2O. Make up to 1 L with dH_2O and adjust pH to 5.3 ± 0.2 with 3 M HCl. Store solution at room temperature and discard stock solutions after 6 months or sooner if turbidity develops.
8. 2× SSC: add 900 mL dH_2O to 100 mL 20× SSC. Adjust pH to 7.0 ± 0.2 with 10 M NaOH. Store solution at room temperature and discard stock solutions after 6 months or sooner if turbidity develops.
9. Sodium bisulfite solution (30%): dissolve 15 g $NaHSO_3$ in 25 mL 2× SSC. Adjust final volume to 50 mL and mix well using magnetic stirrer (see Note 3).
10. 10% Sodium dodecyl sulfate (SDS) (see Note 4): dissolve 10 g SDS powder in 80 mL dH_2O and mix well. Add dH_2O to make a final volume of 100 mL. Store solution at room temperature and discard stock solutions after 6 months or sooner if turbidity develops.

11. Proteinase K buffer: add 0.85 mL 10% SDS to 500 mL PBS. Adjust pH to 7.0 ± 0.2 with 10 M NaOH. Store solution at room temperature and discard stock solutions after 6 months or sooner if turbidity develops.
12. Proteinase K (14–22 mg/mL) (Proteinase K, recombinant, PCR Grade, Roche Diagnostics Gmbh, Mannheim, Germany). The enzyme will be stable if stored at 2–8°C through to the expiration date.
13. Phosphate-buffered saline (PBS): dissolve the contents of one plastic vial in 1 L dH_2O (see Note 5). Store solution at room temperature and discard stock solutions after 6 months or sooner if turbidity develops.
14. 4′-6-Diamidino-2-phenylindole (DAPI) II counterstain: 125 ng/mL in antifade mounting solution (Vysis, Downers Grove, IL).

2.2. Hybridization

1. Hotplate: 80–95°C.
2. Glass coverslips – 12-mm circular (Menzel-Glaser, Braunschweig, Germany).
3. Fixogum (Marabu, Bietigheim-Bissingen, Germany).
4. Humidified chamber (see Note 6).
5. 37°C incubator.
6. Pipette (0–10 μL).
7. Pipette tips.

2.3. Posthybridization Washes

1. Waterbath (73°C).
2. Nonidet P-40 (NP-40) (Calbiochem, San Diego, CA) (see Note 7).
3. 0.4× SSC/0.3%NP-40: mix 20 mL of 20× SSC with 980 mL distilled water. Add 3 mL of NP-40 and mix thoroughly until NP-40 is dissolved. Adjust pH to 7.0 ± 0.2 with NaOH. Store solution at room temperature and discard stock solutions after 6 months or sooner if turbidity develops.
4. 2× SSC/0.1%NP-40: mix 100 mL of 20× SSC with 900 mL distilled water. Add 1 mL of NP-40 and mix thoroughly until NP-40 is dissolved. Adjust pH to 7.0 ± 0.2 with NaOH. Store solution at room temperature and discard stock solutions after 6 months or sooner if turbidity develops.
5. Glass Coverslips – 22 × 50 mm (Menzel-Glaser, Braunschweig, Germany).
6. Pipette (20–200 μL).
7. Pipette tips.

2.4. Analysis

1. Epifluorescence imaging microscope – the following protocol has been developed with the use of a Zeiss Axioplan 2 Imaging microscope with appropriate objectives and filter sets (Zeiss, Jena, Germany).
2. Immersion oil (high grade for fluorescence) (Zeiss, Jena, Germany).
3. Image Capture and Analysis software – the ISIS image acquisition and analysis software (Metasystems, Altlussheim, Germany) has been used for the development of this protocol and for routine FFPE FISH testing.

3. Method

Prior to commencing this protocol, there are several requirements regarding the FFPE diagnostic and control samples which need to be met.

Ensure that: (1) the FFPE tissue has been cut into 2-μm sections and mounted on a positively charged or silanated slide; (2) at least four diagnostic slides are available for testing; (3) at least four normal control slides of the same tissue type are available for testing; (4) the FFPE sections have not been decalcified, and (5) an H&E-stained FFPE section, adjacent to the diagnostic slides, has been provided.

3.1. Slide Preparation

1. Examine the H&E-stained slide to identify the optimal region of tumor for FISH (see Note 8).
2. Mark the corresponding region of tissue on the underside of the unstained FFPE slides with a diamond tip pen.
3. Identify and mark the region to be assayed on the control sections (see Note 9).

3.2. Deparaffinization

1. Deparaffinize the tissue sections (see Note 10) by placing slides in 2 × 50 mL coplin jars containing xylene at room temperature for 10 min each. Agitate occasionally (see Note 11).
2. Prewarm two coplin jars, one containing 50 mL of sodium bisulfite and the other 50 mL of proteinase K buffer solution (without Proteinase K) to 45°C in a waterbath.
3. Place slides in 100% ethanol (2×) at room temperature for 5 min each.
4. Allow slides to air-dry.
5. Transfer slides to a coplin jar containing prewarmed 30% sodium bisulfite and incubate for 20 min at 45°C.
6. Rinse in 2× SSC (2×) at room temperature for 1 min each.
7. Air-dry the slides.

3.3. Slide Digestion

1. Label slides with proposed digest times (see Note 12).
2. Add 200 μL proteinase K (14–22 mg/mL) to the 45°C proteinase K buffer (see Note 13).
3. Place slides in the proteinase K solution and incubate for appropriate time.
4. Rinse slides for 1 min in 2× SSC (2×) at room temperature (see Note 14).
5. Wash slides for 2 min in PBS at room temperature.
6. Dehydrate slides by immersing in the following graded ethanol series at room temperature for 1 min each: 70, 80, 90, and 100%.
7. Air-dry slides.
8. Add 20 μL DAPI counterstain to each slide, apply a 22 × 50 coverslip and view using an epifluorescence microscope.
9. Evaluate the digestion efficiency by:
   (a) Focusing on the nuclear peripheries using a DAPI filter.
   (b) Assessing extracellular tissue/debris with a double/triple band-pass filter (DAPI/FITC/TRITC). All extracellular tissue autofluoresces with a green/gray appearance.
   (c) Refer to Table 1 and Fig. 1 for tissue digestion evaluation and corresponding troubleshooting. At this point, it is

Table 1
Guide to evaluating tissue digestion following treatment with proteinase K

	Appearance	Action to be taken
Over-digested	• Poorly defined cell peripheries • May appear "ghostly"	• Once over-digested, a sample cannot be rescued • Digest a freshly deparaffinized slide with a reduced digest time • Do not discard any slide/s until successful result is achieved
Under-digested	• High intercellular background • Autofluorescent connective tissue will be visible (when viewed using the double/triple band-pass filter) • Poor uptake of DAPI	Digest slide further: (a) Rinse slide in 2× SSC to loosen/remove the coverslip. If the coverslip does not fall off unaided, gently remove manually and wash off excess DAPI (~1 min), followed by a 1-min wash in 100% ethanol (b) Return to section 3.3, step 3 and perform additional digests, increasing the time by 5 or 10 min
Optimal digestion	• Well-defined, intact cell borders and effective uptake of DAPI	Proceed to step 10 of Subheading 3.3

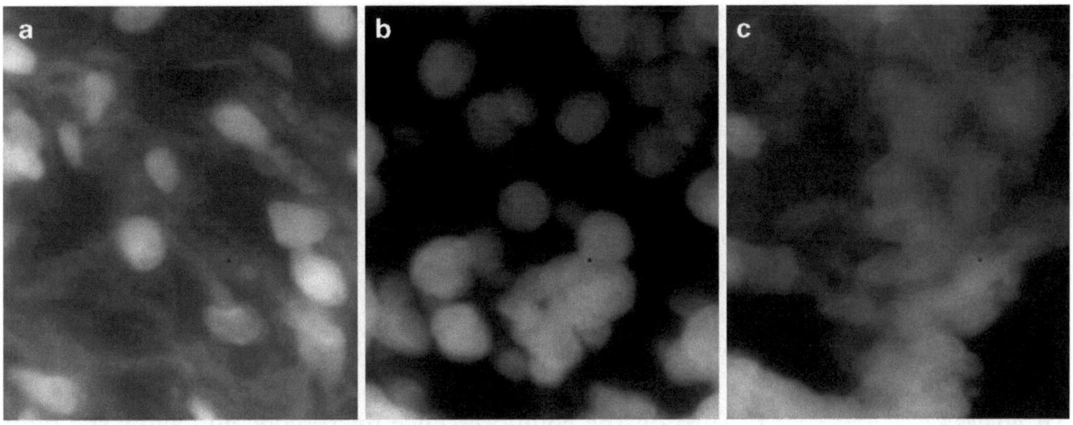

Fig. 1. FFPE digestion evaluation (as viewed using a DAPI filter). (**a**) Tissue under-digestion: note the intercellular connective tissue which overlies the entire sample; this connective tissue will interfere with FISH signal evaluation due to its auto-fluorescent properties; (**b**) adequate digestion highlighted by the well-defined, intact cell peripheries and effective DAPI uptake; (**c**) over-digested FFPE sample with complete disintegration of the cells; once over-digested, FFPE samples are unsuitable for FISH.

no longer necessary to proceed with duplicate slides. Select the slide which presents the best morphology and least background interference before proceeding.

10. Immerse slide/s in 2× SSC at room temperature for ~1 min to loosen/remove the coverslip and wash off excess DAPI.
11. Wash slide in 100% ethanol at room temperature for 1 min (see Note 15).
12. Air-dry slides.

3.4. Hybridization (see Note 16)

1. Preheat hotplate to 80°C (see Note 17).
2. Prewarm a humidified chamber in a 37°C incubator.
3. Prepare probe mixtures according to the manufacturer's protocol (see Note 18).
4. Pipette the appropriate volume of probe mixture (see Note 19) within the premarked region of the slides and immediately apply a 12-mm circular glass coverslip (see Note 20).
5. Seal the edge of the coverslip with Fixogum.
6. Place slides on the hotplate for 10 min to denature both target and probe DNA (co-denaturation).
7. Transfer all slides to the prewarmed humidified chamber (see Note 21).
8. Hybridize overnight or for a minimum of 14 h in a 37°C incubator (see Note 22).

3.5. Posthybridization Washes

1. Prewarm a coplin jar of 0.4× SSC/0.3%NP-40 to 73°C in a waterbath (see Note 23).
2. Carefully remove Fixogum and the coverslips from the slides and wash for 2 min in the 0.4× SSC/0.3%NP-40 solution (see Note 24).
3. Wash the slides in 2× SSC/0.1%NP-40 solution at room temperature for <1 min (see Note 25).
4. Air-dry the slides.
5. Pipette 20 μL of DAPI onto each slide and overlay with a 22 × 50 glass coverslip (see Note 26).

3.6. Analysis

1. Turn on the epifluorescence microscope and allow the bulb to warm up for 5–10 min.
2. Use the 10× objective to locate the focal plane of the predefined area of interest.
3. Apply a drop of immersion oil to the coverslip and switch to the 100× objective.
4. Using the triple and/or individual band-pass filters, check that FISH signals are present within the nuclear boundaries of the cells (see Notes 27 and 28).
5. Assess several areas of nuclei across the entire probed region to account for possible heterogeneity (8).
6. Score a total of 200 cells between two scientists (100 each) (see Note 29) for each FISH assay.
7. Capture a minimum of five fields of view illustrating each normal/abnormal clonal population observed (see Notes 30 and 31).
8. Record the results.

3.7. Interpretation of Results

Due to the nature of FFPE sections, a significant proportion of nuclei will be truncated in any given sample. As a result of sectioning artifact, scoring of FISH will identify a considerable number of nuclei with lost signals. These should be scored, but their inclusion in the final total is dependent on the type of probe being used, the nuclear diameter, and the tissue type. Normal ranges for detection of signal gains or losses will need to be established for each individual laboratory following the scoring and statistical analysis of many nuclei in normal sections (9). Average normal cutoff values of ~10% cells with signal gain and ~50% cells with signal loss have been suggested (5) (see Note 32). Cutoff values for translocation probes (i.e., break-apart, dual fusion, and extra signal) will need to be developed from a negative control data set for each individual probe.

4. Notes

1. Only the tissue which has been preserved in neutral-buffered formalin and paraffin-embedded is suitable for FISH. Tissue fixation methods and thickness of sections other than those specified may affect tissue morphology and/or signal intensity (8).
2. This approach will not be addressed in this chapter. An overview of this method is provided in ref. 9.
3. Sodium bisulfite working solution cannot be stored and must be used on the day it is prepared.
4. SDS is a denaturing agent and has been shown to increase the activity of proteinase K by as much as sevenfold (10).
5. It is advisable to aliquot the PBS solution into 200-mL bottles and then autoclave to ensure sterility.
6. A plastic container (with airtight lid) containing a sponge which has been dampened with 2× SSC will suffice. This will ensure that the slides will not dry out while at the same time maintaining a steady temperature.
7. Note that NP-40 is no longer commercially available. Igepal CA-630 is now offered as a replacement (Sigma-Aldrich, St Louis, MO, USA).
8. When selecting the region/s of interest, ensure that areas of necrosis are avoided.
9. When selecting suitable regions to be used as negative controls, endeavor to identify regions of comparable cell density (to the diagnostic FFPE section) and with similar amounts of connective tissue, so as to minimize any artifact-related variation.
10. It is recommended that at least two patient slides and two control slides be processed for each FISH probe.
11. The incomplete removal of embedding medium will result in an increase in nonspecific fluorescence. If necessary, place the sections in a 60°C oven for 1 h prior to processing through xylene to aid with deparaffinization (8).
12. Certain tissue types are more resistant to Proteinase K digestion than others, and as such digest times must be determined empirically (9). Efficiency of digest can also vary due to factors such as the length of tissue fixation (usually unknown), thickness of the sections, and the proportion of connective tissue in the sample. The optimal digest time will vary from case to case. Generally, however, normal tissue will be more refractory to digestion than tumor tissue derived from the corresponding cell type (5). A suggested starting point to

digest the two tumor sample slides would be at 5 and 10 min, respectively, whereas, a suitable starting point for the control samples would be 10 and 15 min respectively.

13. The final concentration of the proteinase K is to be 0.05–0.09 mg/mL. Also note that proteinase K working solution (i.e., proteinase K buffer + proteinase K) must be used on the day it is prepared. Proteinase K is stable and retains full activity for several hours when incubated at pH 7.0 and 45°C. Ensure that the working solution is at the correct temperature, as an increase in temperature will further activate the enzyme. In fact, proteinase K is 12 times more active at 65°C than at 25°C (11).
14. The length of time slides are immersed in 2 × SSC at this point in the protocol is not critical. They may remain in 2 × SSC at room temperature for up to 1 h.
15. At this point, slides may have a "cloudy" appearance. To improve this, a second wash in 100% ethanol at room temperature is recommended.
16. The fluorophore labels of FISH probes are easily photobleached with excessive exposure to light. To limit this degradation, perform all steps, as is practical, in darkness or reduced light.
17. 80°C is a suggested starting point. Certain samples will be more resistant to denaturation, and therefore an increased hotplate temperature will be required. In our experience, co-denaturation temperatures of up to 95°C may be required.
18. Some probe manufacturers provide ready-to-use probes, while others require the user to prepare the probe mixture from separate probe and hybridization buffer. When preparing a probe mixture, ensure the probe is well mixed before adding to the hybridization buffer.
19. In our experience, 3 μL of probe mixture will suffice when using a 12-mm round coverslip.
20. Ensure the probe mixture spreads evenly under the whole coverslip without any air pockets. If air pockets are present, gently press on coverslip to expel them.
21. If batching of slides is required, transfer slides to a 37°C hotplate following co-denaturation as an interim step prior to placing them in the humidified chamber.
22. Although many probe manufacturers suggest that hybridization incubations of as little as 2 h may provide adequate results, in our experience, incubation times of no less than 14 h were required for reliable and reproducible results. This is especially true for many of the smaller locus-specific or translocation probes.

23. We have found that these wash solutions and conditions are optimal for most probes. If the probe manufacturer's conditions for the posthybridization washes vary from those mentioned above, refer to those recommended by the manufacturer.
24. To ensure that the temperature of the wash solution is maintained, wash no more than four slides simultaneously.
25. A fixed time has not been provided for this wash, as this will depend on the probe size. For example, large centromeric probes will benefit from a wash of ~1 min as this will minimize the background "noise" without compromising the FISH signals themselves. Conversely, smaller locus-specific probes will need no more than 30 s. Any more than this and the FISH signals may become too weak to reliably interpret.
26. The slides should be analyzed within 1 week. Fading of signals will result if slides are exposed to excessive light or high temperatures. Slides should therefore be stored away from light sources at 2–8°C.
27. Troubleshooting suboptimal or failed FISH assays on FFPE sections is inherently difficult due to the sizeable number of steps prior to analysis (5). Variations in specimen preparation, embedding, and sectioning can in turn introduce more variables. Characteristics of particular tissue types and specimen variability further complicate this issue. Refer to Table 2 for a quick reference guide to FFPE troubleshooting.
28. It may be necessary to "focus through" the several focal planes to visualize all of the signals in individual nuclei due to the three-dimensional effect of FFPE sections. Ideally, it is recommended that the following cells are ignored: nuclei that require subjective judgment, nuclei with a weak signal intensity or nonspecific high background, and areas where the nuclear borders are unclear.
29. If there is a discrepancy between the results of the two scorers, it is recommended that a further 100 cells be scored by a third scorer.
30. Images of each of the various signal patterns should be captured by the first scorer. The second scorer can then assess the captured images and if necessary relocate cells to confirm the signal pattern.
31. The Metasystems ISIS software has a functionality called *Focus Stack Capture* which allows for image acquisition over several focal planes (on the *Z* axis). In our experience, this feature is invaluable in dealing with the extreme three-dimensional effect which is characteristic of FFPE sections.

Table 2
Quick reference guide to troubleshooting FFPE FISH

Issue	Possible cause	Solution
No or weak FISH signals	Inappropriate tissue fixation	Ensure that only neutral-buffered FFPE tissue sections have been used
	Insufficient tissue digestion	Ensure that the appropriate digest temperature has been used and that the proteinase K has not passed its expiry date. Further digest, assess, and re-probe same slide
	Inadequate denaturation conditions	Check that the co-denaturation temperature used was at least 80°C for 10 min. Repeat the assay with an increased co-denaturation temperature. Temperatures as high as 95°C are necessary for certain tissue types
	Incorrect hybridization conditions	Ensure that hybridization occurred at 37°C for at least 14 h. Repeat with appropriate temperature and time
	Drying-out of probe during hybridization	Ensure that hybridization chamber is set up correctly, i.e., with 2× SSC to allow for sufficient humidity. Ensure that Fixogum is applied generously to completely seal the probe under the coverslip
	Excessive stringency of posthybridization wash conditions	Ensure that the recommended wash solutions, temperatures, and times are used. If necessary, decrease the time in, or even omit the 2× SSC/0.1% NP-40 wash
	Microscope not set up correctly	Check that an appropriate filter set is in use, a suitable mercury lamp is being used and is not beyond its expected life, the collector lens is not dirty or cracked, and that an appropriate fluorescence microscopy oil is in use
	Signals have faded	Minimize exposure to strong light sources and check probe stock
Region with no or patchy signals	Insufficient probe added	Ensure that the probe volume was sufficient to cover the entire area under the coverslip without any air pockets

Excessive background noise	Inappropriate tissue fixation	Ensure that only neutral-buffered FFPE tissue sections have been used
	Incomplete removal of paraffin	Repeat protocol ensuring that the deparaffinization is complete. If necessary, place the sections in a 60°C oven for 1 h prior to processing through xylene (10)
	Inadequate stringency of posthybridization wash conditions	Ensure that the recommended wash solutions, temperatures, and times are used. If necessary, increase the time for the 2× SSC/0.1%NP-40 wash
Poor nuclear morphology, poor staining, loss of cell borders, or loss of nuclei altogether	Excessive proteinase K treatment	Repeat the protocol with a reduced proteinase K incubation time
	High co-denaturation temperature	Although co-denaturation temperatures of as high as 95°C are occasionally called for, this does have an adverse effect on the morphology of the nuclei. Repeat protocol with a reduced co-denaturation temperature
Green autofluorescence, excessive extracellular material, or poor DAPI staining	Insufficient proteinase K treatment	Repeat protocol with increased incubation time in protease solution. Digest times for subsequent Proteinase K treatment will be dependent on the existing degree of digestion
	FFPE samples sectioned at >2 µm	Thicker sections will require increased proteinase K treatment. They will also exhibit an increased three-dimensional effect due to the multiple layers of nuclei. Ensure sections are 2 µm

32. In order to develop robust cutoff levels for each probe, at least ten data sets should be gathered (with 200 nuclei scored per sample). An abnormal result may be defined as the abnormal signal pattern being observed in the number of cells greater than 3 standard deviations above the mean of results from a negative control group (12, 13).

Acknowledgments

I would like to thank Kathy Somana, Kathleen Rayeroux, Bruce Mercer, Crisoula Batzios, and Lynda Campbell for their assistance in the development and implementation of this technique in our laboratory.

References

1. Schurter MJ, LeBrun DP, Harrison KJ. Improved technique for fluorescence in situ hybridisation analysis of isolated nuclei from archival, B5 or formalin fixed, paraffin wax embedded tissue. Mol Pathol. 2002;55(2): 121–4.
2. Schofield DE, Fletcher JA. Trisomy 12 in pediatric granulosa-stromal cell tumors. Demonstration by a modified method of fluorescence in situ hybridization on paraffin-embedded material. Am J Pathol. 1992;141(6): 1265–9.
3. Wolman SR, Waldman FM, Balazs M. Complementarity of interphase and metaphase chromosome analysis in human renal tumours. Genes Chromosomes Cancer. 1993; 6(1):17–23.
4. Ventura RA, Martin-Subero JI, Jones M, et al. FISH analysis for the detection of lymphoma-associated chromosomal abnormalities in routine paraffin-embedded tissue. J Mol Diagn. 2006;8(2):141–51.
5. Squire JA, Marrano P, Kolomietz E. FISH in Clinical Cytogenetics. In: Beatty BG, Mai S, Squire JA, eds. FISH A Practical Approach. Oxford University Press, New York, 2002. 183–204.
6. Roche Applied Science. DIG Application Manual for Nonradioactive In Situ Hybridisation. 3rd Edition. Roche Applied Science, Penzberg. 2002.
7. Hyytinen E, Visakorpi T, Kallioniemi A, Kallioniemi OP, Isola JJ. Improved technique for analysis of formalin-fixed, paraffin-embedded tumours by fluorescence in situ hybridisation. Cytometry. 1994;16(2):93–9.
8. Dako Cytomation. Histology FISH Accessory Kit Instructions. 2004. 1–16 (Product Insert).
9. Schofield DE, Weremowicz S. Current Protocols in Human Genetics. Wiley, New Jersey, USA. 2008;8.8.1–8.8.8.
10. Hilz H, Wiegers U, Adamietz P. Stimulation of proteinase K action by denaturing agents: application to the isolation of nucleic acids and the degradation of 'masked' proteins. Eur J Biochem. 1975;56(1):103–8.
11. Roche Applied Science. Proteinase K, recombinant, PCR Grade Lot-specific Data Sheet. 2005 (Product Insert).
12. Cogliatti SB, Novak U, Henz S, et al. Diagnosis of Burkitt lymphoma in due time: a practical approach. Br J Haematol. 2006;134(3):294–301.
13. Barry TS, Hwang HC, Goldstein LC, Yaziji H, Hing AW, Gown AM. High Sensitivity and Specificity of FISH in Routine Detection of t(14;18)(q32;q21) and t(11;14)(q13;q32) in Formalin-Fixed, Paraffin-Embedded (FFPE) Tissue sections Using Automated Morphometric Image Analysis. United States & Canadian Academy of Pathology 2005 Annual Meeting, Poster 195.

Chapter 15

The Use of M-FISH and M-BAND to Define Chromosome Abnormalities

Ruth N. MacKinnon and Ilse Chudoba

Abstract

Multicolour fluorescence in situ hybridisation (M-FISH) and multicolour banding (M-BAND) are advanced chromosome painting techniques combining multiple chromosome- or region-specific paints in one step. M-FISH identifies all chromosomes or chromosome arms at once, whereas M-BAND identifies the different regions of a single chromosome. The use of either or both can improve the accuracy of karyotyping and help identify cryptic chromosome rearrangements.

These probes are prepared by pooling multiple chromosome- or chromosome region-specific DNA libraries, each labelled with a unique combination of fluorochromes. Commercial probes are available, avoiding the need for probe preparation. In the protocol described here, a commercial probe is used. Well-spread metaphases are prepared according to standard techniques, followed by alkaline denaturation and application of the denatured probe. After an incubation period, the slides are washed. A fluorescence microscope with filter sets specific to the fluorescent labels is used for analysis, together with specialised image analysis software. The software interprets the combination of fluorochromes to identify each chromosome and produce a false colour image specific for each chromosome or region. The single colour galleries – which show the hybridisation patterns of the individual fluorochromes – are useful to help interpret and confirm the false colour images produced by the software, including ambiguous signals.

Key words: M-FISH, M-BAND, mBAND, mFISH, Multicolour FISH, Molecular cytogenetics, Chromosome identification, Banding

1. Introduction

Traditional chromosome banding techniques are of limited use in the interpretation of complex abnormal karyotypes, as the characteristic banding patterns typical of normal chromosomes can become disordered and are not as readily recognised out of context. Sophisticated high resolution genomic analysis techniques, such as array comparative genomic hybridisation (array CGH) (Chapter 16), single nucleotide polymorphism (SNP) array

Lynda J. Campbell (ed.), *Cancer Cytogenetics: Methods and Protocols*, Methods in Molecular Biology, vol. 730,
DOI 10.1007/978-1-61779-074-4_15, © Springer Science+Business Media, LLC 2011

(Chapter 17) (1–3), and single locus interphase and metaphase fluorescence in situ hybridisation (FISH) (Chapter 3) are high resolution genome analysis techniques that are extremely useful for detecting breakpoints and overall copy number changes. In the case of array techniques, this is done without reference to chromosome organisation or the composition of different cell lines, and so the changes in minor cell lines can be missed and balanced rearrangements are not detected.

Multicolour fluorescence in situ hybridisation (M-FISH) and multicolour banding (M-BAND) are advanced chromosome painting techniques which can reveal gross chromosome structure and rearrangements based on the DNA composition of the chromosomes. They help fill the gap between chromosome banding and high resolution molecular techniques. Combining these techniques gives comprehensive information on genome reorganisation.

The development of M-FISH was based on whole chromosome painting FISH probes. These are pooled DNA extracts of a single chromosome class labelled with single fluorochromes (4, 5). They can be used to identify chromosome abnormalities containing a particular chromosome and a small number of paints labelled with different single fluorochromes can be combined together in one hybridisation. However, the number of readily available fluorochromes is far fewer than the number of human chromosomes, so not all chromosomes can be identified at once in this way. The spectral karyotyping (SKY) (6) and M-FISH (7) procedures were developed separately by two different laboratories to overcome this problem. These approaches make possible the identification of all chromosomes in one hybridisation by combining paints for all 24 different chromosomes. Each chromosome paint is made distinguishable by its unique labelling scheme: a different single fluorochrome or combination of contrasting fluorochromes is used for each chromosome and during the analysis computer-captured images are processed to produce a unique identifying false colour. The use of n different fluorochromes gives 2^n-1 possible different fluorochrome combinations, so all 24 different human chromosomes can be distinguished by using five different fluorochromes. In SKY, a similar probe kit is used that is composed of combinatorially labelled paints and during analysis classification colours (pseudocolours) are assigned to individual chromosome classes which are characterised by their unique colour combinations. However, identification of the different chromosome classes is different as SKY quantifies the spectral composition of each individual pixel within the image, whereas M-FISH is based on determining whether a certain fluorochrome is present or absent.

An advantage of M-FISH over the use of individual whole chromosome paints or traditional banding is the ability to identify all chromosomes in one experiment, which saves both time and specimen and avoids guesswork in choosing chromosome paints to combine. There is also the possibility that unexpected cryptic abnormalities are found. The resolution of M-FISH is, in our experience, about 3 Mb, although in specific instances it may be higher. Advantages over array karyotyping include the preservation of individual chromosome structure and the need for only a small number of cells, as long as metaphases are present. If chromosomes have already been prepared for karyotyping, M-FISH or M-BAND can be readily carried out on this material to help interpret difficult karyotypes, without the need for additional specimen or processing of the specimen.

A more specialised M-FISH technique is arm M-FISH (MetaSystems) (8, 9), where each chromosome arm (except the short arms of the acrocentric and Y chromosomes) is identified. The advantage of arm M-FISH over standard M-FISH is that not only the chromosomal origin of a certain chromosomal fragment can be determined, but also whether it is derived from the p- or the q-arm of this particular chromosome.

M-BAND (10, 11) is based on the same approach as M-FISH but a region-specific M-BAND pattern is produced for a single chromosome. In M-BAND, a series of regional chromosome paints for sequential partially overlapping chromosome regions is produced for a single chromosome. The characteristic pattern of M-BAND is generated by the fluorescence intensity ratios which are determined by quantifying the relative intensities of overlapping fluorochromes. These ratios are translated into a band-specific display colour. M-BAND provides information that allows for a precise analysis of even highly complex inter- and intra-chromosomal aberrations. M-BAND is a proprietary technique available from MetaSystems.

M-FISH is the most direct means of detecting unsuspected cryptic translocations. In cancer cytogenetics, it is also useful in interpreting complex karyotypes. M-BAND can be used to help define translocation breakpoints and to interpret inversions. Like other metaphase FISH techniques, M-FISH and M-BAND retain information on the structure of the individual chromosomes and the composition of different cell lines, as opposed to array-based copy number analysis, in which copy number information is given without reference to chromosome organisation or the composition of different cell lines. This is particularly useful in cancer cytogenetics as M-FISH and M-BAND can help interpret complex karyotypes and different cell lines, and evolution of the different cell lines can be followed.

Commercial M-FISH probes can be purchased, eliminating the need for chromosome library preparation. The protocol below is for use with the MetaSystems XCyte probes, and is based on the manufacturer's instructions. These probes have been prepared by pooling microdissected chromosomes or chromosome regions, amplification by PCR and labelling with one or a combination of up to six fluorochromes. The probes include human repetitive (Cot-1) DNA as a competitor to reduce hybridisation of the probe to repetitive sequences on other chromosomes. M-FISH or SKY probes can also be made using chromosome-specific libraries produced from flow-sorted chromosomes (6). M-FISH and M-BAND experimental protocols are identical.

2. Materials

2.1. Slide Preparation

1. Fixed metaphase suspensions prepared according to standard procedures (see Chapter 5).
2. Clean glass microscope slides.
3. Phase contrast microscope for the examination of spreading quality.

2.2. Chromosome Pre-treatment and Denaturation

1. Water, deionised or distilled.
2. 0.07M NaOH.
3. 20× SSC Saline-sodium Citrate Buffer: 3.0 M NaCl and 0.3 M $C_6H_5Na_3O_7 \cdot 2H_2O$ (tri-Sodium citrate Dihydrate) for preparation of: 0.1× SSC, pH 7.0–7.5 and 2× SSC, pH 7.0–7.5.

   Adjust pH with concentrated NaOH or HCl solutions (see Note 1).
4. Waterbath.
5. Coplin jars.

2.3. Probe Denaturation

1. Probe cocktail: MetaSystems 24XCyte (M-FISH) or XCyte (M-BAND) (see Note 2).
2. Waterbaths at 75 and 37°C and ice, or thermocycler.

2.4. Hybridisation

1. Coverslips.
2. Marabu Fixogum (Beitigheim, Bissingen, Germany) (see Note 3).
3. Sealed humidified chamber (see Note 4).
4. 37°C incubator.

2.5. Washes

1. Tween20 Polyoxyethylene sorbitan-monolaurate Syrup (e.g. Sigma P-1379) (Tween™ is a trademark of ICI America, Inc.) (see Note 5).
2. 1× SSC, pH 7.0–7.5.
3. 4× SSCT, pH 7.0–7.5 (prepared by adding 0.05% Tween 20 to 4× SSC, pH 7.0–7.5) (see Note 5).

4. Phosphate-Buffered Saline Solution (PBS): 120 mM NaCl, 7 mM Na_2HPO_4, 3 mM NaH_2PO_4, and 2.7 mM KCl.

2.6. Mounting and Counterstain

1. DAPI/antifade (available separately from MetaSystems) (see Note 6).
2. Coverslips (24 × 50–60 mm).

2.7. Capture and Analysis

1. Fluorescence microscope with filter sets appropriate to the fluorochromes used.
2. Low fluorescence immersion oil.
3. Capture and analysis station.

3. Methods

The basic procedure is similar to standard FISH with denaturation of probe and chromosomes, incubation together under a coverslip and washing. Metaphase chromosomes are prepared and spread according to standard techniques. These are subjected to alkaline DNA denaturation with sodium hydroxide. Alkaline denaturation, adapted from Fritz et al. (12) and Rieder et al. (13), is used as it causes less damage to chromosome morphology than formamide denaturation. The probe is separately heat-denatured and applied to the denatured chromosomes under a sealed coverslip; then, the slides are incubated overnight at 37°C. Slides are washed and mounted in antifade with a counterstain. The analysis is carried out with a fluorescence microscope with filter sets specific to the fluorochromes used, and captured images are analysed with software specific to the commercial probe, in this case Isis capture and analysis software from MetaSystems. The captured images are processed to produce false colour karyotypes or M-BAND images, and these are interpreted based on the false colours and confirmed by the examination of the single colour galleries produced by each individual fluorochrome.

3.1. Metaphase Preparation

1. Metaphase chromosomes are prepared and slides made according to standard methods (see Chapter 5) (see Note 7).
2. Metaphases should be checked under a phase contrast microscope for morphology and spreading. Good morphology is indicated by well-defined chromosomes clear of cytoplasm, neither phase bright nor pale. Well-spread chromosomes with minimal overlaps are best for the interpretation of M-FISH, whereas more overlaps can be tolerated with M-BAND (see Note 8).
3. The area to be hybridised is marked, for example, with a diamond pencil or by spreading onto pre-marked slides.

3.2. Chromosome Denaturation

The following protocol does not include a pre-treatment step (see Note 9).

1. Incubate in 2× SSC, pH 7.0–7.5, 70°C, 30 min (see Note 10).
2. Remove coplin jar from the water bath and cool at room temperature down to ~37°C (this takes about 20 min).
3. Start probe denaturation (Subheading 3.3).
4. Immerse slides in 0.1× SSC, pH 7.0–7.5, room temperature, 1 min.
5. Denature in 0.07 M NaOH, room temperature for exactly 1 min.
6. Immerse slides in 0.1× SSC, pH 7.0–7.5, 4°C, 1 min.
7. Immerse slides in 2× SSC, pH 7.0–7.5, 4°C, 1 min.
8. Dehydrate through ethanol series: 1 min each in 70, 95, and 100% ethanol.
9. Air dry (see Note 11).

3.3. Probe Denaturation

1. Prepare water baths at 75 and 37°C (see Note 12).
2. Spin the probe stock tube briefly in a microfuge to ensure that the probe is at the bottom of the tube.
3. Pipette the required amount of probe into a microfuge tube:

   4 μL for 12 mm diameter round coverslip

   7 μL for 18 × 18 mm^2 coverslip

   10 μL for 22 × 22 mm^2 coverslip

   12 μL for 24 × 24 mm^2 coverslip or

   24 μL for 24 × 50 mm^2 coverslip (whole slide).
4. Denature at 75°C for 5 min.
5. Place briefly on ice.
6. Incubate at 37°C for 30 min to reduce background by pre-annealing labelled and unlabelled repeat sequences in the probe cocktail.

3.4. Hybridisation

1. Spin the tube briefly in a microfuge to ensure that the denatured probe is at the bottom of the tube.
2. Pipette the probe onto the marked area of the slide containing the metaphases to be probed.
3. Carefully place a coverslip over the probe, avoiding air bubbles, which can be gently pressed out if necessary (see Note 13).
4. Seal around the coverslip using rubber cement.
5. Incubate in the sealed humidified chamber at 37°C overnight (see Note 14).

3.5. Washes

1. Pre-warm 1× SSC, pH 7.0–7.5 to 75°C.
2. Remove rubber cement and coverslips carefully.
3. Wash slides in 1× SSC, pH 7.0–7.5, 75°C, 5 min.
4. Incubate slides in 4× SSCT, pH 7.0–7.5, 5 min.

3.6. Mounting and Counterstain

1. Wash slide briefly in distilled water to avoid crystal formation.
2. Drain fluid off and let air dry.
3. Apply 20 µL of DAPI/antifade and coverslip (see Note 15).
4. Slides are best stored in the dark at –20°C (see Note 16).

3.7. Capture and Analysis

Detailed instructions on image capture and analysis are provided with the software specific to the probe (Isis, MetaSystems).

1. Metaphases can be found by viewing the DAPI counterstain with a low power objective.
2. Locate a spread and view under high power (see Note 17).
3. Assess for hybridisation by viewing different fluorochromes (see Note 18).
4. Ensure that the spectrum symbol is selected before capture and any of the following image processing steps (see Note 19).
5. Select the experiment type (M-FISH, mouse M-FISH, or a specific M-BAND) and capture the DAPI image and each fluorochrome in succession using automatic integration control (see Note 20). Figure 1 shows a screenshot of a captured metaphase in Isis.
6. Define the region of interest (see Note 21).
7. Apply background correction.
8. Apply automatic upper and lower thresholding.
9. Correct pixel shift if necessary. Do not use the automatic register colour function for M-BAND.
10. Maximise the metaphase as desired.
11. Adjust the object threshold and separate the chromosomes (see Note 22).
12. Enter karyotype mode.
13. For M-FISH: apply automatic karyotyping and then make any adjustments to allow for incorrectly classified chromosomes (see Note 23, Fig. 2a–c).
14. The first false colour image is produced using the built-in Classifier (Fig. 2a). The chromosomes in the karyogram may be used to train a new Classifier. Place translocated chromosomes in the buffer and untranslocated chromosomes in their correct position and select Train M-FISH Classifier. This results in a more homogeneous false colour image, by adjusting to the image produced by the conditions in the

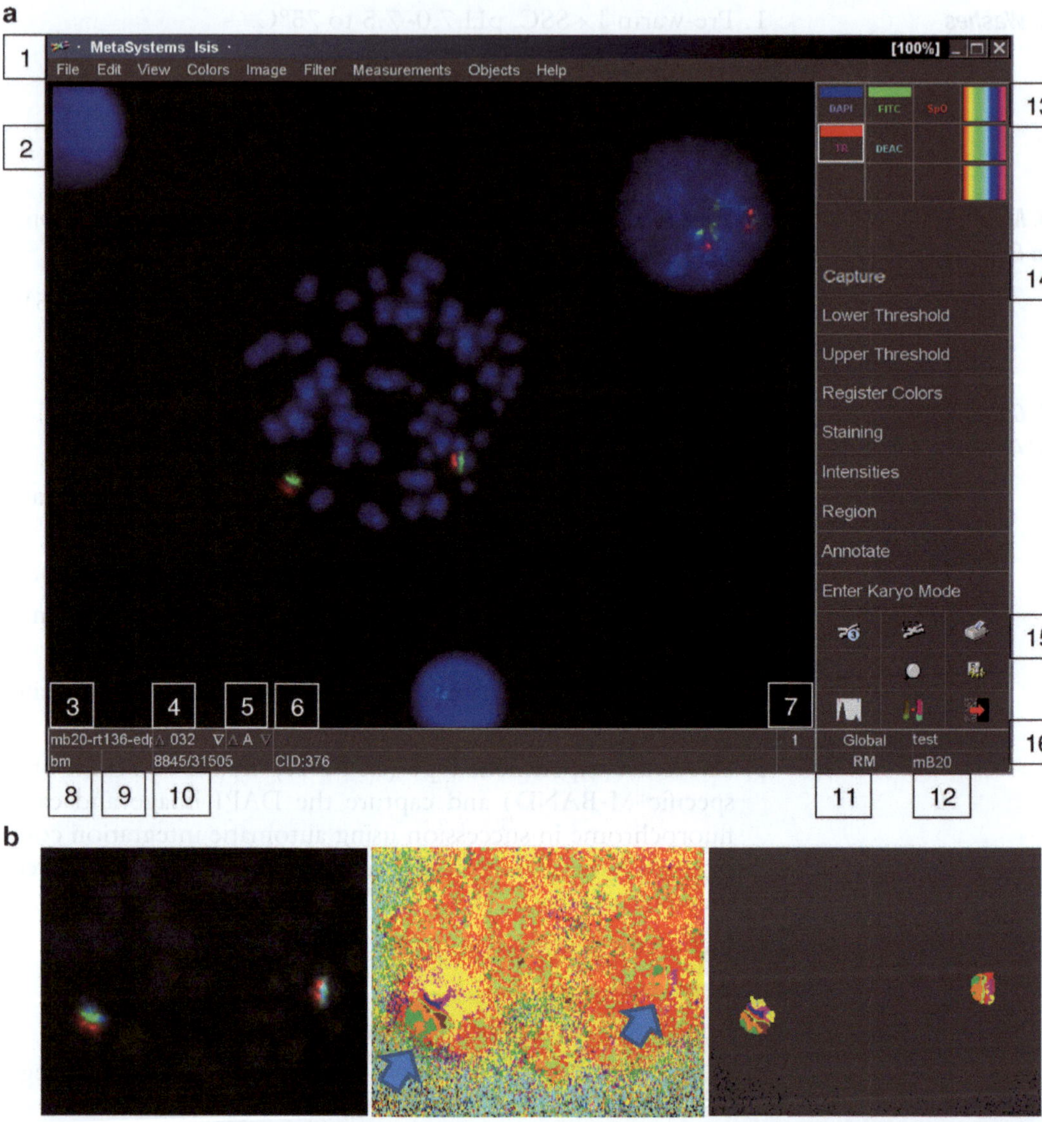

Fig. 1. (**a**) Screenshot of a multicolour banded metaphase labelled with the M-BAND chromosome 20 probe XCyte 20 with sections of the screen labelled. The three highlighted fluorochrome channels DAPI (*blue* counterstain), FITC (*green*), and TR (Texas Red)) are visible in this red-green-blue (RGB) display (interphase nuclei are also visible). (**b**) Partial image of the same metaphase displayed in: *left*, FITC, SpectrumOrange, DEAC (*aqua*); *centre*: false colours with minimum classification intensity set at 0%, i.e. every pixel is assigned a false colour (labelled chromosomes are *arrowed*); *right*, false colours with minimum classification intensity set at 22% so that only the labelled chromosome regions are assigned a false colour. The inbuilt XCyte 20 Classifier is used for this image – the image is unprocessed except for DAPI enhancement to make the chromosomes clearer. Key: (1) menu bar; (2) image area; (3) file name; (4) image number; (5) image field number; (6) comments; (7) object count; (8) culture; (9) slide; (10) co-ordinates; (11) user; (12) technique; (13) fluorochrome channels; (14) command buttons; (15) icons; (16) data directory.

laboratory and can result in correction of incorrect false colours (see Note 24, Fig. 2b).

15. For M-BAND Classifier training: place the abnormal labelled chromosomes at the correct position and the normal

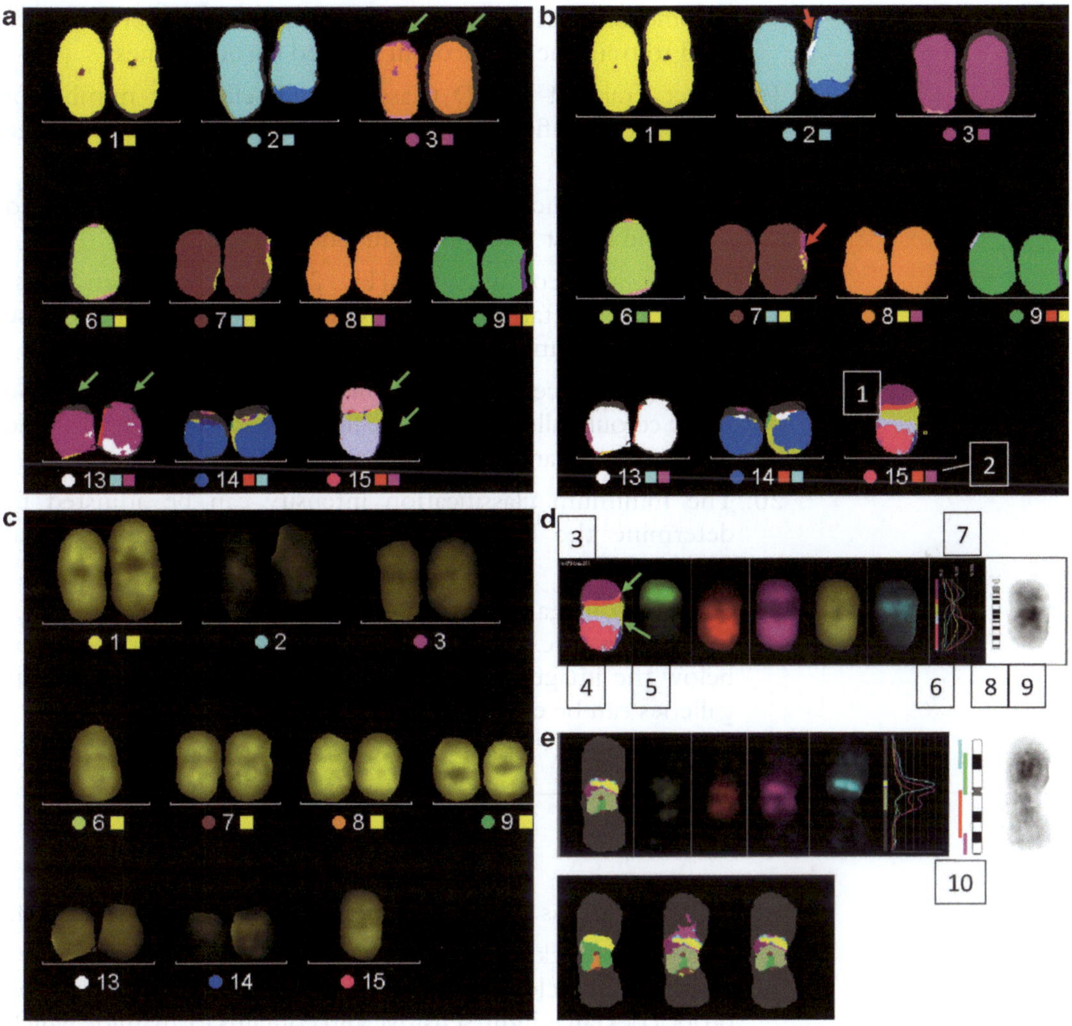

Fig. 2. (**a–d**) Karyotype elements from a metaphase probed with XCyte 24 (M-FISH). The chromosome at position 15 was determined by a combination of G-banding, M-FISH and M-BAND to be a derivative of chromosomes 11, 15, and 20 with inversion of the chromosome 20 section: trc(11;20;15)(11qter → 11p11.2::20p13 → 20p12::20q13.3 → 20q13.2::20q13.2 → 20q12::20q11.2 → 20p11.2::15q10 → 15qter). This metaphase is used as an illustration of problematic analysis, but other metaphases from this hybridisation had less ambiguous labelling. Some chromosomes had a high level of Cy5 background, and true Cy5 signal on chromosome 20 was relatively low. (**a**) In false colour mode before training but with incorrectly coloured chromosomes (*green arrows*) placed in the correct positions after assessing the DAPI banding, individual fluorochromes, and single colour galleries; (**b**) in false colour mode after training (*red arrows* show examples of flaring produced by fluorochromes from neighbouring chromosomes). Although there was no normal chromosome 15, a more accurate false colour pattern was produced for the abnormal chromosome labelled (1) after training with this abnormal chromosome at position 15; (2) Indicates the labelling scheme: the *circle* on the *left* of the chromosome number shows the false colour and the squares on the *right* show the individual fluorochromes combining to produce the false colour; (**c**) Cy5 single fluorochrome image. Some chromosomes have a high background level of Cy5 labelling which explains the misidentified chromosomes in 2a (*green arrows*). (**d**) M-FISH single colour gallery of the chromosome labelled (1) from (**b**). There is an intermediate level of Cy5 labelling along the chromosome, suggesting that the Cy5 label is background. Note that misleading false colours (*green arrows*) are produced at the interfaces of the three main false colour regions; (**e**) (*upper*) XCyte 20 M-BAND image of the same abnormal chromosome confirms that the false colour produced in the central region in 2(d) is true chromosome 20 signal, and shows the M-BAND pattern interpreted as 20p13 → 20p12::20q13.3 → 20q13.2::20q13.2 → 20q12::20q11.2 → 20p11.2; (*lower*) the same chromosome (*left*) trained

chromosomes at the chromosome 1 position, where they are used to train the multicolour banded chromosomes.

16. Select Train M-BAND Classifier and select the number of bands. A Classifier from another metaphase can be used (see Note 25).
17. For M-FISH and M-BAND: DAPI banding can be enhanced using the adjust filter power function.
18. Use the single colour gallery (Fig. 2d, e) to confirm and assist in the interpretation of false colour bands and check false colour classifications (see Note 26).
19. The fluorescence intensity profile (a graph at the right of the single colour gallery) shows the fluorescence intensity of all the fluorochromes and the resulting false colour (see Note 27).
20. The minimum classification intensity can be adjusted to determine the level of labelling accepted as true signal (see Note 28).
21. Single colour galleries can be copied to compare with other metaphases or experiments. Annotations can be made on or below the image, and the final image and/or single colour galleries can be exported and printed.

4. Notes

1. If SSC solutions are to be stored, they should be autoclaved.
2. The probe stock can be split into aliquots to avoid excessive freeze–thaw cycles of the whole mixture. Please note that the probe cocktail is light-sensitive and contains formamide which is a suspected teratogen.
3. Rubber cement purchased from bicycle suppliers as an adhesive for repairing tyre punctures can be used, but sealants such as Fixogum sold expressly for FISH are becoming more readily available.
4. A sealed humidified chamber can be made by placing absorbent material, for example an absorbent wipe, in the bottom of a sealable plastic container and wetting it with water or 2× SSC

Fig. 2. (continued) with the normal chromosome 20 from this metaphase, (*centre* and *right*) trained with a Classifier produced using chromosome 20 from another metaphase. The M-BAND boundaries have been adjusted in the different images by varying the minimum classification intensity. Note that interpretation of the inversion is helped by the examination of the single colour gallery, as fluorochromes that are not normally together are juxtaposed. Key: (3) Classifier name; (4) false colour image; (5) fluorochromes: FITC, SpO, TR, Cy5, DEAC; (6) fluorescence intensity profile; (7) signal intensity relative to strongest signal on this chromosome; (8) idiogram; (9) DAPI-banded chromosome; (10) XCyte 20 labelling scheme showing coverage of each fluorochrome-labelled partial chromosome paint.

until just saturated. This is pre-warmed, the slides are placed on the saturated material and incubated in the sealed container.

5. Igepal CA-630 (which replaces Nonidet P-40, available from Sigma-Aldrich) can be substituted for Tween 20. Tween20 (or Igapal CA-630) is added after autoclaving.
6. It is possible to use DAPI/antifade prepared in the laboratory. However, the DAPI concentration must be low to avoid crosstalk with the DEAC (aqua) fluorochrome.
7. Successful M-FISH depends on the preparation of metaphase chromosomes from an appropriate specimen. Fixed chromosome suspensions are relatively stable and can be stored long term at –80°C. Chromosomes must be well spread and with minimal overlaps, particularly for M-FISH. More overlapping can be tolerated for M-BAND. Slides stored at room temperature should not be older than 2 weeks. If slides are stored for longer, they should be stored at –20°C with desiccant (silica gel) in a sealed slide box, which is warmed to room temperature before opening. Chromosome suspensions or slides can be prepared in one laboratory for FISH in another laboratory, allowing exchange of samples between laboratories. Metaphase spreads may be targeted to an area equal to the size of the coverslip which is to be used, if the specimen is precious. Aging is normally carried out overnight in a desiccator. Alternatively, chromosomes may be "aged" by heat treatment of the slides, at 45°C for 3 h, but only if the slides have been made on the same day as hybridisation. Slides which have been probed previously can be used. To do this, heat-denature the slides with formamide (see Note 29) to more effectively remove the old signal. Be aware that the old signal may not be completely removed, and there is probably deterioration in morphology. Although results are not as good, previously G-banded slides can be used. The immersion oil and mounting medium must be removed by rinsing or soaking in Xylol for at least 10 min, or as long as is needed. The slide is rinsed in fresh Carnoy's fixative (methanol:acetic acid 3:1), dehydrated in 70, 90, and 100% ethanol for 2 min each and air dried. If it is known that G-banded slides will be demounted for M-FISH, the slides can be mounted for a short period in a thin layer of PBS (as thin as is practical to avoid reduction in clarity of the G-banded image), which is easily removed by rinsing with water before destaining in Carnoy's fixative and ethanol dehydration as detailed above.
8. Chromosomes with suboptimal morphology can still be used for M-FISH, but results may not be as good.
9. A pepsin pre-treatment step is not necessary in most cases but can be applied if there is cytoplasmic protein over the

chromosomes which may reduce hybridisation efficiency. Add 500 μL pepsin solution (20 mg/mL Sigma, P-7012 dissolved in water; for different types of pepsin the concentration and the incubation time have to be determined) to 100 mL 0.01 M HCl pre-warmed to 37°C, mix well and immediately incubate the slides: 1–2 min for amniocyte or lymphocyte chromosomes and up to 5 min for bone marrow chromosomes. Wash in PBS, 3 min. Incubate under 100 μL post-fixation solution (1% formaldehyde/50 mM $MgCl_2$ in PBS), coverslipped, at room temperature for 10 min. Wash in PBS, 3 min. Go straight to denaturation or dehydrate through an ethanol series if the procedure is to be paused, and rehydrate before denaturation. Pepsin pre-treatment is not needed for G-banded slides.

10. Heat the waterbaths used for various steps with the coplin jars in them as sudden changes of heat can lead to the coplin jars cracking. Check the pH of solutions that should be pH 7.0–7.5 at room temperature before use. Check the temperature of the solutions as this can be different from the temperature of the waterbath. Best results are obtained if solutions are freshly set up for each experiment.
11. Chromosome denaturation can be checked under phase contrast microscopy. Regular checking familiarises the user with the appearance of successfully denatured chromosomes, allowing difficult cases to be assessed for chromosome denaturation. Denatured chromosomes are light grey and a little blown out. Over-denatured chromosomes are very light grey and their structure is destroyed. Dark chromosomes are not denatured. For chromosomes which are difficult to denature, formamide denaturation can be used (see Note 29) and the denaturation temperature can be increased up to 75°C or no more than 80°C.
12. A thermocycler can be used, set at 75°C for 5 min, 10°C for 30 s, and 37°C for 30 min.
13. Examine the coverslip before use to ensure that it is clean and dust-free so that it fits snugly against the slide – if the probe does not reach the edges of the coverslip, this may be due to dust or dirt under the coverslip; handle with forceps or gloves.
14. Incubations can be carried out for up to 4 days. Make sure that the hybridisation chamber is well sealed to prevent drying out.
15. Only 15–20 μL DAPI/antifade should be used. Too much can interfere with the clarity of the image. Excess can be carefully pressed out and blotted from around the coverslip.
16. A good signal can still be seen at least a month after hybridisation (stored at −20°C). However, fluorescent signals fade over time and so are best analysed as soon as is practical.

17. Different immersion oils should not be mixed. Use a low fluorescence immersion oil. The fluorescence field should be evenly illuminated. Avoid over-exposure of the slides to light by blocking the filter when not viewing the slide.
18. FITC signals are very sensitive to low pH. The processing solutions must be at pH 7.0–7.5 at room temperature.
19. This is most important in order to preserve the correct fluorescent ratio of the raw image.
20. DEAC is programed to capture last to avoid crosstalk with the DAPI counterstain. Obtain a clear image for each fluorochrome. Use the up and down arrows to select the field used to determine exposure time. Any high intensity background should be excluded from this field or the real signal will be reduced in intensity. Ensure a positive signal for each fluorochrome is included in this field (particularly for M-BAND) or exposure times will be too high. Avoid moving or shaking the microscope during capture or successive images will not overlap precisely. Minimise exposure to light: block the filters from illuminating the slide when not viewing; close the diaphragm as far as possible when capturing; work in a darkened room as much as possible; store slides in the dark.
21. For M-BAND, when eliminating unlabelled chromosomes check the red-green-blue (RGB) image for all fluorochromes by clicking on each fluorochrome channel (Fig. 1); otherwise, chromosomes labelled with only a subset of fluorochromes may be missed. A double-click on the fluorochrome channel displays the single fluorochrome only (Fig. 2c).
22. Adjust the object threshold close to the edge of the chromosomes and check that all chromosomes are preserved: the smaller ones are easily lost by over-thresholding. Increasing the object threshold makes separation of the chromosomes easier. Use the separation function while viewing the DAPI image rather than the false colour image. The false colour image can be referred to in difficult separations.
23. Select "show labelling scheme" to view the fluorochrome combinations (Fig. 2b) used to classify the chromosomes and check that the single colour galleries (Fig. 2d) match these.
24. A colour classifier defines the correlation between the fluorochrome combinations of the labelling scheme and a display false colour. It determines how any combination of different fluorescence intensities is classified. For M-FISH probe kits, the number of colour classes (i.e. the number of false colours) is equivalent to the number of fluorochrome combinations (number of chromosome classes). It may be necessary to use an abnormal chromosome if there is no example of a normal chromosome; a small amount of another chromosome in the wrong position will not affect the classifier. Or a classifier

from another patient or metaphase can be used, if there are no examples of a particular normal chromosome. This also applies for M-BAND (see Note 25).

25. M-BAND classifiers have the same structure as M-FISH classifiers. The basic configuration of MetaSystems imaging system (Isis.mFISH) provides a set of M-BAND classifiers, one for each chromosome. The classifiers can be adjusted and trained in order to get the best matching classifier for the chromosomal aberration to be analysed. The user sets the number of bands in each M-BAND Classifier. The labelling pattern of the chromosome or chromosomes in position 1 is used to determine the classification rules and labelled chromosomes in the karyotype are false colour banded by these rules, according to their relative fluorescence intensities. The M-BAND Classifier produced for one metaphase can be saved and used to train labelled chromosomes in other metaphases. This is useful if there are no examples of the normal chromosome to use for training. A good representative classifier from a single metaphase can be used repeatedly. Although a large number of bands can be produced, comparison of banded chromosomes is more robust if there are not too many bands. A useful number of false colour bands is three times the number of fluorochromes. A different Classifier can be produced for each trained multicolour banded metaphase. However, to compare M-BAND patterns from different cells or samples, each metaphase should be trained with the same classifier (see Fig. 2e). Combining M-BAND probes is not recommended as the interpretation is difficult.
26. This is particularly important at the junction of two different chromosomes where an incorrect false colour may result (Fig. 2d), and for M-BAND where the juxtaposition of bands not normally together may produce a misleading false colour. Some chromosomes may need to be checked in the metaphase image to see if a signal is caused by flaring from adjacent chromosomes (Fig. 2b). Small sections of a chromosome may not produce a false colour due to stronger signals from adjacent chromosome regions – sometimes, these are detectable in the single colour gallery or fluorescence intensity profile (14). M-BAND can detect small sections of a chromosome with greater sensitivity as there is no stronger adjacent signal from other chromosomes.
27. The fluorescence intensity profile can be misleading if the chromosome is bent – the chromosome should be straightened or the section of interest aligned vertically.
28. With the minimum classification intensity set at zero, there will be 100% false colour labelling (Fig. 1b). As the minimum classification intensity is reduced, unlabelled regions including

heterochromatic regions such as centromeres and the short arms of acrocentrics become unlabelled (Fig. 2a, b). When analysing M-BAND, the minimum classification intensity should be adjusted so that unlabelled sections are grey (Figs. 1b and 2e). It can be adjusted so that the outermost false colour band on a translocated chromosome corresponds with the edge of the outermost fluorochrome signal. In the same image, different fluorochromes may require a different minimum classification intensity to achieve this.

29. Formamide denaturation (15). Requirements: 70% deionized formamide in 2× SSC, pH7.0 and an ethanol series, 70% (-20°C), 90% and 100%. Pre-warm formamide solution to 70°C for the first one to two slides plus 1°C for every additional slide. Incubate the slides in formamide solution for 3 min then dehydrate through the ethanol series, 3 min each, and air dry.

Acknowledgements

We thank Kathleen Rayeroux for critical reading of the manuscript and provision of material for the figures.

References

1. Redon R, Fitzgerald T, Carter NP. Comparative genomic hybridization: DNA labeling, hybridization and detection. Methods Mol Biol 2009;529:267–78.
2. Horvath A, Stratakis C. Basic molecular techniques for the detection of single nucleotide polymorphisms: genome-wide applications in search for endocrine tumor related genes. Methods Mol Biol 2009;590:143–63.
3. Pinkel D, Segraves R, Sudar D, et al. High resolution analysis of DNA copy number variation using comparative genomic hybridization to microarrays. Nat Genet 1998;20:207–11.
4. Carter NP. Cytogenetic analysis by chromosome painting. Cytometry 1994;18:2–10.
5. Deng HX, Yoshiura K, Dirks RW, et al. Chromosome-band-specific painting: chromosome in situ suppression hybridization using PCR products from a microdissected chromosome band as a probe pool. Hum Genet 1992;89:13–7.
6. Schrock E, du Manoir S, Veldman T, et al. Multicolor spectral karyotyping of human chromosomes. Science 1996;273:494–7.
7. Speicher MR, Gwyn Ballard S, Ward DC. Karyotyping human chromosomes by combinatorial multi-fluor FISH. Nat Genet 1996;12:368–75.
8. Guan XY, Xu J, Anzick SL, Zhang H, Trent JM, Meltzer PS. Hybrid selection of transcribed sequences from microdissected DNA: isolation of genes within amplified region at 20q11-q13.2 in breast cancer. Cancer Res 1996;56:3446–50.
9. Karhu R, Ahlstedt-Soini M, Bittner M, Meltzer P, Trent JM, Isola JJ. Chromosome arm-specific multicolor FISH. Genes Chromosomes Cancer 2001;30:105–9.
10. Liehr T, Starke H, Heller A, et al. Multicolor fluorescence in situ hybridization (FISH) applied to FISH-banding. Cytogenet Genome Res 2006;114:240–4.
11. Chudoba I, Plesch A, Lorch T, Lemke J, Claussen U, Senger G. High resolution multicolor-banding: a new technique for refined FISH analysis of human chromosomes. Cytogenet Cell Genet 1999;84:156–60.
12. Fritz B, Kuster W, Orstavik KH, Naumova A, Spranger J, Rehder H. Pigmentary mosaicism in hypomelanosis of Ito. Further evidence for functional disomy of Xp. Hum Genet 1998;103:441–9.
13. Rieder H, Bonwetsch C, Janssen LA, et al. High rate of chromosome abnormalities

detected by fluorescence in situ hybridization using BCR and ABL probes in adult acute lymphoblastic leukemia. Leukemia 1998;12:1473–81.

14. de Vree P, Simon M, van Dooren M, et al. Application of molecular cytogenetic techniques to clarify apparently balanced complex chromosomal rearrangements in two patients with an abnormal phenotype: case report. Mol Cytogenet 2009;2:15.

15. Harper ME, Saunders GF. Localization of single copy DNA sequences of G-banded human chromosomes by in situ hybridization. Chromosoma 1981;83:431–9.

Chapter 16

Utility of Array Comparative Genomic Hybridization in Cytogenetic Analysis

Rashmi R. Singh, K-John J. Cheung, and Douglas E. Horsman

Abstract

Conventional comparative genomic hybridization (CGH), high-resolution oligonucleotide, and BAC array CGH have modernized the field of cytogenetics to enable access to unbalanced genomic aberrations such as whole or partial chromosomal gains and losses. The basic principle of array CGH involves hybridizing differentially labeled proband/test (e.g., tumor) and normal reference DNA on an array of oligonucleotide or BAC clones instead of normal metaphases as in conventional CGH. The sub-megabase resolution tiling BAC arrays are extremely useful for the analysis of acquired aberrations in cancer genomes. Array CGH can be extremely useful to identify the chromosomal makeup of marker and ring chromosomes, to define/delineate the precise location/bands involved in structural aberrations and the accurate localization of translocation breakpoints in both simple and complex karyotypes either alone or in combination with standard karyotype analysis.

Key words: Array comparative genomic hybridization, Sub-megabase resolution tiling, Unbalanced genomic aberrations, Copy number variation

1. Introduction

Array-based technologies have revolutionized the field of interphase cytogenetics and have been applied to the analysis of inherited/constitutional and somatic chromosomal aberrations (1). Among the different types of technology, array comparative genomic hybridization (CGH) has become one of the most popular techniques for genome-wide screening for structural alterations involving genomic copy number changes (gains/losses) in tumor cells (2–4). The method is based on the principle of conventional CGH where the test/proband (e.g., tumor) and normal (reference) DNA are differentially labeled with fluorochrome dyes such as Cy3-dCTP and Cy5-dCTP, and then hybridized to

Lynda J. Campbell (ed.), *Cancer Cytogenetics: Methods and Protocols*, Methods in Molecular Biology, vol. 730,
DOI 10.1007/978-1-61779-074-4_16, © Springer Science+Business Media, LLC 2011

normal metaphase chromosomes (4). The main difference is that arrays consist of probes covering up to the entire genome spotted onto glass slides as a matrix of DNA hybridization targets. Array CGH, also called molecular karyotyping, can be highly comprehensive, sensitive, relatively fast, and, most importantly, amenable to very high resolution (resolution being dependent on the density of clones spotted on the array) (5, 6).

The human BAC arrays used for array CGH analysis consist of linker–adapter representations of BAC clones that are mapped by DNA fingerprinting, PCR-based BAC end sequencing, and chromosome mapping by fluorescent in situ hybridization (FISH). Nonspecific hybridization especially with highly repetitive sequences such as centromeres, telomeres, and internal high repetitive sequences are avoided by the addition of Cot-1 DNA (7). Different configurations of arrays are available based on the density and spacing of the spotted BAC clones, which result in variable resolution of detection of copy number changes. Some of the available arrays include the commercial Spectro-Genomics arrays with 2,632 BACs distributed at an average spacing of ~1 Mb across the genome (7), tiling arrays such as the sub-megabase resolution tiling (SMRT) arrays spotted with >25,000 BAC clones with an average spacing of ~80 kb (8), and oligonucleotide arrays with oligos ranging in length from 60 to 100 bp spotted at an average spacing of ~45–90 kb (4). Of these, higher density oligo arrays can precisely detect small deletions, duplications, amplifications, and aneuploidies, with a resolution down to ~10 kb. BAC arrays have a lower resolution capability (~1 Mb for Spectral Genomics arrays and ~300 kb for SMRT arrays) (7). A genomic gain or loss is defined by the apparent shift of a minimum three adjacent mapped BACs (see Subheading 2.2). The detection resolution of these arrays is also influenced by tumor heterogeneity, and the aberrant clone size can determine the nature and degree of shifts from baseline. In this context, SMRT arrays are relatively more tolerant of tumor heterogeneity and archived (formalin-fixed, paraffin-embedded) DNA, and thus are very useful to analyze cancer genomes in both clinical and research settings (8).

Apart from detecting chromosomal aberrations, array CGH has the potential to detect copy number variations (CNV). CNVs are inherited copy number alterations that occur in segments ranging from 1 kb to several Mb and are known to vary due to deletions or duplications. So far, 3,654 segmental CNVs have been identified with 800 appearing at a frequency of at least 3% (9). CNVs must be distinguished from acquired copy number alterations that occur in cancer cells. The use of a patient's own constitutional DNA as normal reference can help to eliminate visualization of CNVs, while the use of pooled DNA from a control population can enhance the polymorphic shifts. The density of arrays is known to influence the smallest CNV that can be detected such that higher density arrays can detect CNVs less than 10–50 kb in size.

Array CGH software packages are visualization tools that enable both whole genome and single chromosome views of a single case, facilitating the precise localization of copy number changes. Another tool, called multiple alignment, allows the user to view segmental changes affecting an individual chromosome across a large number of cases. An example of a CNV showing a localized region of gain/loss in the multiple alignment view is shown in Fig. 1 (10).

Despite the apparent utility of array CGH, there is always a degree of uncertainty regarding the performance and data output of arrays, and thus validation of detected genomic aberrations using other methods is essential. FISH is a useful method for validation of copy number gains and losses, using derivatives of the BAC probes from the array, or selected from independent BAC libraries; however, small tandem duplications and very small deletions may be difficult or impossible to validate by this approach

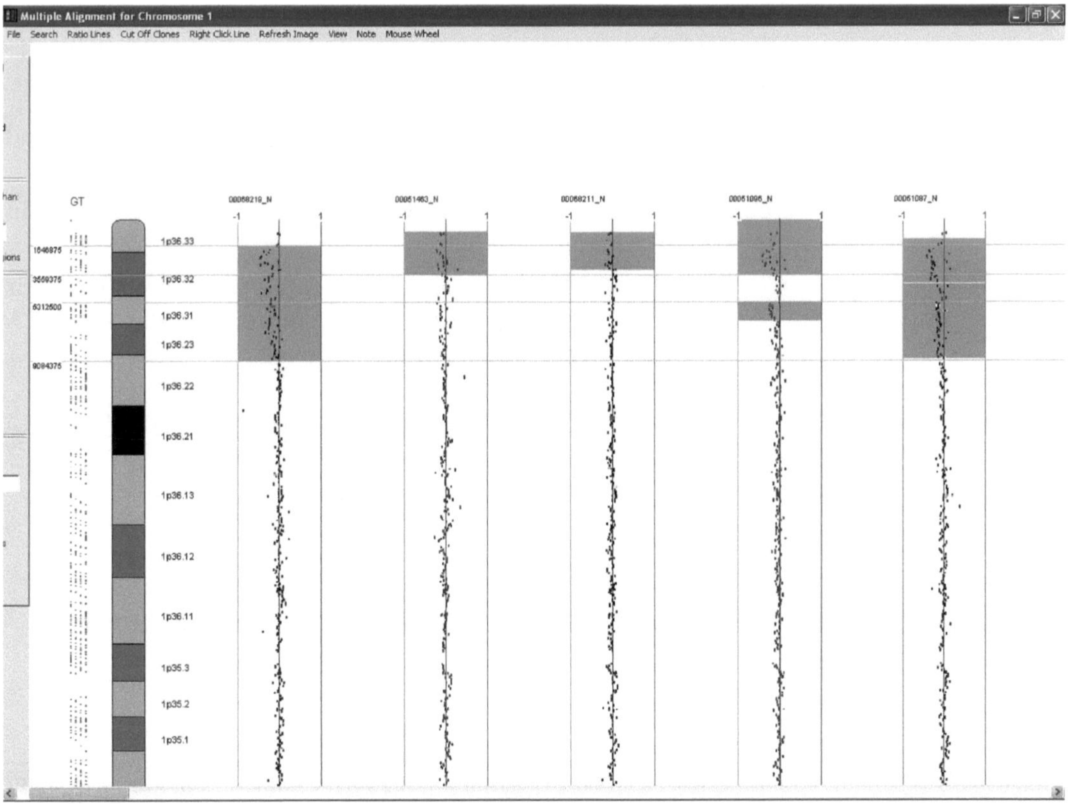

Fig. 1. Shown here are five representative follicular lymphoma (FL) cases being viewed in the SeeGH multiple alignment application illustrating the loss of 1p36 (*highlighted*), a region deleted in approximately 20% of FL specimens (10). The majority of these array CGH-detected changes can be validated by FISH. The *blocks* on the *left side* of the illustration represent the gene track (GT) corresponding to their respective chromosomal bands and BAC clones (hg15 assembly).

(~100 kb for losses and ~500 kb for gains). Smaller rearrangements may be addressed using quantitative-PCR (Q-PCR), multiplex ligation-dependent probe amplification (MLPA), or by using another array platform (4). FISH is the preferred method to validate large aberrations (200–500 kb and larger) and can supplement karyotype analysis.

2. Materials

2.1. Genomic DNA Extraction from Frozen Tumor Tissues

1. Six to eight slices of frozen tumor tissue, each 25 μm thick, or cell pellets in a 1.5 mL Eppendorf tube.
2. Proteinase K digestion buffer: mix Proteinase K (20 mg/mL, Invitrogen) with equal volumes of DNA buffer (=10 mg/mL) (see item 7, the composition of DNA buffer, below). Aliquot 100 μL/tube (adequate for five samples). Store the buffer at –20°C. Refer to Table 1 for reagent calculator for varying tissue sizes.
3. Solution of preprepared phenol (70%)/chloroform/isoamyl alcohol (Invitrogen).
4. Chloroform:isoamyl alcohol (24:1) (Fluka).
5. 10% SDS: weigh 20 g of SDS in 200 mL double distilled water (ddH_2O).
6. 5 M sodium chloride (NaCl): dissolve 29.22 g of NaCl in 100 mL ddH_2O.
7. TE buffer (DNA buffer): mix 2 mL of preprepared 1 M Tris (pH 8.0) (Tris–HCl) and 2 mL of preprepared 0.5 M EDTA (Sigma) along with 10 mL of sterile water.

2.2. DNA Quantitation

1. Perform Nanodrop spectrophotometry on DNA at $OD_{260:280}$ ratios; acceptable readings should be between 1.7 and 1.9.

Table 1
A reagent calculator for varying sample sizes for DNA extraction from tumor and normal tissues

Tissue (mg)	5	10	15	20	40	60	80	100
Total volume (μL)	400	800	1,200	1,600	3,200	4,800	6,400	8,000
DNA buffer (μL)	340	680	1,020	1,360	2,720	4,080	5,400	6,800
Proteinase K (μL)	20	40	60	80	160	240	320	400
10% SDS (μL)	40	80	120	160	320	480	640	800

2. If there is sufficient DNA, run 2–5 μL on an agarose gel with lambda size markers on either side to quantitate the DNA.
3. Store at −20°C until needed.
4. Optional RNase treatment of DNA.

2.3. Labeling of Genomic DNA for Array CGH

2.3.1. Specimen Requirements

1. Sex matched genomic reference DNA (200–400 ng) consisting of male and female from a reputable source (Novagen) (stored at −20°C).
2. Genomic DNA from the proband/tumor (200–400 ng) is extracted using the suitable methods (stored at −20°C) (see Note 1).
3. All DNA must be quantitated by Nanodrop spectrophotometry at $OD_{260:280}$ ratios before use.

2.3.2. Compositions and Preparation of Labeling Reagents

1. 5× Random Octamers/primer buffer (7 μg final concentration) (Alpha DNA, NNNNNNNN): Calculate as follows:

$$\frac{\text{Number of micrograms of primer}}{14} = \text{number of microliters of water.}$$

   Add an equal volume of 10× Klenow buffer. Store at −20°C until use.
2. 10× dNTP (Promega) mix (1 mL): mix 20 μL each of dATP (100 μM stock), dGTP (100 μM stock), dTTP (100 μM stock), 12 μL dCTP (100 μM stock), and 928 μL ddH_2O to a total volume of 1 mL. Mix reagents together by pipetting gently up and down. Store at −20°C until use.
3. Dilute both reference (control DNA: Novagen, male genomic DNA and female genomic DNA) and tumor DNA into the range of 20–400 ng/μL. Store the original undiluted DNA at −20°C until use.
4. Sheared herring sperm DNA (Invitrogen): pass the salmon sperm DNA through a 25-G needle attached to a 1-mL syringe five times. Denature for 10 min at 100°C. Transfer onto ice immediately. Store original DNA at −20°C until use.
5. Yeast tRNA (25 mg) (Invitrogen): add 1 mL ddH_2O to make a concentration of 25 mg/mL. Store at −20°C until use.
6. Cy3-dCTP (1 mM) (Amersham), Cy5-dCTP (1 mM) (Amersham), Klenow and 10× Klenow buffer (Promega), Cot-1 DNA (Invitrogen). Store at −20°C until use.
7. Wash buffer: 0.1× SSC, 0.1% SDS; pH 7.0 (1 L): Mix 5 mL of 20× SSC (stocks adjusted to 7.0) and 5 mL of 20% SDS (stocks adjusted to 7.0), and make up the volume to 1 L with 990 mL dH_2O. The pH of the solution should be 7.0. Store at room temperature.

8. 0.1× SSC; pH 7.0 (1L): mix 5 mL of 20× SSC (stocks adjusted to 7.0) with 995 mL of dH_2O. The pH of the solution should be 7.0. Store at room temperature.
9. ddH_2O, stored at room temperature.
10. DIG-easy hybridization solution (Roche), stored at room temperature.
11. Supplies: 0.2-mL PCR tubes, Microcon filters YM-30, 1.5 mL tubes, 1,000 μL pipettes, 200 μL pipettes, 10 μL pipettes, pipette tips, 24 mm×60 mm glass cover slips (Fisher Scientific), 50 mL tubes.

3. Methods

3.1. Extraction of Genomic DNA from Tissue Samples

3.1.1. Digestion of Tissue Samples

1. Place slices of frozen tissue or frozen cell pellets into 1.5 mL Eppendorf tubes. Refer to Table 1 for reagent calculator for varying sample sizes for DNA extraction.
2. Add 360 μL DNA buffer, 20 μL of Proteinase K (10 mg/mL) and 40 μL 10% SDS.
3. Incubate overnight (or for 3–6 h if thin sections) at 45°C, shake occasionally.
4. Heat-inactivate Proteinase K at 85°C for 15 min. Allow to cool down.
5. Add 1 μL RNase A (10 mg/mL).
6. Incubate for 30 min at 37°C.

3.1.2. Phenol–Chloroform Extraction

1. Add 400 μL of room temperature PCI (phenol/chloroform/isoamyl alcohol); shake by hand or orbital mixer for 5 min.
2. Centrifuge at 2,580 *g* for 5 min.
3. Transfer upper aqueous layer to new tube.
4. Repeat steps 1–3.
5. Add 400 μL of chloroform/isoamyl alcohol (24:1); shake by hand or orbital mixer for 5 min.
6. Centrifuge at 2,580 *g* for 5 min.
7. Transfer upper aqueous layer to a new tube.

3.1.3. Ethanol Precipitation

1. Add 1,000 μL 100% ethanol and 26 μL 5 M NaCl.
2. Transfer the tubes to −20°C for 30 min (DNA precipitation step can be left in freezer overnight).
3. Centrifuge at 2,580 *g* for 10 min.
4. Remove supernatant; add 500 μL of 70% ethanol onto DNA pellet.

5. Centrifuge at 2,580 *g* for 10 min.
6. Remove ethanol and air-dry DNA for ~15 min.
7. Add 50 μL of sterile water to DNA. Leave at 37°C for 24 h to fully dissolve.
8. Perform Nanodrop spectrophotometry on DNA at $OD_{260:280}$ ratios.
9. If there is sufficient sample, quantitate the DNA further (see Subheading 2.2).
10. Store at −20°C until needed.

3.2. Preparation of Labeled Probes for Array CGH

3.2.1. Labeling of DNA

1. For *each slide*, make reaction mixes in 0.2-mL tubes. Add the appropriate contents to tubes 1 and 2 as indicated in the Table 2.
2. Boil the contents of tubes 1 and 2 for 10 min at 100°C.
3. Transfer tubes immediately onto ice.
4. Add 5 μL of the 10× dNTP mix to each tube (2 mM for dATP, dGTP, and dTTP; 1.2 mM for dCTP) (see Note 2).
5. Add 2 μL Cy3-dCTP (2 mM) to the sample DNA – tube 1 (see Note 2).
6. Add 2 μL Cy5-dCTP (2 mM) to the reference DNA – tube 2 (see Note 2).
7. Add 2 μL dH_2O to each tube.

Table 2
Instructions for differential labeling of both test (tumor) and normal reference DNA for CGH analysis

Reagent	Fluorescent label	Tube 1	Tube 2
Sample DNA (tumor)	Cy3	*x*μL (200–400 ng)[a]	–
Reference DNA (patient PB/control)	Cy5 (more unstable)	–	*x*μL (200–400 ng)[a]
5× Random octamers	–	5 μL (7 μg total)[b]	5 μL (7 μg total)[b]
ddH_2O	–	yμL $[40-(x+20)]$	yμL $[40-(x+20)]$
Final volume	–	40 μL	40 μL

[a]200 ng of each DNA sample is routinely used

[b]Final concentration of the random octamers in the 5× Klenow buffer is 7 μg/μL (see Subheading 2.3.2 for preparation of random octamers)

8. Add 1 μL (40 U) Klenow to all tubes (1 and 2). Total volume = 50 μL. (Check the final reaction volume is 40 μL adjusting the water volume as in Table 2) (see Note 2).
9. Incubate overnight in a PCR thermocycler at 37°C (~18 h).

3.2.2. Removal of Unincorporated Nucleotides: Using Microcon YM-30 Columns

1. Insert a Microcon column narrow end down into the 1.5-mL tube provided by Millipore (see Note 3).
2. Pool reaction tubes together (reference and sample DNA) with 100 μL Cot-1 DNA (1 μg/μL) onto the same column.
3. Add the sample to the column (avoid touching the membrane) and spin at 13,000 *g* for 12 min at room temperature.
4. Add 200 μL ddH_2O to membrane and repeat spin to wash.
5. Make a hybridization mix as follows (this is for one slide; multiply volumes depending on the number of slides):
   (a) 40 μL DIG-easy hybridization buffer
   (b) 5 μL (10 mg/mL) sheared herring/salmon sperm DNA
   (c) 5 μL (25 mg/mL) yeast tRNA
6. Add 40 μL DIG-easy hybridization mix to the membrane.
7. Leave for 5 min at room temperature.
8. Invert the Microcon YM-30 column in a new 1.5-mL tube and spin at 1,000 *g* for 3 min.

3.2.3. Calculation of Nucleotide Incorporation (Optional)

1. Remove 1.5 μL from the labeled reaction and place it on the NanoDrop spectrophotometer.
2. Use an appropriate solution to blank the spectrophotometer.
3. Readings below 3.0 pmol/μL in either channel have shown variable results and the DNA should be labeled again.

3.2.4. Probe Prehybridization

1. Denature the probe at 85°C for 10 min.
2. Allow the Cot-1 DNA to anneal to the DNA probe for 1 h at 45°C.

3.2.5. Hybridization

1. Maintain the probe at 45°C (if necessary, pulse-spin the tube to pool any condensed probe) (see Note 4).
2. Pipette 43 μL of the probe solution onto a 24 mm × 60 mm coverslip prewarmed to 45°C.
3. Gently lower the slide over the coverslip (try to avoid generating bubbles).
4. Place the slide into a hybridization cassette (see Note 5).
5. Add 10 μL of water into the lower groove of the cassette (see Note 5).

6. Close the hybridization cassette ensuring a good seal is made by the gasket.
7. Incubate for 36–40 h at 45°C.

3.2.6. Washing

1. Prewarm the wash buffer (0.1× SSC with 0.1% SDS) to 45°C (see Note 6).
2. Remove the slide from the hybridization cassette.
3. Soak slide in wash buffer until the coverslip slides off (usually 2–5 min).
4. Transfer slide immediately into a fresh coplin jar containing wash buffer.
5. Wash the slide in wash buffer for 5 min with agitation. Quickly change the wash buffer and repeat the wash step another two times.
6. Rinse the slide with three changes of 0.1× SSC at room temperature, leaving the slide in the final 0.1× SSC so it will not dry out.
7. Dry the slide with a filtered air stream, or centrifuge the slide in a clean, dust-free 50 mL Falcon tube at 700 *g* for 3 min.
8. Store the slides in a dark, dust-free box until ready to scan. If the slides are not scanned within 2 h, vacuum seal the slide box and leave the box at room temperature until ready.

3.3. Analysis of Array CGH

3.3.1. Slide Scanning

1. Turn the ArrayWoRx scanner (Applied Precision, WA) on before loading the slide.
2. Open the program ArrayWoRx and activate the camera to adjust the intensities of both Cy3 and Cy5 dyes to give the reading between 30 and 40%. Save this setting.
3. Select normal scan to begin scanning.

3.3.2. Spot Finding and Annotations

1. Start SoftWoRx and load your project consisting of the scanned data.
2. Every slide consists of two files, one from each dye channel. Select both and load image.
3. Begin alignment of the grid template to the color spots before exporting the aligned spots for annotations.
4. Save the exported normalized file under "HA data export" in tab-delimited format.

3.3.3. Visualization of Array CGH Data

1. Start SeeGH (download from http://flintbox.ca/technology.asp?Page=706) and fill in the required information (see Fig. 2) described below in the "New Data" window for the new normalized file:
   (a) File Name: *Path to your normalized data*

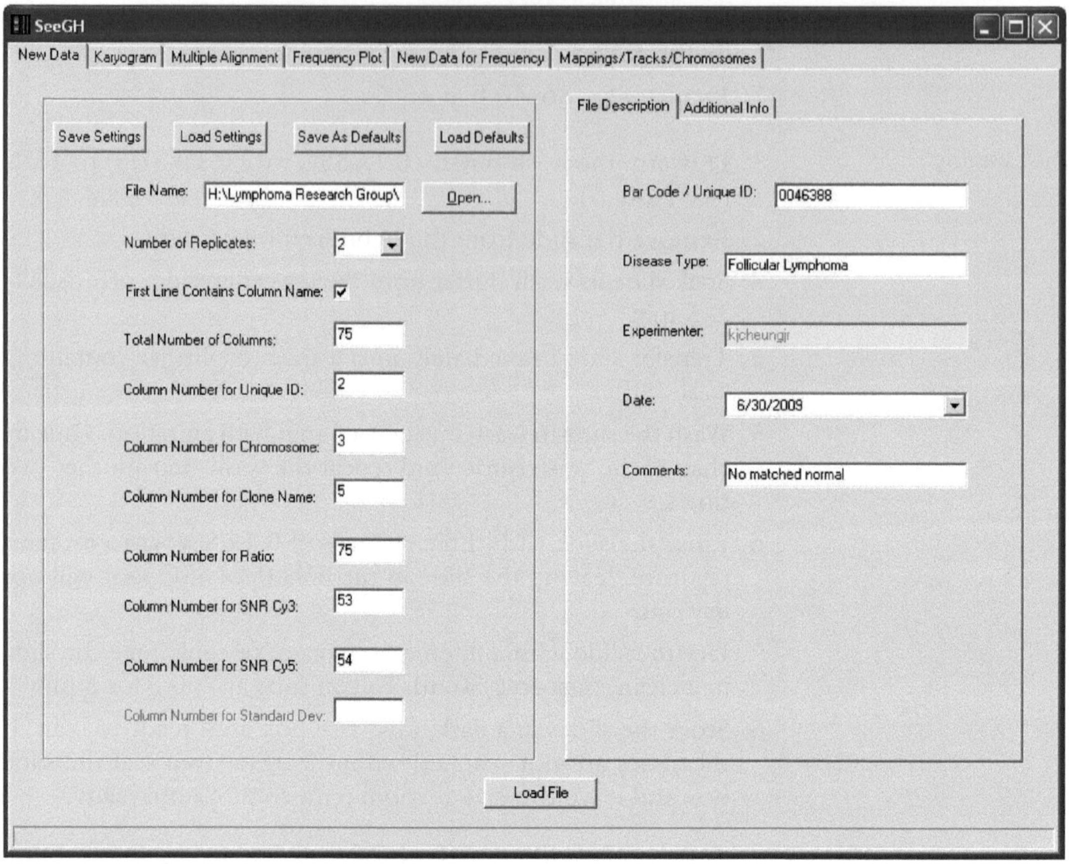

Fig. 2. A screenshot showing the interface of SeeGH under the "New Data" tab. The field identifiers correspond to descriptions in the text.

(b) Number of replicates: *2*

(c) Total number of columns: *75*

(d) Column number for unique ID: *2*

(e) Column number for chromosome: *3*

(f) Column number for clone name: *5*

(g) Column number for ratio: *75*

(h) Column number for SNR Cy3: *53*

(i) Column number for SNR Cy5: *54*

2. Fill in the remaining fields in the "File Description" and "Additional Info" windows before loading the file.

3. The whole genome view of all 24 chromosomes and data points representing BAC clones are displayed (see Fig. 3). Parameters can now be adjusted, for example, to filter out the clones that hybridized poorly:

(a) Hide clones with SD > *0.1*

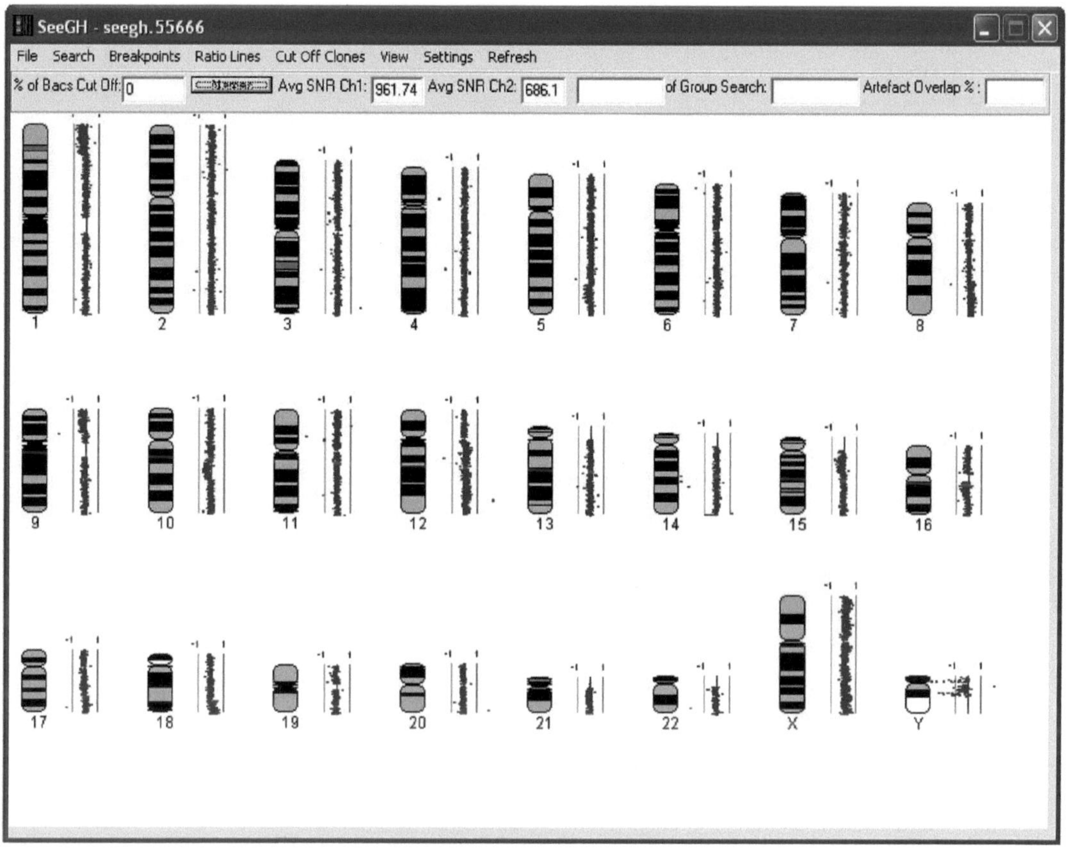

Fig. 3. SeeGH "Genomic View" window. Reconstructed whole genomic array CGH profile from approximately 27,000 array elements. Shown here is the profile of the Tat-1 B-cell lymphoma cell line.

(b) Hide clones in Ch 1 with SNR < *3*

(c) Hide clones in Ch 2 with SNR < *3*

4. Update these parameters by clicking the "Refresh" button.
5. Select a chromosome of interest and zoom in by right clicking the mouse for a detailed look at a particular region (see Fig. 4). The assignment of a state to a segment (gain, amplification, single, or homozygous loss) is done using the threshold value, calculated by the software as the ratio of the fluorochrome intensities of the normal and tumor DNA.
6. Detailed information of BAC clones can be retrieved from the side panel by placing the mouse over the BACs (see Fig. 4).
7. To access the existing files that have already been uploaded into SeeGH, start SeeGH and search the files in the "Karyogram" window (see Fig. 5).
8. To compare data of a particular chromosome among different case files, search and add files in the "Multiple Alignment"

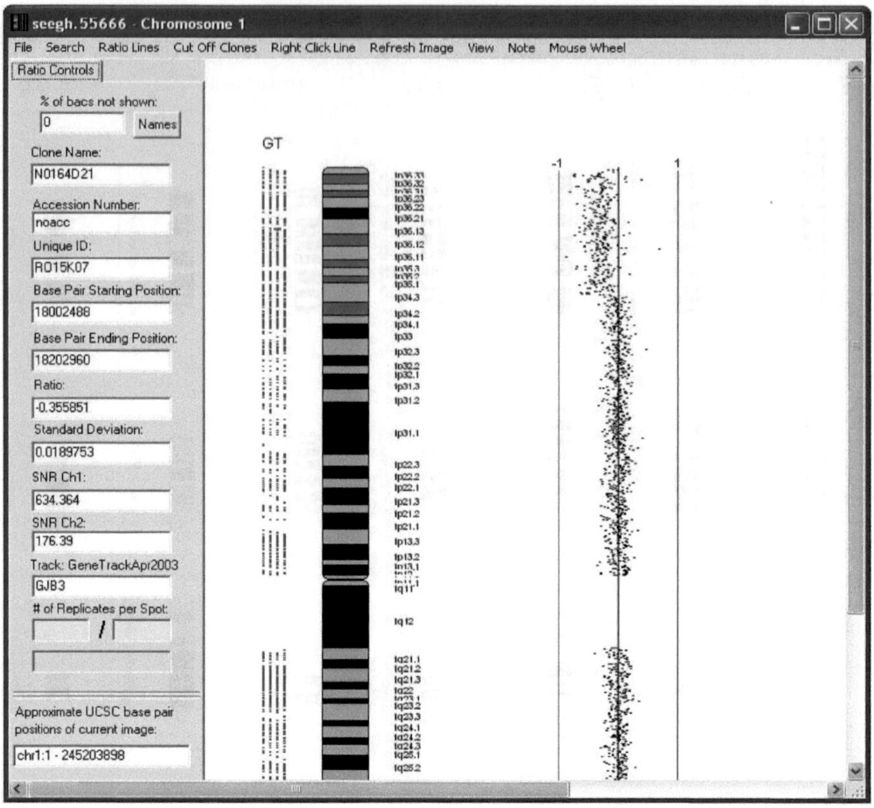

Fig. 4. SeeGH "Chromosome View" window showing a portion of chromosome 1. Each *dot* represents a BAC clone. Loss is indicated by a shift to the left of center and gain by a shift to the right of center. *Vertical lines* are −1 and +1 scale bars of $\log_2$ ratios.

window, then select the chromosome of interest for visualization (see Fig. 6a and b).

4. Notes

1. For efficient labeled DNA probe synthesis, it is essential to ensure that the sample (tumor) and reference (control) DNA are free from contaminants such as phenol, RNA, salts, etc., that may interfere with the labeling step and result in insufficient labeled probes, poor hybridization, and high background signal. For example, the resuspension of DNA in Tris–EDTA (TE) instead of water is not recommended as a

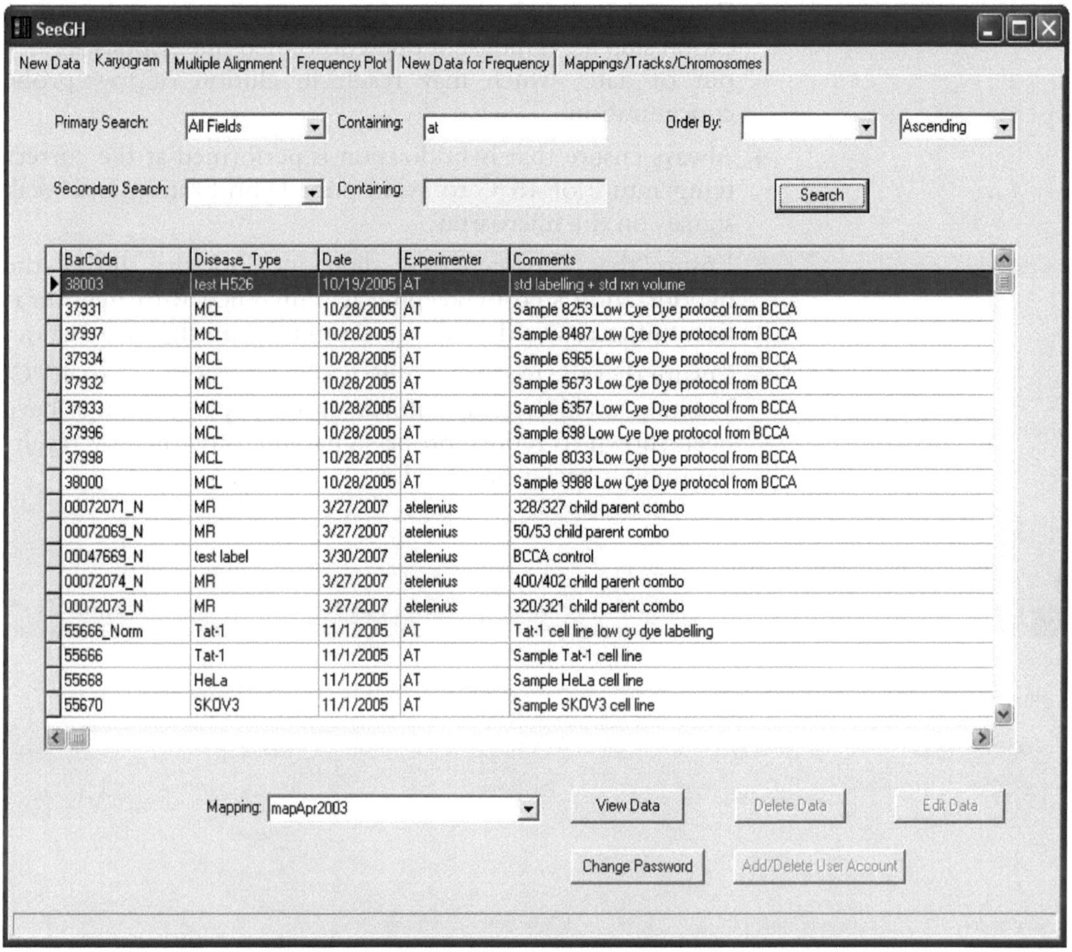

Fig. 5. SeeGH "Existing Data" window. Data that has already been uploaded into the program can be accessed by selecting the "Karyogram" tab. The displayed list can be limited by searching for data sets with specific search criteria. After selecting a particular file, the data can be viewed or deleted. Deleting a data set removes all tables from the database, while viewing opens the "Genomic View" for that data set.

high salt concentration can inhibit the labeling reaction. We also recommend assaying DNA quality using the Nanodrop spectrophotometer to measure the DNA quantity, i.e., 260:280 and 260:230 ratios.

2. Ensure that the concentrations of the template DNA, the dNTPs, Klenow enzyme, and the Cy3 and Cy5 dyes are in good working order and at appropriate working concentrations, which otherwise may result in low probe concentration, insufficient labeled probes, and weak hybridization signals on the microarray.

3. Ensure that the Microcon columns used for the probe purification step for removal of unincorporated nucleotides are not out of date, which may result in elution of low probe concentration.
4. Always ensure that hybridization is performed at the correct temperature of 45°C to avoid poor hybridization and weak signals on the microarray.
5. Ensure that the microarray slide does not dry during the hybridization step by keeping it humidified with hybridization buffer to avoid weak signal intensity on the microarray.
6. Ensure that the washing buffer is always prewarmed to 45°C. Washing at high temperatures or improper washings have been shown to reduce probe signal intensity, thereby resulting in high background noise.

a

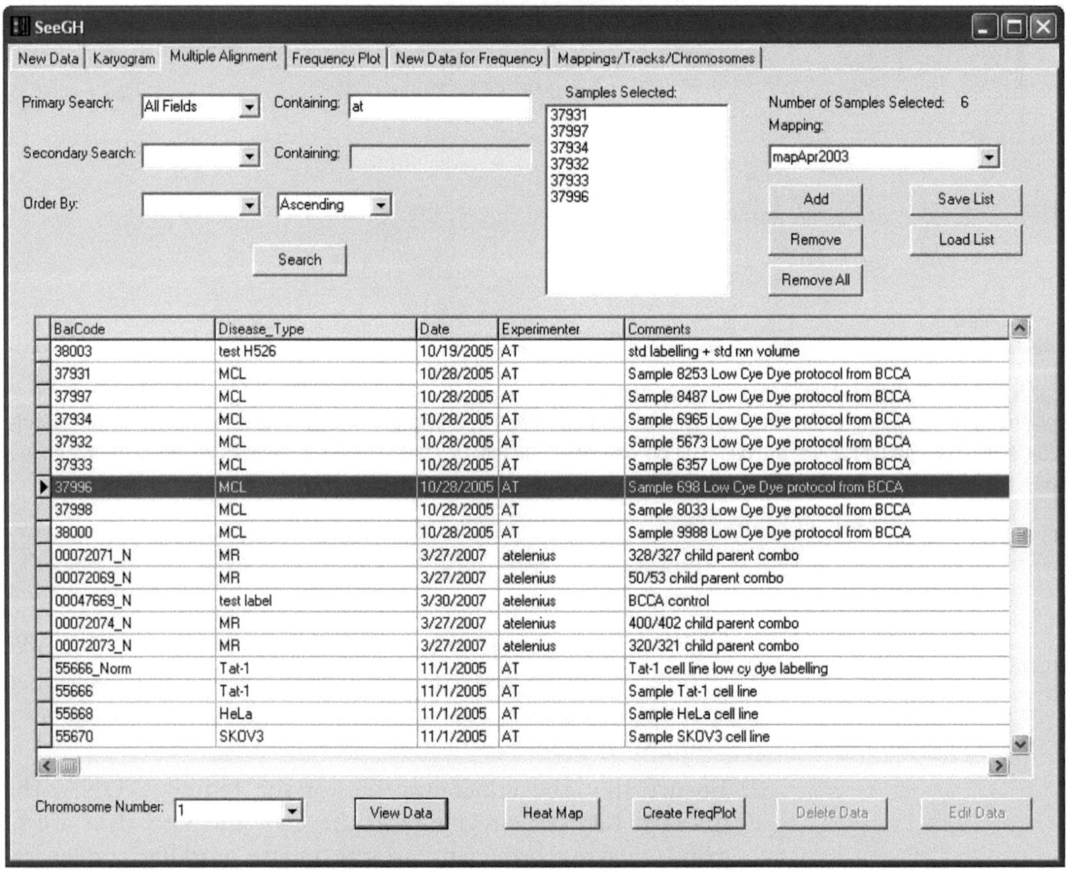

Fig. 6. The use of multiple alignment function to visualize a particular chromosomal region across a number of samples. (**a**) Samples along with the mapping build options and chromosome number can be selected to compare multiple samples at once. (**b**) A representative screenshot of four cases selected for comparison of chromosome 1.

b

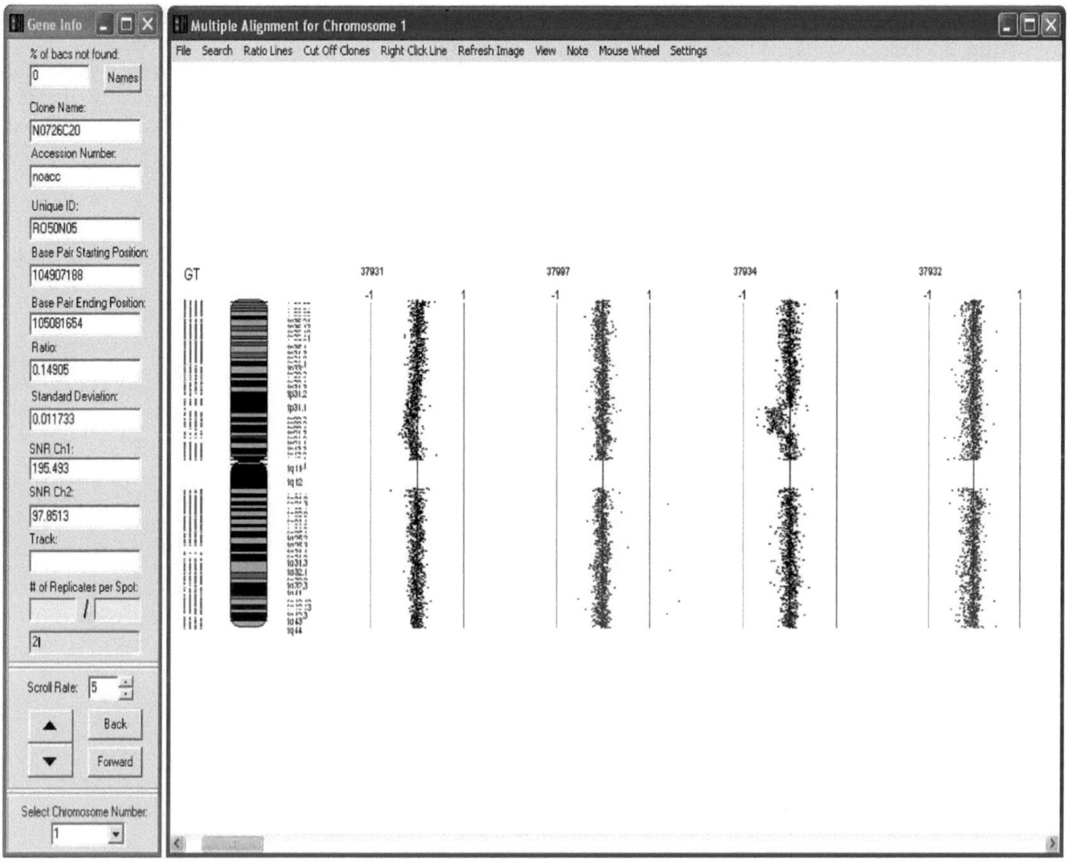

Fig. 6. (continued)

References

1. Maciejewski JP, Mufti GJ. (2008) Whole genome scanning as a cytogenetic tool in hematologic malignancies. *Blood* **112**, 965–974.
2. Pinkel D, Segraves R, Sudar D, Clark S, Poole I, Kowbel D et al. (1998) High resolution analysis of DNA copy number variation using comparative genomic hybridization to microarrays. *Nat Genet* **20**, 207–211.
3. Abdel-Rahman WM, Ollikainen M, Kariola R, Järvinen HJ, Mecklin JP, Nyström-Lahti M et al. (2005) Comprehensive characterization of HNPCC-related colorectal cancers reveals striking molecular features in families with no germline mismatch repair gene mutations. *Oncogene* **24**, 1542–1551.
4. De Lellis L, Curia MC, Aceto GM, Toracchio S, Colucci G, Russo A et al. (2007) Analysis of extended genomic rearrangements in oncological research. *Ann Oncol Suppl* **6**: vi, 173–178.
5. Bruder CE, Hirvelä C, Tapia-Paez I, Fransson I, Segraves R, Hamilton G et al. (2001) High resolution deletion analysis of constitutional DNA from neurofibromatosis type 2 (NF2) patients using microarray-CGH. *Hum Mol Genet* **10**, 271–282.
6. Shaffer LG, Bejjani BA. (2004) A cytogeneticist's perspective on genomic microarrays. *Hum Reprod Update* **10**, 221–226. Review.
7. Salman M, Jhanwar SC, Ostrer H. (2004) Will the new cytogenetics replace the old cytogenetics? *Clin Genet* **66**, 265–275. Review.
8. Garnis C, Coe BP, Lam SL, MacAulay C, Lam WL. (2005) High-resolution array CGH increases heterogeneity tolerance in

the analysis of clinical samples. *Genomics* **85**, 790–793.

9. Wong K, DeLeeuw RJ, Dosanjh NS, Kimm LR, Cheng K, Horsman DE et al. (2007) A comprehensive analysis of common copy number-variations in the human genome. *Am J Hum Genet* **80**, 91–104.

10. Cheung KJ, Shah SP, Steidl C, Johnson N, Relander T, Telenius A et al. (2009) Genome-wide profiling of follicular lymphoma by array comparative genomic hybridization reveals prognostically significant DNA copy number imbalances. *Blood* **113**, 137–148.

Chapter 17

Single Nucleotide Polymorphism Microarray Analysis of Genetic Alterations in Cancer

Charles G. Mullighan

Abstract

The identification of structural genetic alterations, including DNA amplifications, deletions, and loss of heterozygosity (LOH), using single nucleotide polymorphism (SNP) microarrays has provided important insights into the pathogenesis of a number of hematologic malignancies. Currently available SNP arrays comprise over a million SNP and copy number oligonucleotide probes that interrogate the genome at sub-kilobase resolution. The accurate detection of DNA copy number abnormalities and LOH is critically dependent on the use of high-quality DNA, the use of matched reference samples wherever possible, optimal normalization of raw microarray data, and computational algorithms to detect copy number alterations sensitively and robustly. This chapter provides methods and guidelines for preparing samples, processing and analyzing data, and validation of novel lesions. Specific examples are provided for Affymetrix SNP arrays in acute lymphoblastic leukemia.

Key words: Single nucleotide polymorphism microarray, SNP, Normalization, Loss of heterozygosity, Leukemia

1. Introduction

Chromosomal alterations such as aneuploidy and translocations are hallmarks of many hematologic malignancies including acute leukemia (1, 2). However, these alterations rarely alone result in leukemia, suggesting that additional genetic alterations contribute to leukemogenesis. Until recently, identification of submicroscopic genetic alterations such as deletions and gains of DNA relied on low-resolution genomic approaches or techniques examining individual genes. These approaches have identified several recurring genetic alterations, but were unable to survey the whole genome at high resolution.

Lynda J. Campbell (ed.), *Cancer Cytogenetics: Methods and Protocols*, Methods in Molecular Biology, vol. 730, DOI 10.1007/978-1-61779-074-4_17, © Springer Science+Business Media, LLC 2011

The completion of the human genome project has facilitated the development of microarray technology to profile genetic alterations in cancer genomes at high resolution. Several complementary microarray platforms are available to identify gains and losses of DNA at varying resolution. The earliest arrays comprised several thousand probes obtained from bacterial artificial chromosome (BAC) libraries (3). BAC arrays can provide tiling coverage of the genome with high signal-to-noise performance, but the probes are large (frequently over 100 kb in size) and consequently array resolution, and the ability to detect focal genetic alterations, which are a hallmark of leukemia, is low. Oligonucleotide arrays use short (20–100 nucleotides) probes, and current designs incorporate hundreds of thousands to millions of probes that cover whole genomes at high resolution, or provide tiling coverage of specific regions, or indeed whole genomes.

Oligonucleotide arrays are commonly two-color designs in which test and reference DNA samples are labeled with different fluorochromes and co-hybridized to the same microarray (e.g. Agilent). Single nucleotide polymorphism (SNP) microarrays are a specific type of oligonucleotide array, in which the array probes are specific for alternative alleles at SNPs (4, 5). SNP arrays were initially designed as a tool to genotype large numbers of SNPs to facilitate genome-wide association studies (GWAS). However, SNP arrays rely upon the strength of hybridization of a test sample to each SNP probe in order to perform genotyping. Measurement of the strength of hybridization to each probe can be used as a measure of the content of nucleic acid in each sample – in this case, DNA copy number. Thus, a sample with a deletion at a specific locus will show reduced signal intensity at the SNP probes within the region of deletion; conversely, samples with a gain or amplification will show increased signal. The SNP genotyping information may also be used to detect loss of heterozygosity (LOH) in tumor samples. LOH in tumor samples may arise from change in copy number state (most commonly deletion, but also amplification), or may be non-deletional, or "copy-neutral." Such copy-neutral LOH (CN-LOH, also known as acquired uniparental isodisomy) may reflect reduplication of a mutated or epigenetically modified tumor suppressor gene (6–8). The ability to detect changes in both copy number state and CN-LOH has led to the widespread use of SNP arrays to detect somatic (tumor-acquired) structural DNA changes in cancer genomes (6). This has been particularly fruitful in acute leukemia, where SNP arrays have been used to identify a number of new targets of genetic alteration, novel prognostic markers, and novel targets for therapeutic intervention (reviewed in Refs. 7, 9).

The most widely used SNP microarrays are manufactured by Affymetrix (Santa Clara, CA) (5) and Illumina (San Diego, CA) (10). Both platforms have been extensively used and validated for genotyping and detection of DNA copy number abnormalities and LOH. The resolution of SNP arrays has increased dramatically in the last 5 years, from approximately 10,000 SNP markers per array to over 1 million (Illumina) and 1.8 million (the Affymetrix SNP 6.0 array) markers. The laboratory of Mullighan, the author, has extensive experience with the Affymetrix arrays, and the remainder of this chapter will discuss methods and applications for this platform. However, many of the technical and computational issues and solutions are generalizable to both platforms.

1.1. Scope of Chapter

SNP arrays can rapidly generate a tremendous amount of high-resolution data on large numbers of cancer samples. However, accurate and comprehensive identification of DNA copy number alterations and LOH requires careful sample preparation, optimal data preprocessing and normalization, and sensitive and robust computational downstream data analysis to identify genomic lesions. Key aspects of array workflow, potential problems and pitfalls, and proposed solutions will be described. Following sample extraction, DNA processing, array hybridization, and scanning are performed in accordance with the manufacturer's protocols (11), and these aspects will not be discussed further here. SNP arrays also enable genome-wide association studies (GWAS) in cancer, but these are beyond the scope of this review.

1.2. Sample Preparation

High purity, non-degraded DNA is required to obtain high-quality data from SNP microarrays. Ideally, samples that are enriched for tumor cells should be studied. Array-CGH and SNP microarrays have limited sensitivity, and generally at least 30% tumor cells in a sample are required to detect a genetic alteration. Column-based DNA extraction methods (e.g. the DNA Blood Mini Kit, Qiagen, Valencia, CA), salting out (Gentra Puregene, Qiagen) or organic extraction (e.g. using phenol–chloroform) both yield high-molecular weight, pure DNA that performs well on SNP array analysis (see Notes 1 and 2).

2. Materials

2.1. DNA Extractions

2.1.1. General Equipment

1. 1.5-mL Microcentrifuge tubes.
2. Solvent-resistant marking pen for labeling tubes.
3. Nuclease-free, molecular biology grade water.
4. Molecular biology grade agarose.
5. Electrophoresis running buffer, e.g. 0.5× TBE, 1× TAE.
6. DNA stain for visualizing extracted DNA (e.g. ethidium bromide, SYBR Safe).

7. Tabletop microcentrifuge capable of centrifuging 1.5-mL tubes at 16,000 *g*.
8. Waterbath for incubating microcentrifuge tubes at 37 and 55°C.
9. Pipettes (e.g. 10, 200, and 1,000 μL) and pipette tips with filter barriers.
10. Gloves.
11. Glassware, bottles, and flasks for preparing agarose gels and electrophoresis running buffer.
12. Equipment for DNA quantitation: e.g. spectrophotometer (e.g. Nanodrop) or fluorimeter for PicoGreen assay (e.g. Nanodrop fluorimeter, Qubit, Invitrogen).
13. Microwave oven for preparing agarose solution.
14. Agarose gel electrophoresis tanks.
15. Power supply.
16. Gel imaging station with UV transilluminator and photographic equipment or image capture camera.

2.1.2. Equipment for DNA Extraction by Phenol–Chloroform

1. DNA lysis buffer: 10 mM Tris, 10 mM NaCl, 10 mM EDTA.
2. 20% Sodium dodecyl sulfate solution.
3. Proteinase K solution.
4. 5 M Sodium chloride solution.
5. Glycogen solution (0.2 mg/mL).
6. Buffer-saturated phenol.
7. Phenol:chloroform:isoamyl alcohol 24:25:1 pH 8.0.
8. Ethanol: 95–100% and 70% (freshly prepared).
9. RNAse A.
10. DNA hydration solution (e.g. Tris–EDTA pH 8.0).
11. –80° C Freezer or dry ice for chilling ethanol.
12. Fume hood.
13. Dedicated disposal containers for disposal of organic solvents, pipette tips, and microcentrifuge tubes.

2.1.3. Alternative DNA Extraction Methods

1. Column-based DNA extraction kits, e.g. DNA blood mini (Qiagen).

2.2. PCR Methods

1. Nuclease-free water (DNAse/RNAse free).
2. PCR primers (purified by HPLC or oligonucleotide purification columns, e.g. Glen-Pak DNA Purification Cartridges, Glen Research).
3. Fluorescently labeled probes for gene of interest (e.g. custom probes, Applied Biosystems); alternatively, SYBR green.

4. Control primer/probes (e.g. RNAse P (*RPPH1*), Applied Biosystems or custom control probe).
5. Quantitative PCR primer design software, e.g. Primer Express (Applied Biosystems).
6. Real-time PCR instrument (e.g. 7900HT, Applied Biosystems).

3. Methods

3.1. DNA Extraction Using Phenol–Chloroform

1. Pre-label 1.5-mL tubes with sample identifier using a marker resistant to solvents.
2. Pellet up to five million cells per tube at 10,000 *g* for 1 min. Do not exceed five million cells per extraction. Wash once in phosphate-buffered saline and pellet.
3. Resuspend in 480 μL DNA lysis buffer.
4. Add 12.5 μL 20% SDS and 10 μL proteinase K. Cap securely and invert 20–30 times.
5. Incubate at 37°C overnight in a waterbath, or for 2 h at 55°C.
6. Chill 100% ethanol at –80°C, or briefly on dry ice.
7. Add 5 μL RNAse A (Qiagen). Invert 50 times, and incubate at 37°C for 10 min.
8. Add 10 μL 5 M NaCl. Add 1 μL 0.2 mg/mL glycogen.
9. In a fume hood, add 500 μL buffer-saturated phenol, and mix by vigorous inversion.
10. Spin 16,000 *g* for 5 min.
11. With a 1,000-μL pipettor or transfer pipette, carefully aspirate the top aqueous phase into a new 1.5-mL tube. The aqueous phase may be viscous after this first organic extraction.
12. Add 500 μL phenol:chloroform:isoamyl alcohol mix and centrifuge as above.
13. Aspirate aqueous phase into a new tube. If still very viscous, repeat this PCI extraction one more time.
14. Add 1 mL cold ethanol. Invert 50 times. A DNA aggregate should be visible. The DNA may be transferred by pipetting to a new 1.5-mL Eppendorf tube containing 1 mL 70% ethanol, or if material is limiting or a DNA aggregate is not visible, centrifuge for 10 min at 20,000 *g*.
15. Wash once with 70% ethanol. Scrape tube on rack to dislodge the pellet. Invert several times to wash the pellet. If pellet has not already been transferred to a fresh tube for ethanol washing, transfer now. This will minimize any contamination with organic solvents, which can result in spuriously high A260 readings and overestimation of DNA concentration.

16. Spin at 16,000–20,000 *g* for 5 min to pellet.
17. Remove ethanol in two stages: aspirate most of the ethanol with a 1,000-µL pipette. Spin briefly to 16,000 *g* and then remove residual ethanol with a 100- to 200-µL pipette.
18. Air dry for 10 min, but do not overdry.
19. Resuspend in 200-µL DNA hydration solution (e.g. TE pH 8.0 or equivalent) and incubate at 55°C for 2 h. Gently pipette to loosen the pellet and incubate overnight at 37°C.
20. Assess quality by spectrophotometry (e.g. Nanodrop) or PicoGreen, and assess integrity by electrophoresis, either by 0.8% agarose gel, or using an Agilent DNA chip. On gel electrophoresis, a single band above 20 kb should be seen (see Note 3).
21. Then store at 4°C, or −20/−80°C for long-term storage.

3.2. Performing SNP Arrays

3.2.1. Array Formats and Resolution

Currently available SNP arrays that are widely used include the 250K Nsp and 250K Sty arrays (12); the SNP 5.0 array, a composite of the two 250K arrays, each of which genotypes approximately 250,000 SNPs; and the SNP 6.0 array that comprises over 906,600 SNP probes and over 946,600 non-SNP copy number probes (13). The 500K arrays provide an average resolution of approximately 5 kb between markers, and the SNP 6.0 arrays less than 1 kb between SNP/copy number markers. This author has extensive experience with the SNP 6.0 platform in analyzing leukemia genomes, and both SNP and copy number probes may be analyzed concurrently to generate high-quality data (14, 15). Both platforms involve digestion of sample DNA with restriction enzymes (*Nsp*I and *Sty*I), followed by ligation of adapters, single primer PCR amplification, fragmentation and end-labeling, followed by hybridization to the SNP microarrays, and subsequent scanning of the arrays. These steps are performed as described by the manufacturer (16, 17), ideally by core laboratories and individuals with training and experience in the conduct of SNP array genotyping, and are not described here. The methods and procedures described below focus on optimal data processing and validation of copy number alterations.

3.2.2. Comment on Sample Submission and Batch Effects

SNP arrays are single channel arrays in which processed DNA from a single sample is hybridized to a single array, contrary to many array-CGH platforms in which test and reference samples are co-hybridized to the same array. Two points regarding batch effects are emphasized.

3.2.2.1. Sample Number for Genotyping

Current genotyping algorithms for the SNP 6.0 arrays (e.g. the Birdseed v2 algorithm implemented in Affymetrix Genotyping Console) require at least 44 samples for optimal genotyping. Where possible, samples should be batched accordingly. If fewer samples are studied, external references can be obtained to facilitate genotyping. SNP genotypes are not required to perform copy number analysis, but are required for LOH inference, and allele-specific copy number analysis.

3.2.2.2. Copy Number Inference

Detection of copy number alterations requires comparison of the SNP array probe signal intensities of each tumor sample to reference samples. The SNP array probes are short, and raw SNP data are noisy, and subject to significant inter-batch variability ("batch effect") even for batches run in the same core facility. To optimize the quality of copy number inference, it is strongly recommended to include at least five to six reference samples (ideally normal samples corresponding to the tumor samples) in each batch.

3.3. Data Processing

Steps in the processing and analysis of SNP array data are shown in Fig. 1.

3.3.1. Software for Analysis of SNP Array Data

A number of free and commercial solutions for the analysis of SNP array data in cancer are available (8). These include the copy number analysis tool (CNAT) in Genotyping Console (Affymetrix) (18), copy number analysis with regression and tree (CARAT) (19), dChip (20, 21), Partek (22), Copy Number Analyzer for Affymetrix GeneChip Mapping arrays (CNAG) (23–25), the Genome Imbalance Map (GIM) (26), *IT*erative and *A*lternative norma*LI*zation and *C*opy number calling for Affymetrix *S*NP arrays (ITALICS) (27), Nexus (28), PennCNV (29), and PLASQ (30). The ideal tool and workflow will enable the rapid processing of large numbers of paired and unpaired samples, optimal array normalization, correction for predictable technical artifacts (e.g. GC content correction and batch effects) sensitive and accurate computational calling of (allele-specific) DNA copy number alterations and LOH (for both samples with paired normal data and unpaired samples), graphical display of results, and the ability to

SNP array analysis workflow:

Generation of SNP call files (Genotyping console)

CEL/CHP file processing, probe summarization (dChip)

Normalization (Reference normalization)

Generate probe intensity / SNP call matrix (dChip)

Viewing of copy number / LOH data (dChip)

Lesion (CNV/LOH) calling (dChip, circular binary segmentation, DNAcopy)

Visualization of called lesions (dChip, IGV, UCSC browser)

Downstream applications
- Statistical assessment of recurring regions of copy number change (GISTIC)
- integration with gene expression profiling data
- analysis of associations with phenotypes, outcome

Fig. 1. SNP array analysis workflow used by the author.

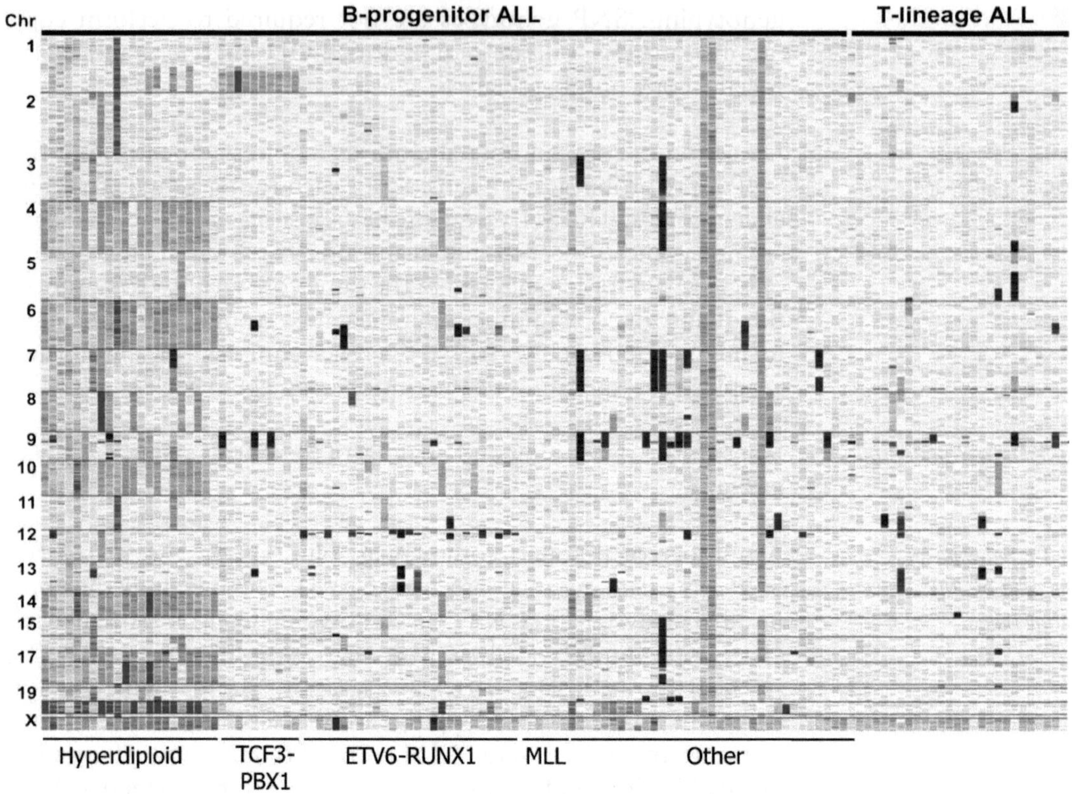

Fig. 2. dChip copy number heatmap of DNA copy number alterations in diagnostic tumor samples obtained from 126 pediatric ALL patients. Each sample is a *column* and SNP data is arranged in *rows*. Median smoothed (window 5 markers) data are shown. Heatmaps are shown as a color scale in dChip, with gains *red* and losses *blue*. Here, this has been shown as gray scale.

export results for downstream analyses. Arguably, no single tool possesses all these capabilities. The workflow described below performs each of these tasks, and utilizes free software (primarily dChip for initial data handling and viewing of results), and custom algorithms written for array normalization. This workflow has been validated for the analysis of thousands of 50K, 250K, and SNP 6.0 SNP arrays in a range of hematologic malignancies, including acute lymphoblastic leukemia (ALL) (14, 15, 31–33) (Fig. 2), acute myeloid leukemia (34), chronic myeloid leukemia (CML), and juvenile myelomonocytic leukemia (35, 36).

3.3.2. Data Preprocessing

3.3.2.1. Generation of SNP Genotypes

SNP array CEL files are loaded into Genotyping Console (Affymetrix) and SNP genotyping performed using the Birdseed v2 algorithm and default parameters. For 50K and 250K arrays, data are processed using GTYPE (Affymetrix), and genotyping is performed using the BRLMM algorithm. This procedure will generate quality control metrics, including contrast QC (for SNP 6.0 data), and SNP call rates to evaluate array quality. SNP calls are saved in CHP files, and may also be exported as SNP call TXT files for subsequent analysis.

3.3.2.2. Normalization and Probe Level Summarization

Prior to performing copy number analysis, it is necessary to normalize raw microarray data, and to summarize the hybridization intensities for the probes for each SNP or copy number probe set (probe level summarization). Many normalization algorithms are based upon those developed for the analysis of gene expression microarray data, and commonly use probe intensity data across each array to align arrays (e.g. median centering). These approaches are problematic for SNP array analysis of cancer genomes, which are commonly aneuploid. The goal of normalization is to ensure that copy number states are inferred correctly, and that data from multiple arrays may be compared. However, if the normalization procedure uses data from probes located on aneuploid chromosomes, the arrays may be incorrectly centered, resulting in erroneous copy number inferences. Secondly, many normalization procedures "borrow" information across multiple arrays to guide normalization and center the arrays. If samples are from multiple batches, or are of low quality, this can have a detrimental effect on downstream copy number analyses. These issues are described in detail elsewhere (31, 37).

To avoid these problems, we developed a normalization algorithm, "reference normalization" that uses only data from probes from diploid chromosomes to guide normalization of all probes on the array (31, 37). In this procedure, SNP CEL and CHP (or SNP call TXT) files are first read into dChip, and probe summarization performed using the dChip MBEI routine. Normalization is *not* performed in dChip. Processed, non-normalized data is then exported as a single flat file. This file comprises a column listing the SNP or copy number probe ID, and two columns per sample: one for probe intensity data and the other for SNP call. All files processed in dChip (up to 1,000 arrays in one session) may be exported into a single flat file. Normalization is then performed in R (38). Selection of diploid chromosomes for normalization in each sample may be performed manually (e.g. using existing cytogenetic information) or computationally, in which the frequency of heterozygous SNP calls for each chromosome is used to identify chromosomes likely to be aneuploid. Following normalization, the normalized data are saved in flat file with identical structure to the input file that may be read back into dChip for downstream analysis, or used for lesion calling by other algorithms (e.g. circular binary segmentation (CBS), see below). Source code, instructions, and sample data are publicly available (39).

The Reference Normalization algorithm has several advantages over alternative approaches. Data are optimally normalized, and the diploid state correctly assigned, resulting in accurate copy number inferences (Fig. 3). Normalization is performed for each

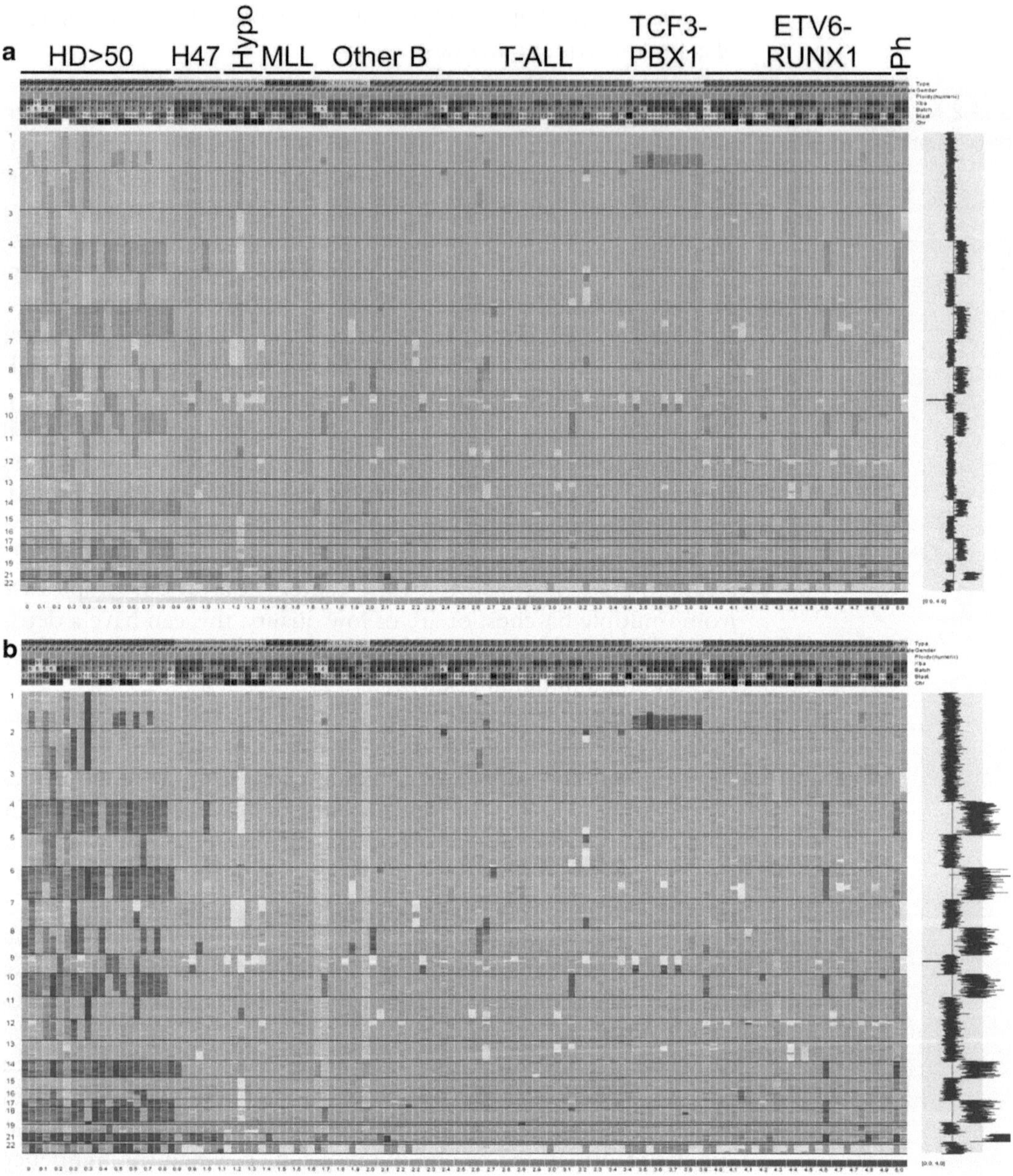

Fig. 3. Normalization methods in SNP array data. Panel (**a**) shows a dChip copy number heatmap converted to grayscale for median centered data. This results in inaccurate copy number inferences in aneuploid samples, notably for high hyperdiploid (HD>50) samples, where whole chromosomal gains are called weakly, and diploid chromosomes are called as having a copy number state intermediate between 1 (heterozygous loss) and 2 (diploid). Moreover, several near haploid samples are not identified. (**b**) The same data set following normalization using the reference normalization algorithm, which uses only those probe sets from chromosomes known or predicted by computational means to be diploid to guide normalization of the entire array. Copy number states are more accurately identified after normalization using this algorithm.

array independently, without borrowing information across a set of arrays, reducing artifacts from inter-batch effects and low-quality samples. Furthermore, as samples may be normalized individually, the procedure may be performed on multiple CPUs to increase speed and may be performed immediately after the generation of CEL files, without waiting for accumulation of a large number of arrays.

Other algorithms have implemented alternative approaches to normalize SNP array data (e.g. CRMA) (40–42), or to manually adjust the diploid baseline following normalization (e.g. dChip, CNAG).

3.3.2.3. Inference of DNA Copy Number Alterations and Loss of Heterozygosity

Identification of genetic alterations has two key components: the use of computational approaches to identify abnormalities in an automated fashion, and visual inspection of correctly normalized data (e.g. in dChip). Neither approach alone is satisfactory. Visual inspection of copy number or LOH data (e.g. as line plots or heatmaps) will readily identify gross genetic alterations, but may miss focal lesions, particularly with very high resolution data (Fig. 4). Computational approaches to identify lesions are generally highly sensitive, but even with high-quality data, they are likely to identify putative alterations that are due to technical artifacts (e.g. batch effects and noise). Careful visual inspection of all lesions, preferably with secondary validation of key lesions is required for optimal lesion calling.

The approach used by the author is to import normalized data into dChip for visualization, and detect DNA copy number alterations using CBS (43, 44). An advantage of dChip is flexibility and scalability. A single dChip session can load up to 1,000 arrays simultaneously, allowing inspection of data across multiple samples, tumors, and batches simultaneously. This is useful to distinguish lesions from artifact and to assess batch effects. dChip can display both raw and inferred copy number data (e.g. by median smoothing or Hidden Markov Model), can perform both paired and unpaired analysis of copy number and LOH (within the same session), and can export data for viewing in other tools (e.g. the Integrative Genomics Viewer (IGV) (45), and the UCSC Genome Browser (46)). dChip is also a highly useful platform to examine the results of automated lesion calling algorithms such as CBS.

CBS was developed to detect copy number alterations in array-CGH data, and may also be used to analyze SNP array (single channel data) by comparing the probe intensity data for each tumor to a corresponding normal sample or a pool of reference samples. CBS partitions the data into regions (segments) of identical copy number and assumes that a proportion of probes in a region of copy number alteration will be outliers due to the noise inherent in raw data. This approach is preferable to other approaches such as simply stipulating a threshold of raw copy number data to identify lesions, which will result in many false-positive calls of copy number alteration. CBS calculates the significance of change points in copy number state, and results of CBS may be filtered as desired

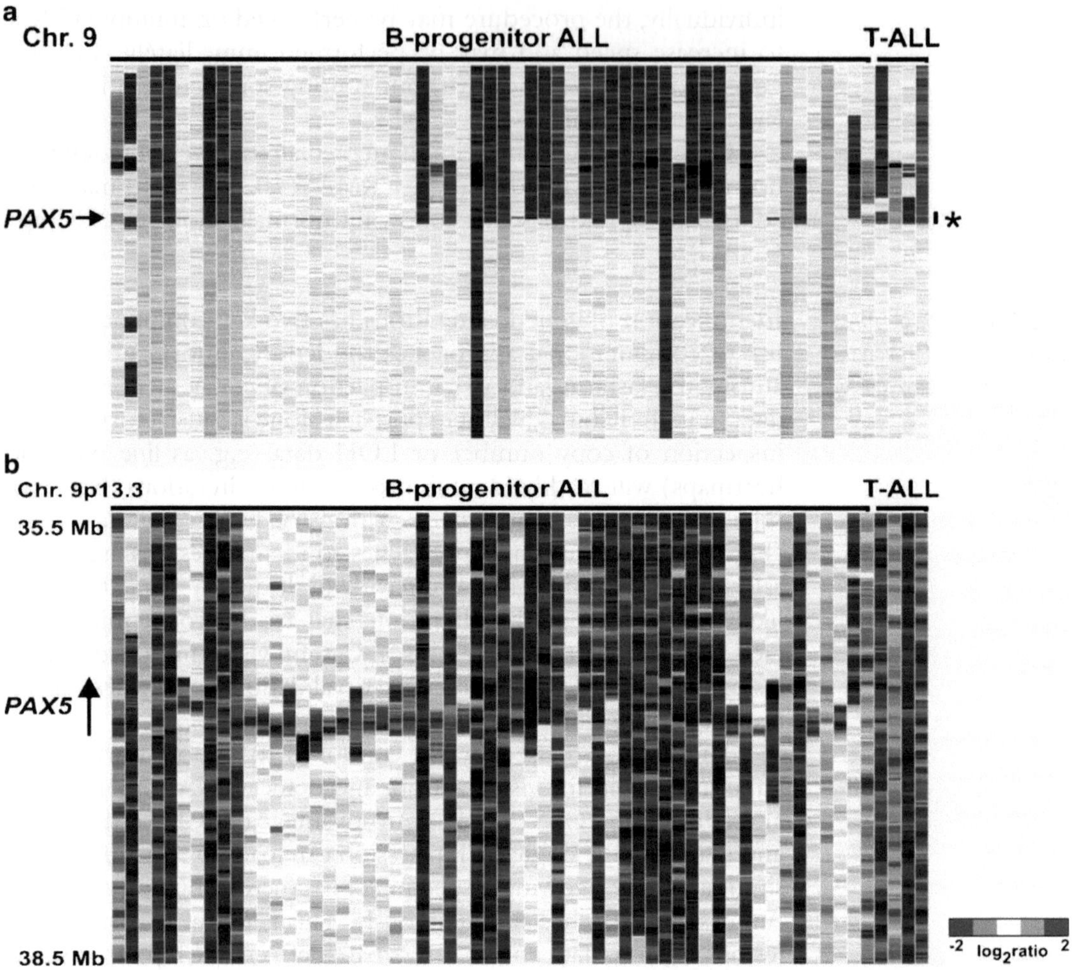

Fig. 4. *PAX5* deletions in ALL, illustrating the potential for visual inspection of data to miss key lesions. Panel (**a**) shows 62 cases with deletions involving *PAX5*. Data has been shown as grayscale, with deletions as *dark gray-black*. Many cases are focal and are not evident on inspection of chromosome-wide data. Panel (**b**) shows the same data around the *PAX5* locus. Computational methods (e.g. segmentation) are required to robustly identify all copy number alterations from high-resolution SNP array data.

(e.g. to include only segments defined by a minimum number of probes, or above or below a defined copy number threshold). For SNP 6.0 data, we typically examine segments comprising at least eight markers with copy number greater than 2.3 or less than 1.7 (on an absolute copy number scale where zero is homozygous deletion, one is heterozygous deletion, two is diploid, and greater than two is gain). An example of CBS data for a pediatric ALL sample is shown in Table 1.

Inference of regions of LOH may be performed using a variety of algorithms and software platforms (e.g. Genotyping Console, dChip, Partek). We routinely use the Hidden Markov Model (HMM) implemented in dChip. This facilitates the direct comparison of LOH data with corresponding copy number data, and dChip

Table 1
Sample outputs from circular binary segmentation for a case of pediatric ALL

Segment	Chromosomes	Cytoband	Start	End	Markers	Segment mean	Segment CN	Size (kb)	# Genes	First 10 genes in segment
10	1	q31.2	189082439	189173550	85	−0.7701	1.17	91.111	0	
12	1	q31.3	192847341	192957563	67	−1.1401	0.91	110.222	1	/KCNT2
57	3	p13	72579940	72860038	204	−0.7851	1.16	280.098	0	
72	4	q13.1	59929717	60478173	352	−0.7256	1.21	548.456	0	
136	9	p21.3	21723500	21889343	106	−0.7313	1.2	165.843	1	/MTAP
137	9	p21.3	21890112	21995382	82	−1.5032	0.71	105.27	2	/CDKN2A/CDKN2B
138	9	p21.3; p21.2	21999960	25819002	2,758	−0.7735	1.17	3819.042	3	/DMRTA1/ELAVL2/ TUSC1
140	9	p21.1	28592289	29281768	522	−0.7254	1.21	689.479	1	/LRRN6C
144	9	p13.2	36920536	37019569	95	−0.7671	1.18	99.033	1	/PAX5
148	9	q32	114292170	114530528	154	−0.8017	1.15	238.358	3	/DFNB31/ATP6V1G1/ C9orf91
161	10	q25.1; q25.2	111758900	111858222	29	−1.0353	0.98	68.282	1	/ADD3
182	12	p13.1	14423954	14550272	54	−0.6715	1.26	126.318	2	/ATF7IP/FLJ22662
184	12	p11.21	31731994	31796366	38	−0.8137	1.14	64.372	1	/LOC196394
187	12	q14.3	63870049	63891303	13	−0.902	1.07	21.254	1	/LEMD3
226	15	q12	23586514	23655099	76	−0.4353	1.48	68.585	1	/ATP10A
246	17	q23.3	59774925	59823735	31	−0.4423	1.47	48.81	1	/PECAM1

(continued)

Table 1 (continued)

Segment	Chromosomes	Cytoband	Start	End	Markers	Segment mean	Segment CN	Size (kb)	# Genes	First 10 genes in segment
249	18	q11.2	17517695	17541946	10	-0.6729	1.25	24.251	1	/ABHD3
251	18	q12.1	29687573	29803069	48	-0.7219	1.21	115.496	1	/NOL4
266	21	q11.2;q21.1	13354643	39548890	1,322	0.3587	2.56	2217.862	9	/ANKRD21/LOC441956/C21orf81/LIPI/RBM11/ABCC13/STCH/SAMSN1/NRIP1
276	21	q22.3	42109653	46921373	3,113	0.409	2.66	4811.72	89	/PRDM15/C21orf25/ZNF295/C21orf121/UMODL1/C21orf128/ABCG1/TFF3/TFF2/TFF1
291	23	p22.33	108465	154907376	1,496	-0.6979	1.23	4054.802	22	/PLCXD1/GTPBP6/PPP2R3B/SHOX/CSF2RA/IL3RA/SLC25A6/CXYorf2/ASMTL/P2RY8

The CBS algorithm exports a list of segments, genomic coordinates, and segment copy number, which may then be annotated with additional information, including chromosomal cytoband, and number and identifiers for genes in the region of copy number alteration

CN copy number

also incorporates the ability to perform LOH inference of unpaired tumor samples by comparing and filtering putative regions of tumor-acquired LOH against a pool of reference samples. Regions of homozygosity also present above a (user-defined) proportion of reference samples are considered to be inherited regions of homozygosity, and are removed (47). It should be emphasized that LOH inference from SNP genotype data needs to be performed with great care. LOH inference by HMM (or other algorithms) in unpaired tumor samples will typically identify many putative regions of LOH, many of which will be found to be inherited when tumors are directly compared to matched normal samples (Fig. 5). Thus, LOH inference of individual tumor samples, without reference to control samples, is strongly discouraged. Ideally, LOH should be performed by directly comparing tumor SNP genotypes to those obtained from normal tissue from the same individual (Fig. 5a, b). When matched normal samples are not available, it is important to filter raw calls of LOH by comparing to pools of reference samples (Fig. 5c, d). This can be performed within dChip (47).

3.3.2.4. Distinguishing Tumor-Acquired from Inherited Copy Number Variants

A critical issue in detecting copy number alterations in tumor samples is distinguishing tumor-acquired from inherited copy number variations (copy number polymorphisms, CNPs). A substantial proportion of the normal human genome exhibits DNA copy number polymorphism (48–53). In paired analysis of copy number alterations, tumors are directly compared to their corresponding normal samples and this will only identify tumor-acquired variations. In unpaired copy number analysis, in which each tumor is compared to a pool of reference samples, both inherited and tumor-acquired alterations will be identified. It is essential to filter these results subsequently to distinguish tumor-acquired from inherited copy number variants. Results of BAC, oligonucleotide, and SNP array CNP studies are available online in the Database of Genomic Variants (49, 54) and in the UCSC genome browser (46, 55).

Several issues should be considered when comparing results of SNP array studies in cancer to these online databases. The CNP studies cited above used methods of variable resolution, and a common finding when comparing results of high-resolution SNP array studies to these online databases is that putative tumor-acquired CNVs may show only partial overlap with (a) published CNP(s). It is important to define criteria for filtering lists of putative CNVs. For example, an overlap of 80% of the putative CNV and a CNP may be used. Recent studies have performed SNP 6.0 CNP analysis on the Hapmap samples and these should also be examined (13). It is also emphasized that even after careful filtering, CNVs identified in unpaired analyses may represent uncommon private CNPs that have not been previously identified. Thus, reports of novel CNVs in cancer-involving genes not previously identified

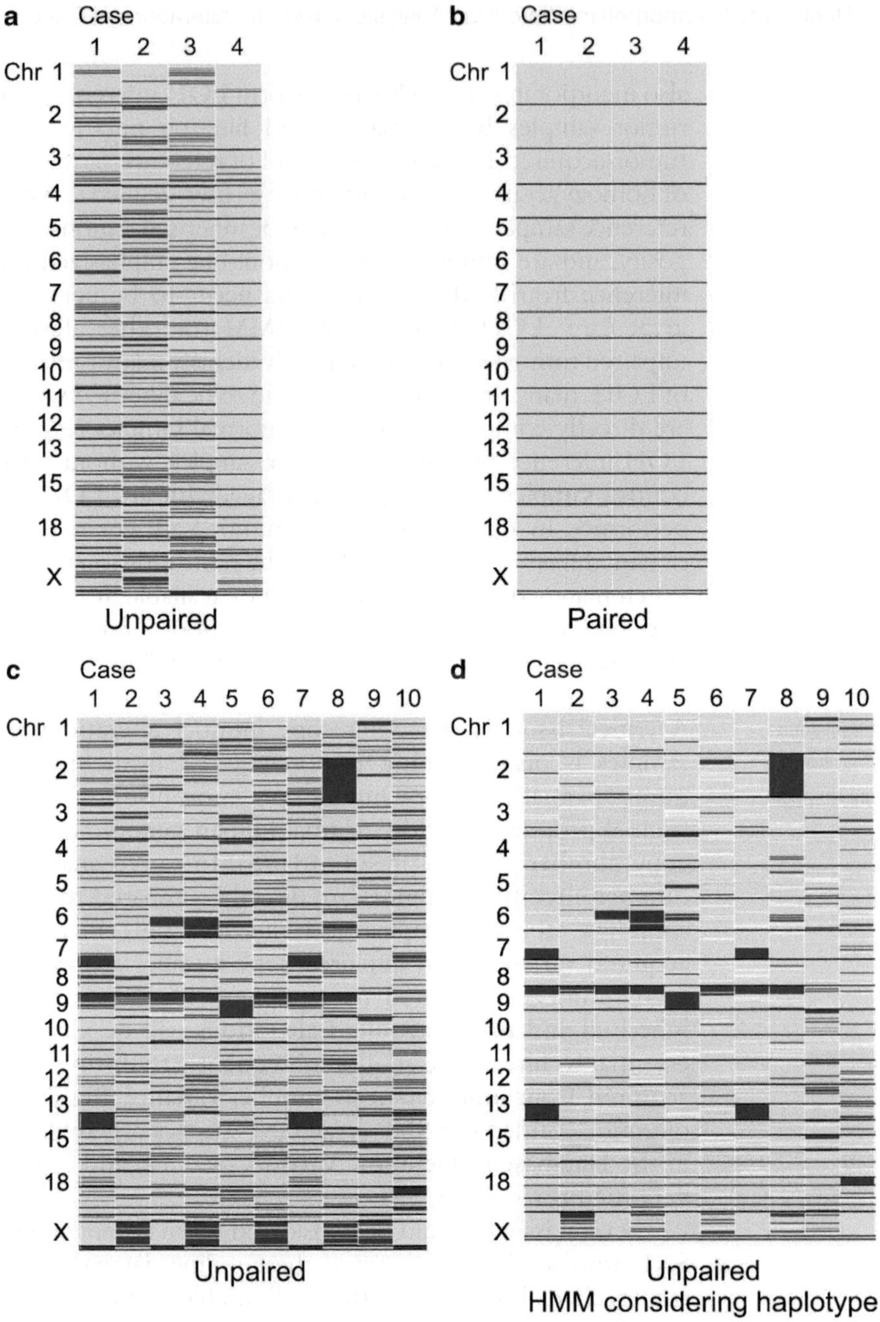

Fig. 5. Loss of heterozygosity (LOH) inference from SNP array data. Panels (**a** and **b**) depict unpaired and paired LOH inference in dChip for four chronic phase CML samples. LOH data have been converted to grayscale, and regions of LOH are shown as *black horizontal lines/bars*. The Hidden Markov Model algorithm identifies multiple putative regions of LOH in each sample when LOH inference is performed unpaired, without reference to any control samples (**a**). When paired LOH inference is performed for the same samples (**b**), in which SNP calls are directly compared between the tumor samples and batched normal DNA, none of the regions are tumor-acquired. Panels (**c** and **d**) illustrated LOH inference for samples that lack a matched normal reference samples. Ten cell line samples are shown. Panel (**c**) shows unpaired LOH inference performed in dChip by HMM. Several large regions of LOH are identified (chromosomes 2, 6, 7, 9, 14), but numerous putative focal regions of LOH are also identified. Panel (**d**) shows the same samples, where unpaired LOH inference by HMM has again been performed, followed by filtering of LOH regions using SNP data from a pool of reference samples (HMM considering haplotype implemented in dChip (47)). This considerably reduces the number of putative regions of LOH.

from paired CNV studies should be interpreted with great caution and preferably complemented by paired analysis in at least a subset of samples.

3.3.3. Downstream Analysis and Visualization

Tools such as dChip are useful in simultaneously viewing data for large numbers of samples, and incorporate a limited amount of reference genome data (e.g. the location of Refseq genes). Several other free tools are available that allow visualization of raw and/or inferred copy number data and the simultaneous visualization of public or user-provided data and annotations. The IGV (45) is a free tool that allows the simultaneous display of genomic data from multiple samples (Fig. 6). Data from multiple platforms (e.g. copy number, gene expression, and methylation) may be displayed simultaneously. The UCSC genome browser (46, 55) allows the uploading of genomic data as custom tracks from a variety of different analytical platforms, including dChip (Fig. 7). Advantages of this approach include the ability to upload raw (uninferred) as well as inferred copy number data, to view a large number of genomic features simultaneously, and to save custom visualizations as sessions that can be reloaded or shared with colleagues for future viewing. The author has found this tool particularly useful for precisely defining the extents of copy number alterations for downstream studies, such as developing lesion-specific qualitative and quantitative genomic PCR assays.

Additional tools are available to compute the significance of recurring regions of copy number alterations, such as Genomic Identification of Significant Targets in Cancer (GISTIC) (56, 57), to examine associations between DNA copy number alterations and outcome (33), and to perform cross-platform analyses, such as the integration of SNP array and gene expression data, but these are beyond the scope of this review.

3.4. Validation of Copy Number Alterations

It is often desirable to confirm DNA copy number alterations, particularly for novel and/or recurrent lesions. Suitable procedures are fluorescence in situ hybridization (FISH), which may be performed on interphase or metaphase preparations of leukemic cells and quantitative genomic PCR. FISH may be performed using commercial probes or custom-labeled BAC or fosmid probes. This is a useful technique for relatively large alterations (greater than 50–100 kb in size), and may provide information about clonality and structural rearrangements, depending on the assay design and labeling strategy. A detailed discussion of FISH is outside the scope of this review.

Quantitative genomic PCR may be used to confirm lesions of any size, and has been extensively used by the author (14, 15, 31–33).

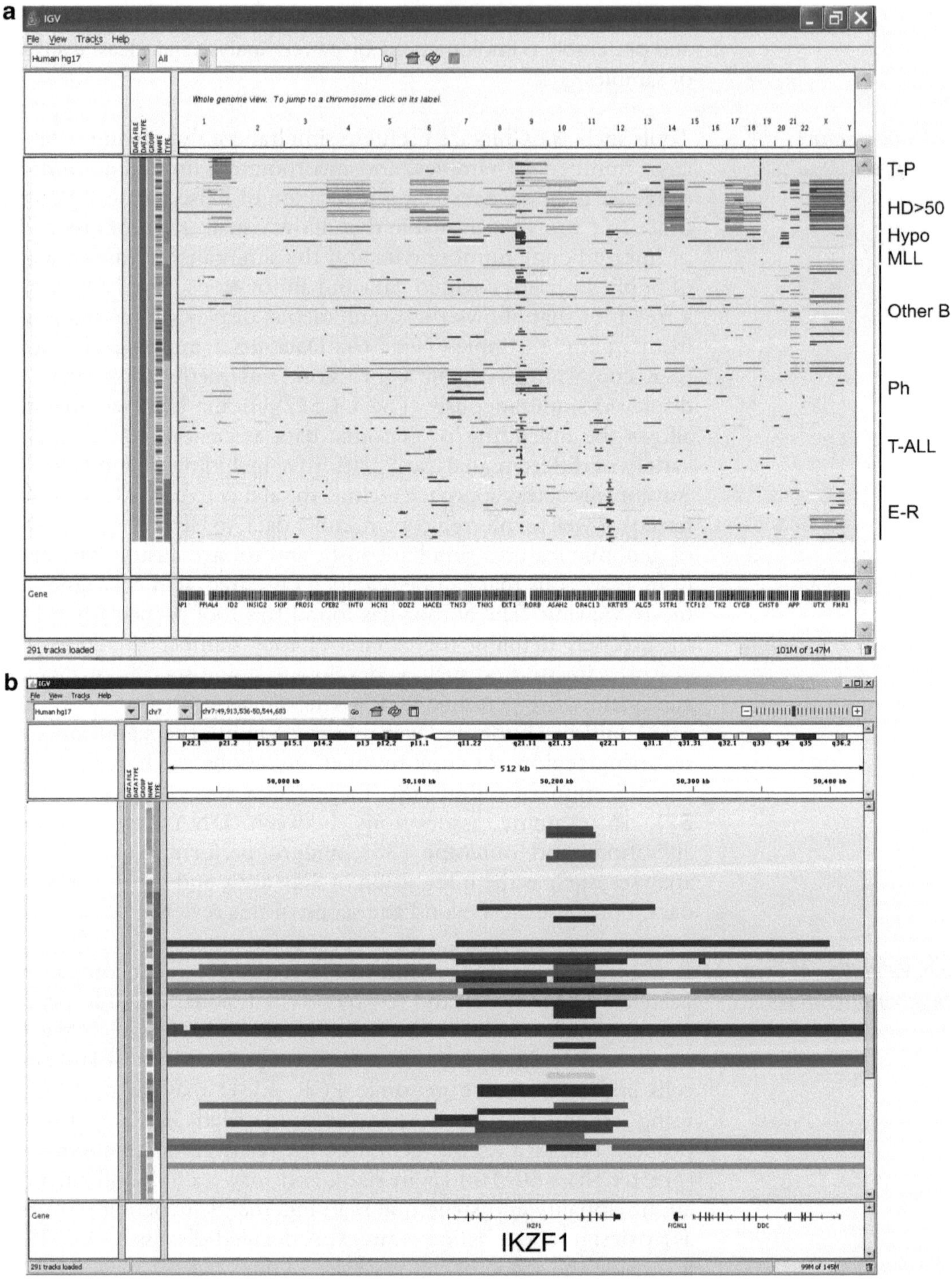

Fig. 6. The Integrative Genomics Viewer (IGV) (45). Panel (**a**) shows curated, segmented copy number data for over 290 ALL samples. Regions of copy number alterations are shown as *gray-black*. Each sample is a *row*. Panel (**b**) shows focal deletions at *IKZF1* (IKAROS) in IGV. E-R, ETV6-RUNX1; HD>50, high hyperdiploidy with greater than 50 chromosomes; Hypo, hypodiploid; Ph, BCR-ABL1; T-P, TCF3-PBX1.

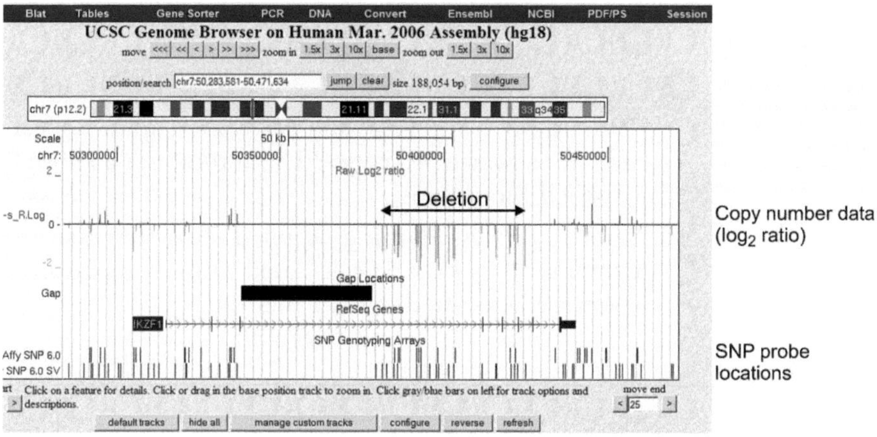

Fig. 7. Display of copy number data in the UCSC genome browser. Raw log2 ratio copy number data for chromosome 7 from an ALL case with deletion of *IKZF1* were exported from dChip and loaded into the UCSC genome browser as a custom track. This allows detailed inspection of the copy number data at locus, and correlation with available reference genome tracks (including the location of Affymetrix SNP probes).

3.5. Quantitative Genomic PCR for Confirmation of DNA Copy Number Abnormalities

1. Design PCR primers and probes using Primer Express (Applied Biosystems, Valencia, CA). We use MGB-labeled probes and default design parameters. Primer/probe sets lacking hairpins and with minimal self-complementarity are desirable. PCR primers should be purified (HPLC or oligonucleotide purification columns, e.g. Glen-Pak DNA Purification Cartridges, Glen Research) (see Note 4).
2. Preparation of DNA: dilute all DNA to optimized concentration (samples are usually 20–50 ng/10 μL; standards 150 ng/10 μL).
3. Make a tenfold serial dilution of standard DNA to at least 10^{-4}.
4. Oligonucleotides are diluted to stock concentrations 0.3 μg/μL, and are then diluted (typically to 300 nM) just before being added to the PCR reaction master mix. Probes (stock concentration 100 μM) are diluted 1:40 just before use.
5. Reaction composition for a typical test reaction is DEPC-treated water (9 μL), real-time PCR master mix (25 μL), forward and reverse primers (2 μL; each at 300 nM/2 μL), and probe (2 μL of 1:40 dilution). Total volume: 40 μL.
6. Reaction composition for a commercial internal housekeeping gene (e.g. RNAse P, Applied Biosystems) (see Note 5). DEPC water (12.5 μL), PCR mastermix (25 μL), and primer/probe mix (2.5 μL). Total volume: 40 μL.
7. It is recommended to prepare the mastermix for at least two extra reactions.

8. Dispense mastermix (40 μL) to PCR plate and then add DNA (10 μL) to plate, resulting in a final reaction volume of 50 μL.
9. Thermal cycling parameters are 50°C for 2 min, 95°C for 10 min, 95°C for 15 s, and 60°C for 1 min. Repeat last two steps for a total of 40 cycles. The author uses a 7900HT system (Applied Biosystems).
10. Data is processed using Applied Biosystems software, and copy numbers may be calculated using comparison of slopes (preferred), or by the ΔΔCt method.

4. Notes

1. Protocols for the Qiagen DNA blood mini kit and the Gentra Puregene kit can be obtained from the manufacturer's website (58, 59). For samples with low cell numbers, organic extraction is recommended as this maximizes the DNA yield. Column-based methods retain some of the sample DNA on the column during final elution. In general, approximately 2×10^5 leukemic cells are required to provide enough DNA for SNP array processing (500 ng) and quality control (100–200 ng).
2. For limiting samples, there is considerable interest in using whole genome amplified (WGA) DNA as input for SNP array analysis. A number of WGA kits are available, many using strand displacement (e.g. Phi 29 DNA polymerase) (60). It is this author's experience that WGA DNA performs extremely well for less-demanding applications (e.g. PCR and resequencing), and results in generally accurate SNP genotype calls from SNP arrays (61). However, WGA is associated with amplification bias, and suboptimal results on copy number analysis, especially for the detection of focal variants (62). Similarly, DNA extracted from formalin-fixed, paraffin-embedded material generally exhibits substantial fragmentation and may be used for SNP array genotyping (63), but generally results in a lower signal-to-noise ratio than that obtained with non-degraded DNA. Alternative platforms, such as Molecular Inversion Probe Technology, have emerged as a promising approach to generate high-quality DNA copy number data from FFPE material (64). WGA and FFPE DNA should thus be used with caution for SNP array analysis.
3. For DNA samples that have been extracted using column-based methods (in which organic solvents such as phenol are not used) quantitation by spectrophotometry usually provides an accurate estimate of DNA concentration. For samples extracted using phenol, trace amounts of phenol may result in

spuriously high-absorbance readings and overestimation of concentration. Thus, samples should be carefully washed with ethanol, preferably in a fresh microcentrifuge tube, to remove all traces of phenol. Fluorescence measurement (e.g. PicoGreen) is less convenient than spectrophotometry, but provides accurate measurement of dsDNA concentration.

4. While nonspecific DNA intercalating agents may be used for quantitative PCR (e.g. SYBR green), we find that custom fluorescent probes provide more reproducible data, in part as the range of DNA copies (zero to two or more) is much narrower than that frequently observed for RNA quantitation.
5. The control amplicon must be chosen carefully. RNase P is a single exon gene, and works well for genomic quantitative PCR; however, the genomic locus (*RPPH1*) is located at 14q11.2 and this chromosome is aneuploid in some tumors (e.g. ALL). Other control reactions may be used, or custom control reactions may be designed. At the least, SNP array copy number data should be checked to exclude copy number alterations at the *RPPH1* (or alternative) locus.

5. Key Points

1. SNP arrays provide an exceptionally powerful platform for genomic profiling of tumors. SNP arrays enable analysis of DNA copy number alterations (deletions and gains), inherited copy number polymorphisms, and SNP genotype information that may be used to identify LOH and assess associations between inherited genotype and clinical phenotypes.
2. Optimal analysis of SNP array data includes utilizing high-quality, non-degraded DNA; SNP array analysis of matched tumor and normal samples; running samples within the same batch; and using tools that appropriately normalize data and perform rigorous paired and unpaired copy number/LOH analysis and computational lesion calling.
3. Raw SNP data is noisy. Careful array preprocessing, normalization, and lesion calling are required for obtaining accurate results.
4. Many normalization algorithms are not suited to the analysis of cancer genomes, which are frequently aneuploid, and result in erroneous copy number inferences.
5. Where possible, both copy number and LOH inference should be performed as a pairwise analysis between the tumor and the corresponding normal genome. If matched normal

tissue is unavailable, copy number and LOH calls must be compared to pools of normal samples and online databases to distinguish tumor-acquired from inherited DNA copy number alterations and LOH.

6. Novel lesions should be validated by confirmatory techniques, such as FISH or genomic quantitative PCR.
7. Useful tools for processing and visualizing data include dChip, the IGV, the UCSC Genome Browser, and GISTIC.

Acknowledgments

Work described in this chapter was supported National Cancer Institute Cancer Center Support Grant P30 CA021765, the American Lebanese Syrian Associated Charities of St. Jude Children's Research Hospital. The author is supported by the National Health and Medical Research Council (Australian), the American Society of Hematology, the American Association for Cancer Research and is a Pew Scholar in the Biomedical Sciences. The author thanks the many collaborators at St Jude Children's Research Hospital who have contributed to this study, notably James Downing and Sheila Shurtleff (Pathology), Stan Pounds (Biostatistics), Jing Ma (Hartwell Center for Bioinformatics and Biotechnology), Xiaoping Su and Letha Phillips (Pathology), who have supported these studies, and developed and performed the bioinformatic and laboratory procedures described; John Morris, Emily Walker and Geoff Neale of the Clinical Applications of Core Technology laboratory of the Hartwell Center for Bioinformatics and Biotechnology at St Jude, for performing SNP arrays; and Cheng Li, for helpful discussions and development of dChip.

References

1. Pui CH, Robison LL, Look AT. Acute lymphoblastic leukaemia. Lancet 2008;371:1030–43.
2. Harrison CJ. Cytogenetics of paediatric and adolescent acute lymphoblastic leukaemia. Br J Haematol 2009;144:147–56.
3. Albertson DG, Pinkel D. Genomic microarrays in human genetic disease and cancer. Hum Mol Genet 2003;12(Spec No 2):R145–52.
4. Carter NP. Methods and strategies for analyzing copy number variation using DNA microarrays. Nat Genet 2007;39:S16–21.
5. Matsuzaki H, Dong S, Loi H, et al. Genotyping over 100,000 SNPs on a pair of oligonucleotide arrays. Nat Methods 2004;1:109–11.
6. Dutt A, Beroukhim R. Single nucleotide polymorphism array analysis of cancer. Curr Opin Oncol 2007;19:43–9.
7. Mullighan CG, Downing JR. Genome-wide profiling of genetic alterations in acute lymphoblastic leukemia: recent insights and future directions. Leukemia 2009;23(7):1209–18.
8. LaFramboise T. Single nucleotide polymorphism arrays: a decade of biological, computational and technological advances. Nucleic Acids Res 2009;37:4181–93.
9. Mullighan CG, Downing JR. Global genomic characterization of acute lymphoblastic leukemia. Semin Hematol 2009;46:3–15.

10. Gunderson KL, Steemers FJ, Lee G, Mendoza LG, Chee MS. A genome-wide scalable SNP genotyping assay using microarray technology. Nat Genet 2005;37:549–54.
11. Affymetrix Technical Manuals. (Accessed at http://www.affymetrix.com/support/technical/manuals.affx.)
12. Genome-Wide Human SNP Array 5.0. (Accessed at http://www.affymetrix.com/support/technical/datasheets/500k_datasheet.pdf.)
13. McCarroll SA, Kuruvilla FG, Korn JM, et al. Integrated detection and population-genetic analysis of SNPs and copy number variation. Nat Genet 2008;40:1166–74.
14. Mullighan CG, Phillips LA, Su X, et al. Genomic analysis of the clonal origins of relapsed acute lymphoblastic leukemia. Science 2008;322:1377–80.
15. Mullighan CG, Collins-Underwood JR, Phillips LAA, et al. Rearrangement of *CRLF2* in B-progenitor and Down syndrome associated acute lymphoblastic leukemia. Nat Genet 2009;41:1243–6.
16. Genome-Wide Human SNP Nsp/Sty 6.0 User Guide. (Accessed at http://www.affymetrix.com/support/downloads/manuals/genomewidesnp6_manual.pdf.)
17. Genome-Wide Human SNP Nsp/Sty Assay 5.0. (Accessed at http://www.affymetrix.com/support/downloads/manuals/genome_wide_snp_5_0_manual.pdf.)
18. Huang J, Wei W, Zhang J, et al. Whole genome DNA copy number changes identified by high density oligonucleotide arrays. Hum Genomics 2004;1:287–99.
19. Huang J, Wei W, Chen J, et al. CARAT: a novel method for allelic detection of DNA copy number changes using high density oligonucleotide arrays. BMC Bioinform 2006;7:83.
20. Lin M, Wei LJ, Sellers WR, Lieberfarb M, Wong WH, Li C. dChipSNP: significance curve and clustering of SNP-array-based loss-of-heterozygosity data. Bioinformatics 2004;20:1233–40.
21. dChip. (Accessed at http://biosun1.harvard.edu/complab/dchip/.)
22. Partek. (Accessed at http://www.partek.org.)
23. Nannya Y, Sanada M, Nakazaki K, et al. A robust algorithm for copy number detection using high-density oligonucleotide single nucleotide polymorphism genotyping arrays. Cancer Res 2005;65:6071–9.
24. CNAG. (Accessed at http://www.genome.umin.jp/CNAGtop2.html.)
25. Yamamoto G, Nannya Y, Kato M, et al. Highly sensitive method for genomewide detection of allelic composition in nonpaired, primary tumor specimens by use of affymetrix single-nucleotide-polymorphism genotyping microarrays. Am J Hum Genet 2007;81:114–26.
26. Ishikawa S, Komura D, Tsuji S, et al. Allelic dosage analysis with genotyping microarrays. Biochem Biophys Res Commun 2005;333:1309–14.
27. Rigaill G, Hupe P, Almeida A, et al. ITALICS: an algorithm for normalization and DNA copy number calling for Affymetrix SNP arrays. Bioinformatics 2008;24:768–74.
28. Nexus. (Accessed at http://www.biodiscovery.com/index/nexus.)
29. Wang K, Li M, Hadley D, et al. PennCNV: an integrated hidden Markov model designed for high-resolution copy number variation detection in whole-genome SNP genotyping data. Genome Res 2007;17:1665–74.
30. LaFramboise T, Weir BA, Zhao X, et al. Allele-specific amplification in cancer revealed by SNP array analysis. PLoS Comput Biol 2005;1:e65.
31. Mullighan CG, Goorha S, Radtke I, et al. Genome-wide analysis of genetic alterations in acute lymphoblastic leukaemia. Nature 2007;446:758–64.
32. Mullighan CG, Miller CB, Radtke I, et al. BCR-ABL1 lymphoblastic leukaemia is characterized by the deletion of Ikaros. Nature 2008;453:110–4.
33. Mullighan CG, Su X, Zhang J, et al. Deletion of IKZF1 and Prognosis in Acute Lymphoblastic Leukemia. N Engl J Med 2009;360:470–80.
34. Radtke I, Mullighan CG, Ishii M, et al. Genomic analysis reveals few genetic alterations in pediatric acute myeloid leukemia. Proc Natl Acad Sci USA 2009;106:12944–9.
35. Flotho C, Coustan-Smith E, Pei D, et al. A set of genes that regulate cell proliferation predicts treatment outcome in childhood acute lymphoblastic leukemia. Blood 2007;110:1271–7.
36. Loh ML, Sakai DS, Flotho C, et al. Mutations in CBL occur frequently in juvenile myelomonocytic leukemia. Blood 2009;114:1859–63.
37. Pounds S, Cheng C, Mullighan C, Raimondi SC, Shurtleff S, Downing JR. Reference alignment of SNP microarray signals for copy number analysis of tumors. Bioinformatics 2009;25:315–21.
38. R Development Core Team. R: A language and environment for statistical computing. 2009. (Accessed at http://www.R-project.org.)

39. Reference Marker Normalization. (Accessed at http://www.stjuderesearch.org/depts/biostats/refnorm/.)
40. Bengtsson H, Irizarry R, Carvalho B, Speed TP. Estimation and assessment of raw copy numbers at the single locus level. Bioinformatics 2008;24:759–67.
41. Bengtsson H, Ray A, Spellman P, Speed TP. A single-sample method for normalizing and combining full-resolution copy numbers from multiple platforms, labs and analysis methods. Bioinformatics 2009;25:861–7.
42. Bengtsson H, Wirapati P, Speed TP. A single-array preprocessing method for estimating full-resolution raw copy numbers from all Affymetrix genotyping arrays including GenomeWideSNP 5 & 6. Bioinformatics 2009;25:2149–56.
43. Olshen AB, Venkatraman ES, Lucito R, Wigler M. Circular binary segmentation for the analysis of array-based DNA copy number data. Biostatistics 2004;5:557–72.
44. Venkatraman ES, Olshen AB. A faster circular binary segmentation algorithm for the analysis of array CGH data. Bioinformatics 2007;23:657–63.
45. The Integrative Genomics Viewer. (Accessed at http://www.broadinstitute.org/igv/.)
46. Kent WJ, Sugnet CW, Furey TS, et al. The human genome browser at UCSC. Genome Res 2002;12:996–1006.
47. Beroukhim R, Lin M, Park Y, et al. Inferring loss-of-heterozygosity from unpaired tumors using high-density oligonucleotide SNP arrays. PLoS Comput Biol 2006;2:e41.
48. Conrad DF, Andrews TD, Carter NP, Hurles ME, Pritchard JK. A high-resolution survey of deletion polymorphism in the human genome. Nat Genet 2006;38:75–81.
49. Iafrate AJ, Feuk L, Rivera MN, et al. Detection of large-scale variation in the human genome. Nat Genet 2004;36:949–51.
50. Redon R, Ishikawa S, Fitch KR, et al. Global variation in copy number in the human genome. Nature 2006;444:444–54.
51. McCarroll SA, Hadnott TN, Perry GH, et al. Common deletion polymorphisms in the human genome. Nat Genet 2006;38:86–92.
52. Komura D, Shen F, Ishikawa S, et al. Genome-wide detection of human copy number variations using high-density DNA oligonucleotide arrays. Genome Res 2006;16:1575–84.
53. McCarroll SA. Extending genome-wide association studies to copy-number variation. Hum Mol Genet 2008;17:R135–42.
54. The Database of Genomic Variants. (Accessed at http://projects.tcag.ca/variation/.)
55. UCSC Genome Browser. (Accessed at http://www.genome.ucsc.edu/cgi-bin/hgGateway.)
56. Beroukhim R, Getz G, Nghiemphu L, et al. Assessing the significance of chromosomal aberrations in cancer: methodology and application to glioma. Proc Natl Acad Sci USA 2007;104:20007–12.
57. Gene Pattern. (Accessed at http://genepattern.broadinstitute.org/gp/pages/index.jsf.)
58. DNA Blood Mini Kit Technical Manual. (Accessed at http://www1.qiagen.com/products/genomicdnastabilizationpurification/qiaampsystem/qiaampdnabloodminikit.aspx.)
59. Gentra Puregene Blood Kit Technical Manual. (Accessed at http://www1.qiagen.com/Products/GenomicDnaStabilizationPurification/GentraPuregeneBloodKit.aspx.)
60. Lovmar L, Syvanen AC. Multiple displacement amplification to create a long-lasting source of DNA for genetic studies. Hum Mutat 2006;27:603–14.
61. Xing J, Watkins WS, Zhang Y, Witherspoon DJ, Jorde LB. High fidelity of whole-genome amplified DNA on high-density single nucleotide polymorphism arrays. Genomics 2008; 92:452–6.
62. Pugh TJ, Delaney AD, Farnoud N, et al. Impact of whole genome amplification on analysis of copy number variants. Nucleic Acids Res 2008;36:e80.
63. Oosting J, Lips EH, van Eijk R, et al. High-resolution copy number analysis of paraffin-embedded archival tissue using SNP BeadArrays. Genome Res 2007;17:368–76.
64. Wang Y, Carlton VE, Karlin-Neumann G, et al. High quality copy number and genotype data from FFPE samples using Molecular Inversion Probe (MIP) microarrays. BMC Med Genomics 2009;2:8.

Chapter 18

Cytogenetic Analysis and Reporting

Lynda J. Campbell

Abstract

Use of appropriate methods to produce analysable metaphase spreads and high-quality fluorescence in situ hybridization (FISH) and array results is critical to successful cytogenetic analysis of haematological malignancies, but the analysis and reporting of the findings of these tests are equally important as it is the final report that clinicians rely upon to inform patient management. A methodical approach to reporting with the use of universally recognized nomenclature will ensure that the diagnostic and prognostic information is presented in a form that is interpretable by both clinicians and other cytogeneticists.

Key words: Karyotype, Nomenclature, ISCN, Prognosis score

1. Introduction

Cytogenetic analysis is dependent upon the disease being studied, the number of cultures established, and whether fluorescence in situ hybridization (FISH) testing is incorporated into the analysis. The following outlines are reasonably generic and should apply to most analyses.

Reporting also depends upon the disorder being studied and, to some extent, to whom the report is directed, although the report should be interpretable without extensive prior knowledge of the patient's condition. Close liaison with the referring clinician or pathology laboratory is of utmost importance to clarify the diagnosis as it is impossible to either ensure that the correct cultures have been employed or to write a meaningful report without some knowledge of the underlying condition. Whilst reporting of cytogenetic analysis varies considerably depending upon the policy of the laboratory, there are a number of constant features that should be present and these are outlined below.

Lynda J. Campbell (ed.), *Cancer Cytogenetics: Methods and Protocols*, Methods in Molecular Biology, vol. 730,
DOI 10.1007/978-1-61779-074-4_18, © Springer Science+Business Media, LLC 2011

2. Cytogenetic Analysis

2.1. Karyotyping

Cytogenetic analysis is defined as the examination, either via microscope or an image analysis system, of the individual chromosomes within a cell. A count of the chromosomes must be made and then a careful comparison, band for band, of the chromosome homologs so that any variation from the norm and from each other may be observed. The greater the band resolution the longer this process may take, depending upon the level of skill of the analyser. Cancer chromosomes do not, in general, achieve band resolutions greater than the 550 level, and the majority are closer to or below the 300 band level (see Note 1).

2.2. Number of Cells Analysed, Counted, and Scanned

Cytogenetic analysis requirements vary according to the clinical question being asked. For a new acute leukaemia, it is standard practise to analyse at least 20 metaphase spreads, preferably from two separate cultures. In reality, it may be necessary to analyse many more than 20 if there are complex abnormalities and multiple cell lines to elucidate. Alternatively, when analysing a case post-therapy with a known abnormality identified previously, it is sufficient to analyse only a handful of cells completely but to scan many more for the presence of the abnormality identified at diagnosis.

A normal karyotype is usually defined as at least 20 metaphase spreads fully analysed with no chromosome abnormalities observed. If fewer metaphase spreads are available for analysis or if the banding quality is too poor to allow a complete analysis, the report should be suitably qualified.

Frequently, the 20th cell analysed will contain an abnormality that is well-recognized in the disorder that is being investigated and so it behoves the cytogeneticist to continue to analyse sufficient cells so that a clone may be identified (see Note 2). Alternatively, rather than scanning large numbers of cells, it may be appropriate to use a suitable FISH probe to determine whether the abnormal cell represents a clone. The prognostic significance of the abnormality will determine how exhaustively evidence of its clonality is sought.

The only two settings where a single abnormal cell may be included in the karyotype are (1) where it represents evidence of an abnormal karyotype that has been shown previously to be clonal; for example, a patient with chronic myeloid leukaemia and the standard Philadelphia 9;22 translocation identified at diagnosis might have the following karyotype some months later after therapy: 46,XY,t(9;22)(q34;q11.2)[1]/46,XY[39], indicating evidence of residual disease; and (2) when a single abnormal cell has been identified with, for example, an additional chromosome 8, FISH studies have identified three chromosome 8 centromere signals in interphase cells: 47,XX,+8[1]/46,XX[39]. nuc ish(DZ82x3)[50/200].

The cultures that have been established for a sample may contain different proportions of normal and abnormal cells. This is particularly true for the chronic lymphoid malignancies. Chronic lymphoid malignancy cells are difficult to culture and to induce to divide. Mitogens are generally required for chronic lymphocytic leukaemia (CLL) (see Chapter 9) and so it is likely that an overnight unstimulated culture of CLL cells will result in a normal karyotype, reflecting the residual normal cells in the marrow rather than the CLL genotype. If, however, the stimulated culture contains mainly abnormal cells, it would be appropriate to take the karyotype from the stimulated culture and ignore the normal cells obtained from what proved to be an inappropriate culture for that disease type in that patient.

Although often unsuccessful in producing an abnormal karyotype in the setting of CLL or other chronic lymphoproliferative disorders, a short-term culture may still be useful for FISH as it may provide a more typical representation of the percentage of marrow or peripheral blood infiltration with malignant cells than a long-term culture, even though the leukaemia or lymphoma cells could not be induced to divide in such a short time frame.

It is often difficult to decide how far to investigate the complexity of some cancer karyotypes. Ultimately, this must be a pragmatic decision balancing the desire to discern the intricate nature of the abnormalities and often the multiple clones that have evolved within an individual's leukaemia and the necessity of timely and clinically useful reporting. Where possible, the individual clones should be identified as it simplifies follow-up when clonal evolution occurs or when one of many previously identified clones becomes paramount with subsequent relapse or progression of the patient's disease. It may give a clue to previous disorders if a simple karyotype can be shown to have then evolved through a series of genetic events to one of considerable complexity. It is also possible to discern the patterns of evolution that identify those abnormalities that are more likely to be primary or secondary to the process of leukaemogenesis.

3. Cytogenetic Reporting

3.1. Cytogenetic Nomenclature

Cytogenetic analysis reporting has evolved to include a number of elements. The most important to the cytogeneticist is the karyotype – the string of descriptors that is able to be interpreted internationally and provides a written picture of the chromosome abnormalities detected. The latest ISCN or International Standing Committee for Human Cytogenetic Nomenclature publication is regarded as the primer for nomenclature and is internationally recognized as the standard by which karyotypes should be written (1). However, as cytogenetic and FISH techniques have expanded over the years so too has the ISCN evolved. For cytogeneticists, the importance of using the correct nomenclature cannot be over-estimated, but it is also critical to remember that cytogenetic nomenclature is a language that is an arcane one spoken only by other cytogeneticists. Attempts to instruct non-cytogeneticists in the intricacies of cytogenetic nomenclature are usually met with failure. Most haematologists measure the prognostic seriousness of the karyotype by the number of lines covered by the karyotype string. It is generally a case of "never mind the quality, feel the width"; hence, the importance of the report.

3.2. Description of the Karyotype

The cytogenetics report should include a description of the individual chromosome abnormalities and an assessment of their diagnostic or prognostic value. It is common practise to include an evaluation of the level of banding achieved (see Fig. 1). This was of more particular importance in constitutional karyotypes when the identification of microdeletion syndromes relied on conventional cytogenetic analysis than it is today with the advent of more sensitive and reliable methods such as FISH and array comparative genomic hybridization. However, it is a useful reminder of the limitations of G-banded karyotypes for the detection of subtle structural abnormalities and so has a place in reporting of cancer cytogenetics to alert the clinician that there may be unidentified abnormalities if the level of banding is low. It is also a useful internal quality measure for laboratories to monitor their performance even though the number of factors that contribute to a poor level of banding is legion.

The description of the karyotype based on the ISCN must start with the modal number – the number of chromosomes in the abnormal cell. This often requires a range to be provided as malignancies may include genetic variations between individual cells, not all falling neatly into separate cell lines. The ISCN dictates that the range be presented as 56~57, although not all pathology IT systems are capable of reproducing the ~ symbol. The modal number is followed by a description of the sex chromosomes and then any abnormalities of the autosomes (chromosomes 1–22) in numerical order and with numerical abnormalities of each chromosome preceding structural abnormalities of that chromosome in the string.

Structural abnormalities must be described according to the specific abnormality, whether it is a balanced translocation, an inversion, a deletion, or an unbalanced translocation involving known or unknown chromosomes. One of the most important decisions that the cytogeneticist must make is to determine the point at which each individual chromosome breaks – the

Band Level	G Banding Quality (bands per haploid set)	Chromosome Examples
< 150	• Unequivocal chromosome pairing not possible.	
150 to 250	• Chromosomes can be unequivocally identified from landmark bands.	
	• Three distinct bands on 10q (10q23 separate from 10q25).	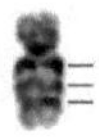
	• 13q13 visible • 22q12 visible	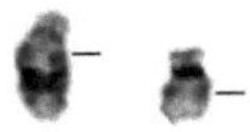
300 to 350	• Two distinct bands in middle of 2p. (2p14-16 separate from 2p22) • Foot band on 5q resolves (5q32 separate from 5q34). • Foot band on 7q visible (7q33-35). • 3p14 separate from 3p12. • Top dark band on 14q resolves into three. (14q12, 14q21, 14q23) • Dark band on 20q resolves (20q12 distinct from 20q13.2).	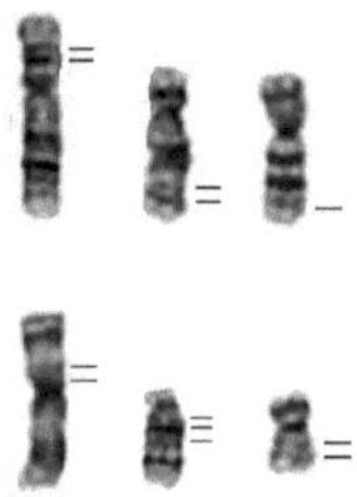
400 to 550	• Two distinct dark bands on each of 8p and 9p. • Three distinct dark bands in middle of 5q (5q14, 5q21, 5q23). • Four distinct dark bands on 18q. • Foot band on 7q resolves (7q33 and 7q35). • Three distinct dark bands on 11p (11p14, 11p15.2, 11p15.4). • 22q13.2 becomes visible.	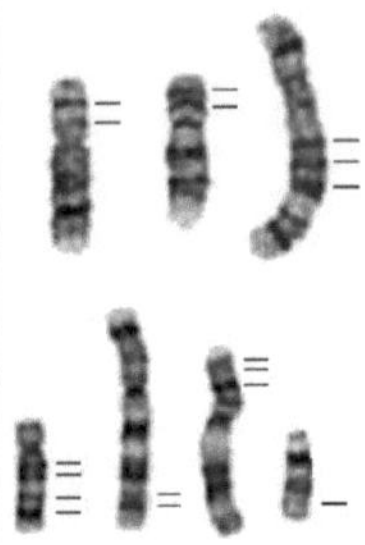
850	• 6q16 splits. • 6q24, 6q25.2 and 6q26 appear as three distinct dark bands. • 11p14 splits. • 15q12 is distinct. • 20p12 splits.	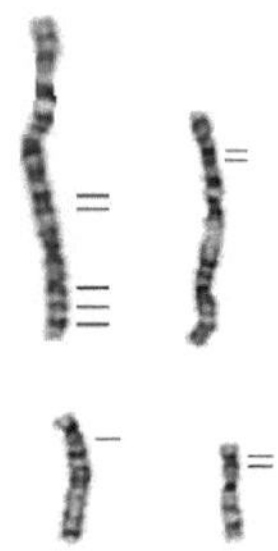

Fig. 1. G-banding levels (*left-hand column*) may be identified by recognizing chromosome landmarks (*centre column*). G-banded chromosome examples of these landmarks are shown in the *right-hand column*: chromosomes 10, 13, and 22 are used to identify banding levels 150–250; chromosomes 2, 5, 7, 3, 14, and 20 are used to identify banding levels 300–350; chromosomes 8, 9, 5, 18, 7, 11, and 22 are used to identify banding levels 400–550; chromosomes 6, 11, 15, and 20 are used to identify 850 banding level.

breakpoint – in the translocation, inversion, or deletion. The accuracy of this determination will depend on the level of banding achieved. The breakpoints should be assessed on the abnormal karyotype with the longest chromosomes and the highest band resolution, even if these are not typical of the banding level overall. It is also reasonable to use prior knowledge of well-recognized abnormalities to define breakpoints in the absence of high banding levels. For example, the balanced 15;17 translocation that is considered characteristic of acute promyelocytic leukaemia (APL) has been described with a variety of breakpoints assigned to the chromosome 17 as it is frequently difficult to determine based on G-banding alone. However, given that the *RARA* gene has been localised to the 17q21 band it is logical that, rather than providing the breakpoint as a range of possibilities such as q12-21, the breakpoint be given as 17q21.

Deletions are often described as "terminal." For example, a deletion of the long arm of one chromosome 7 may be described as del(7)(q21), implying that all chromosomal material beyond the 7q21 band has been lost from the cell. However, our knowledge of the basic biology of the chromosome and the functional necessity of there being a telomere on the end of each chromosome makes the concept of a terminal deletion nonsensical. In reality, del(7)(q21) must be del(7)(q21q36), 7q36 being the last band on the long arm of chromosome 7, but by convention we write it with the last band that we can identify by conventional cytogenetics and assume that although we cannot see the terminal segment of the chromosome, it must be there. In contrast, deletions of the long arm of chromosome 5 are usually written as "interstitial" as it is obvious from the G-banded chromosome that there is retention of the most distal part of chromosome 5, for example, del(5)(q13q33), indicating that although the segment of chromosome 5 between the bands 5q13 and 5q33 is lost, there is retention of the distal segment of the chromosome between 5q33 and the 5q telomere.

The report should be a useful document to the clinician and it should be able to be interpreted without the need to refer to previous reports. For example, an analysis of a bone marrow sample taken from a patient with APL following induction therapy that has the following karyotype: 46,XX[40] may be reported as "No cytogenetic abnormality was detected." This is an accurate description but much less helpful to the clinician than "All metaphase spreads analysed had a normal female karyotype with no evidence of the t(15;17) that was identified at diagnosis. This is cytogenetic evidence of remission." It is particularly useful to include some mention of the abnormalities previously identified as there is no certainty that the clinician in receipt of this report will have immediate access to all previous reports.

3.3. Assigning Prognosis Based on the Karyotype

For many haematological malignancies, the karyotype with or without FISH testing provides prognostic information. The inclusion of a comment regarding prognosis is helpful to the clinician, particularly when assignment of prognosis is reliant upon an interpretation of the karyotype.

The karyotype is particularly informative with regard to prognosis for patients with acute myeloid leukaemia (AML) and myelodysplastic syndromes (MDS). Prognosis categories in AML rely on the finding of individual abnormalities such as inversion 16, t(15;17), or t(8;21) to define "good prognosis" and deletions of chromosome 5, abnormalities of 3q or t(6;9) (to name only a few abnormalities) that relegate an AML to the "poor prognosis" category. These abnormalities are readily discernible by a knowledgeable haematologist, although it is appropriate that the cytogeneticist identify such abnormalities as indicators of good or poor prognosis in the report.

More problematic, however, are the definitions of "complex" chromosome abnormalities. The Medical Research Council (MRC) trials in the United Kingdom define a complex karyotype as one that contains five or more unrelated chromosome abnormalities (2), whereas the South-West Oncology Group (SWOG) and Eastern Co-operative Oncology Group (ECOG) define complexity as three or more unrelated chromosome abnormalities (3). Similarly in MDS, the International Prognostic Scoring System (IPSS) (4) and the WHO revised scoring systems (5) for patients with MDS defined a complex karyotype as one containing three or more abnormalities.

3.4. Counting Chromosome Aberrations

A recent study by the International Working Group on MDS Cytogenetics (IWGMC) highlighted a lack of consistency in scoring the number of abnormalities per karyotype (6). They concluded that standardized international karyotype counting practises were urgently needed and proposed the following set of guidelines.

As a general rule, one aberration should be counted for each item between commas in the nomenclature string. More specifically:

1. Count one aberration for each numerical change (including –Y), balanced translocation, and simple structural change (see Note 3).
2. Count one aberration for each complex structural change; for example, a derivative chromosome showing both a deletion and an unbalanced translocation would still be considered as only one aberration.
3. Count zero for a constitutional aberration, but if in doubt, count one aberration.
4. When multiple clones are present, add all independent aberrations, but count a single (specific) change only once.

5. Count one aberration for tetraploidy.
6. For the IPSS cytogenetic risk, until it is revised, all chromosome 7 abnormalities, even those involving only the p arm, are considered to define a "poor" prognosis karyotype score.

The IWGMC also proposed that cytogeneticists be more pro-active in providing a prognosis score when reporting cytogenetic analyses of MDS and AML cases. Currently, the prognosis score derived from the karyotype is likely to be assigned by a haematologist or data manager, neither of whom is best placed to interpret the karyotype.

3.5. FISH Nomenclature and Reporting

FISH has become such an integral part of cytogenetics that there are seven separate chapters in this volume describing different aspects of the use of FISH in the investigation of different types of malignancies, and FISH testing is mentioned at some point in most other chapters. However, its reporting has been problematic from the beginning as cytogeneticists have struggled to incorporate FISH nomenclature into standard karyotypes. The reason for the difficulties is obvious as the two systems, G-banding and FISH, are based on completely separate modes of observation. G-banding relies on the ability to detect variations in a black and white pattern along the length of each chromosome – it relies on spatial variations, whereas FISH uses fluorescent colour to represent gene number and location.

The ISCN reflects that difficulty; the FISH section has shown greater changes with each successive version than any other section. FISH nomenclature first appeared in the 1995 edition; the In Situ Hybridization chapter consisted of 11 pages, with no mention of breakapart probes, dual fusion probes, or the issue of combining several probe results into one karyotype. At that time, the use of panels of probes was not common practise but now, particularly for patients with acute lymphoblastic leukaemia, CLL, and plasma cell myeloma, panels of probes are used routinely to identify markers of prognosis.

Cytogeneticists are regularly faced with combining G-banding, metaphase FISH, and interphase FISH results into a single string, and the 2009 ISCN has recommended that they flow one after the other, separated by a full stop. For example, a new CML case might be described as follows: 46,XY,t(9;22)(q34;q11.2)[20].ish t(9;22)(ABL1+,BCR+;BCR+,ABL1+).nuc ish(ABL1,BCR)×3(ABL1 con BCR×2)[185/200].

The name of the FISH probe or probes used must be included in the report as the interpretation relies on a precise knowledge of the size and labelling of each probe. The HUGO approved gene names should be used in the karyotype. However, the probe manufacturers are sometimes slow to catch up with changes in the official gene name and it may be that the manufacturer's name for the gene and the gene name in the karyotype string do not correspond.

It behoves the cytogeneticist to make it clear in the report that the two names refer to the same gene. The ISCN has removed the necessity of including the chromosome breakpoints to which each probe hybridizes in the karyotype string. This was a pragmatic decision to reduce the length of the karyotype but has resulted in a string that is meaningless without the knowledge of chromosomal gene localization or, more importantly, a comprehensive report that indicates to the clinician the precise location of each probe hybridization.

The most recent edition of the ISCN includes examples of all of the currently available probe types, but the nomenclature is still open to interpretation. It behoves those reporting FISH analyses to read the ISCN carefully, apply the nomenclature as accurately as possible and, in instances where the ISCN does not cover a particular abnormality, use their best judgement to describe the abnormality within the bounds of the current nomenclature and alert members of the ISCN standing committee of recurrent problems.

4. Notes

1. Definition of banding levels: the ISCN contains idiograms that represent schematic representations of chromosomes corresponding to approximately 300, 400, 550, 700, and 850 bands. These levels represent the resolution achieved by banding and are determined by the time in the cell cycle at which the cell is arrested. They are based on the number of bands visible in a haploid chromosome set. The earlier in the cell cycle, the longer the chromosomes and the higher the banding level achieved. The number of bands visible also depends on the banding technique used. Prophase and prometaphase chromosomes achieve banding at the 700 and 850 band level but good cancer karyotypes rarely achieve 550-band levels.
2. A clone is defined as the loss of an individual chromosome in three or more cells or gain of an individual chromosome or the acquisition of a structural abnormality in two or more cells.
3. A single aberration may involve more than one chromosome; for example, a balanced translocation such as t(9;22) is considered a single aberration.

Acknowledgments

Many thanks to Joanne White, Kathleen Rayeroux, and Adrian Zordan of the Victorian Cancer Cytogenetics Service for the creation of Fig. 1.

References

1. Shaffer LG, Slovak ML, Campbell LJ (eds). ISCN (2009): An International System for Human Cytogenetic Nomenclature. Basel: S. Karger; 2009.
2. Grimwade D, Walker H, Oliver F, et al. The importance of diagnostic cytogenetics on outcome in AML: Analysis of 1,612 patients entered into the MRC AML 10 trial. Blood 1998;92:2322–33.
3. Slovak ML, Kopecky KJ, Cassileth PA, et al. Karyotypic analysis predicts outcome of preremission and postremission therapy in adult acute myeloid leukemia: A Southwest Oncology Group/Eastern Cooperative Oncology Group study. Blood 2000;96:4075–83.
4. Greenberg P, Cox C, LeBeau MM, et al. International scoring system for evaluating prognosis in myelodysplastic syndromes. Blood 1997;89:2079–88.
5. Bernasconi P, Klersy C, Boni M, et al. World Health Organization classification in combination with cytogenetic markers improves the prognostic stratification of patients with de novo primary myelodysplastic syndromes. British Journal of Haematology 2007;137:193–205.
6. Chun K, Hagemeijer A, Iqbal A, Slovak ML. Implementation of standardized international karyotype scoring practices is needed to provide uniform and systematic evaluation for patients with myelodysplastic syndrome using IPSS criteria: An International Working Group on MDS Cytogenetics Study. Leukemia Research 2010;34:160–5.

INDEX

Lynda J. Campbell (ed.), *Cancer Cytogenetics: Methods and Protocols*, Methods in Molecular Biology, vol. 730, DOI 10.1007/978-1-61779-074-4, © Springer Science+Business Media, LLC 2011

U

W

Z

MIX
Papier aus verantwortungsvollen Quellen
Paper from responsible sources
FSC® C105338

If you have any concerns about our products,
you can contact us on
ProductSafety@springernature.com

In case Publisher is established outside the EU,
the EU authorized representative is:
Springer Nature Customer Service Center GmbH
Europaplatz 3, 69115 Heidelberg, Germany

Printed by Libri Plureos GmbH
in Hamburg, Germany